T0296564

THERMODYNAMICS
FOR ENGINEERS

THERMODYNAMICS
FOR ENGINEERS

BY THE LATE

J. A. EWING, K.C.B.

SECOND EDITION

CAMBRIDGE
AT THE UNIVERSITY PRESS
1946

CAMBRIDGE
UNIVERSITY PRESS

University Printing House, Cambridge CB2 8BS, United Kingdom

Cambridge University Press is part of the University of Cambridge.

It furthers the University's mission by disseminating knowledge in the pursuit of education, learning and research at the highest international levels of excellence.

www.cambridge.org
Information on this title: www.cambridge.org/9781107594760

© Cambridge University Press 1946

First edition 1920
Second edition 1936
Reprinted by photo-lithography 1946
First paperback edition 2015

A catalogue record for this publication is available from the British Library

ISBN 978-1-107-59476-0 Paperback

Additional resources for this publication at www.cambridge.org/9781107594760

PREFACE TO THE FIRST EDITION

ALTHOUGH written primarily for engineers, it is hoped that this book may be of service to students of physics and others who wish to acquire a working knowledge of elementary thermodynamics from the physical standpoint.

In presenting the fundamental notions of thermodynamics, the writer has adopted a method which his experience as a teacher encourages him to think useful. The notions are first introduced in a non-mathematical form; the reader is made familiar with them as physical realities, and learns to apply them to practical problems; then, and not till then, he studies the mathematical relations between them. This method appears to have two advantages: it prevents the non-mathematical student from becoming bewildered on the threshold, and it saves the mathematical student from any risk of failing to realize the meaning of the symbols with which he plays. When the non-mathematical student comes to face the mathematical relations, which he must do if he is to pass beyond the rudiments of the subject, he finds it comparatively easy to build on the foundation of physical concepts he has already laid: there is perhaps no better way to learn the meaning and use of partial differential coefficients than by applying them to thermodynamic ideas, once these ideas are clearly apprehended.

Accordingly the plan of the book is to begin with the elementary notions and their interpretation in practice, and to defer the study of general thermodynamic relations till near the end. Finally these relations are illustrated by applying them to characteristic equations of fluids, and in particular to steam, following Callendar's method.

The chapter on Internal-Combustion Engines gives occasion for introducing some results of experiments on the internal energy and specific heats of gases, and this matter is dealt with further in an appendix* which attempts an elementary account of the molecular theory.

In any exposition of the first principles of thermodynamics it is important to choose a way of dealing with temperature such that students may be led by simple and logical steps to understand the thermodynamic scale. The course followed here is first to imagine

* Now Chapter VII.

an ideal gas which serves as thermometric substance, and also as the working substance in a Carnot engine. This gives a perfect-gas scale by reference to which the efficiency of any Carnot cycle is provisionally expressed, and from that the step to the thermo-dynamic scale is easy.

The writer is indebted to Professor Callendar and his publisher, Mr Edward Arnold, for permission to include a much abbreviated version of his *Steam Tables*. By the recent publication of complete tables, Professor Callendar has added substantially to the many obligations under which he has put all students of thermodynamics. The writer would also thank Mr J. B. Peace, of the Cambridge University Press, for various suggestions and for the interest he has taken in bringing out the book; and also Dr E. M. Horsburgh, of the Mathematical Department of this University, for his great kindness in reading the proofs.

J. A. E.

The University, Edinburgh
March 1920

NOTE TO THE SECOND EDITION

SIR ALFRED EWING died on January 7th, 1935: he was busy with the revision of this book till near his last days. He asked me to see the book through the press for him, a task gladly undertaken in tribute to his memory.

The revision was practically done: there remained steam tables to compile and proofs to revise. Had Sir Alfred been spared a little longer, he would have added a chapter on Radiation; apart from that, most of his plan of revision was complete. No attempt has been made to supply that chapter, so that the book may still be truly "Ewing's Thermodynamics."

Steam tables based on the latest 1934 International Steam-Table Conference values have been prepared to replace the tables in the former edition. Ewing realized that recent steam measurements were not in sufficiently good agreement with existing steam-table values and that the latter would eventually be superseded by tables in better agreement. He therefore wished that any provisional tables inserted in the Appendix to replace the former ones should be based on the most recent values. I am grateful to my collaborator, Mr G. S. Callendar, for the preparation of these tables.

Because he was not satisfied with the validity of the tables in the former edition, nor with that of the Callendar equations in the high-pressure regions, Ewing deleted the numerical examples in the text (Chapter III), which had been based on the figures in those tables, and he also shortened the sections (Chapter IX) dealing with the Callendar theory. I have put back the numerical examples, using the figures in the new tables, and feel sure this would have been in accordance with his wishes.

The chapters on the Steam-Engine and on Jets and Turbines were read by Mr H. L. Guy, and that on the Internal-Combustion Engine by Dr F. Ll. Smith. I have made a few small additions and alterations in accordance with their suggestions, for which I am very grateful. The last chapters are mainly taken by Ewing from his articles in the *Dictionary of Applied Physics*, and thanks are due to the publishers, Messrs Macmillan, for permission to use them.

Sir Alfred Ewing had very definite views as to the symbols for the thermodynamical quantities most used by engineers. The symbols he has adopted are in accord with those agreed by the

International Union of Pure and Applied Physics, except that ψ has been used for Free Energy and F for Electromotive Force, while G (Gibbs' Function) is taken as $-G$, the positive value $TS - I$ being more convenient for users of steam tables.

It will be noticed that the Centigrade scale is used throughout; almost the last words I had in conversation with Sir Alfred Ewing were in deprecation of the continued use of the Fahrenheit scale, which he hoped would rapidly go into disuse. Is it too much to hope that engineers will gradually adopt the metric system in their dealings with steam and oil?

A. EGERTON

OXFORD

May 1935

CONTENTS

CHAPTER I

FIRST PRINCIPLES

CHAPTER II

PROPERTIES OF FLUIDS

CHAPTER III

THEORY OF THE STEAM-ENGINE AND OTHER VAPOUR ENGINES

CHAPTER IV

THEORY OF REFRIGERATION

CHAPTER V

JETS AND TURBINES

CHAPTER VI

INTERNAL-COMBUSTION ENGINES AND PROPERTIES OF GASES

CHAPTER VII

MOLECULAR THEORY OF GASES

CHAPTER VIII

GENERAL THERMODYNAMIC RELATIONS

CHAPTER IX

APPLICATIONS TO PARTICULAR FLUIDS

CHAPTER I

FIRST PRINCIPLES

1. The Science of Thermodynamics treats of the relation of heat to mechanical work. In its engineering aspect it is chiefly concerned with the process of getting work done through the agency of heat. Any machine for doing this is called a Heat-Engine. It is also concerned with the process of removing heat from bodies that are already colder than their surroundings. Any machine for doing this is called a Refrigerating Machine.

It is convenient to study the thermodynamic action of heat-engines and refrigerating machines together, because one is the reverse of the other, and by considering both we arrive more easily at an understanding of the whole subject.

2. Heat-Engine and Heat-Pump. In a Heat-Engine heat is supplied, generally by the combustion of fuel, at a high temperature, and the engine discharges heat at a lower temperature. Thus in a steam-engine heat is taken in at the temperature of the boiler and discharged at the temperature of the condenser. In any kind of heat-engine the heat is let down, within the engine, from a high level of temperature to a lower level of temperature, and it is by so letting heat down that the engine is able to do work, as a water-wheel is able to do work by letting water down from a high level to a lower level. But there is this important difference, that some of the heat disappears in the process of being let down: it is converted into the work which the engine does.

In a Refrigerating Machine work has to be spent upon the machine to enable it to take in heat at a low level of temperature, and discharge heat at a higher level of temperature, just as work would have to be spent upon a water-wheel if it were used as a means of raising water by reversing its action, in such a way that the buckets were filled at a low level and emptied at a higher level, so that it should serve as a pump. It would be quite correct to speak of a refrigerating machine as a heat-pump. But again there is an important difference between the refrigerating machine and the reversed water-wheel: the refrigerating machine is a heat-pump which discharges more heat than it takes in, for the work which is spent in driving the machine is converted into heat, which has to be discharged at the higher level of temperature in addition to the heat that is taken in at the low temperature.

3. **Efficiency of a Heat-Engine.** From the point of view of practical thermodynamics the object of a heat-engine is to get work done with the least possible expenditure of fuel. In other words the ratio of the work done to the heat taken in should be as large as is practicable. This ratio is called the Efficiency of the engine as a heat-engine. The theory of heat-engines deals with the conditions that affect efficiency, and with the limit of efficiency that can be reached when the conditions are most favourable.

4. **Coefficient of Performance of a Refrigerating Machine.** In a refrigerating machine the object is to get heat removed from the cold body and pumped up to a higher level of temperature at which it can be discharged, and what is wanted is that this should be done with the least possible expenditure of work. The ratio of the heat taken in by the machine from the cold body to the work that is spent in driving the machine is called the Coefficient of Performance. The theory of refrigeration deals with the conditions that will allow this ratio to be as large as possible.

5. **Working Substance.** In the action of a heat-engine or of a refrigerating machine there is always a working substance which forms the vehicle by which heat passes through the machine. It is because the working substance has a capacity for taking in heat that it can act as a vehicle for conveying heat from one level of temperature to another. In this process its volume changes, and it is by means of changes of volume on the part of the working substance that the machine does work, if it is a heat-engine, or has work spent upon it, if it is a refrigerating machine. Accordingly, an important part of the science of thermodynamics deals with the properties of substances in relation to heat, and the connection between such properties in any substance. The substances with which we are chiefly concerned are fluids in the gaseous or liquid states. They include air and other gases, water and water-vapour, and also some fluids more easily vaporized than water, such as ammonia and carbonic acid, which are used as the working substance in certain refrigerating machines. Each fluid has of course its own characteristics; but many of the relations between its properties are of a general kind and may be studied without limitation

to individual fluids. It will be seen, as we go on, that much of what has to be said applies equally, whatever fluid serves for working substance, and that in any one fluid the various properties are connected with one another in a way that is true for all fluids. The study of the thermodynamic relationships between the various properties of a fluid is useful, not only because of the direct light it throws on the action of heat-engines, but also because it enables a practically complete knowledge of the properties of a fluid in detail to be inferred from a comparatively small number of experimental data. We shall see later, for example, how such relationships have been made use of in calculating modern tables of the properties of steam from the results of careful measurements, made in the laboratory, of a few fundamental quantities.

6. **Operation of the Working Substance in a Heat-Engine.** In general the working substance is a fluid which operates by changing its volume, exerting pressure as it does so. But it is easy to imagine a heat-engine having a solid body for working substance, say a long rod of metal arranged to act as the pawl of a ratchet-wheel with closely pitched teeth. Let the rod be heated so that it lengthens sufficiently to drive the wheel forward through the space of one tooth. Then let the rod be cooled, say by applying cold water, the ratchet-wheel being meanwhile held from returning by a separate click or detent. The rod on cooling will retract so as to engage itself with the next succeeding tooth, which may then be driven forward by heating the rod again, and so on. To make it evident that such an engine would do work we have only to suppose that the ratchet-wheel carries round with it a drum by which a weight is wound up. The device forms a complete heat-engine, in which the working substance is a solid rod, doing work in this case not through changes of volume but through changes of length. While its length is increasing it is exerting force in the direction of its length. It receives heat by being brought into contact with some source of heat at a comparatively high temperature; it transforms a small part of this heat into work; and it rejects the remainder to what we may call a receiver of heat, which is kept at a comparatively low temperature. The greater part of the heat may be said simply to pass through the engine, from the source to the receiver, *becoming degraded as regards temperature in the process.* This is typical of the action of all heat-engines: they convert some heat into work only by letting down a much larger quantity of heat

from a high temperature to a relatively low temperature. The engine we have just imagined would not be at all *efficient*; the fraction of the heat supplied to it which it could convert into work would be very small. Much greater efficiency can be obtained by using a fluid for working substance and by making it act so that its own expansion of volume not only does work but also causes it to fall in temperature before it begins to reject heat to the cold receiver.

7. **Cycle of Operations of the Working Substance.** Generally in the action of a heat-engine or of a refrigerating machine the working substance returns periodically to the same state of temperature, pressure, volume and physical condition in all respects. Each time this has occurred the substance is said to have passed through a complete cycle of operations. For example, in a condensing steam-engine, water taken from the hot-well is pumped into the boiler; it then passes into the cylinder as steam, then from the cylinder into the condenser, and finally from the condenser back to the hot-well; it completes the cycle by returning to the same condition in all respects as at first, and is ready to go through the cycle again. In other less obvious cases a little consideration shows that the cycle is completed although the same portion of working substance does not go through it again: thus in a non-condensing steam-engine the steam which has passed through the engine is discharged into the atmosphere, where it cools to the temperature of the feed-water, while a fresh portion of feed-water is delivered to the engine to go through the cycle in its turn.

In the theory of heat-engines it is of the first importance to consider as a whole the cycle of operations performed by the working substance. If we stop short of the completion of the cycle matters are complicated by the fact that the substance is in a state different from its initial state. On the other hand, if the cycle is complete we know that whatever heat or other energy the substance contained within itself to begin with is there still, for the state of the substance is the same in all respects, and consequently any work that it has done must have been done at the expense of heat which it has taken in during the cycle. The total amount of energy it has parted with must be equal to the amount it has received, during the cycle, for its stock of internal energy is the same at the end as at the beginning. We can at once apply the principle of the Conservation of Energy and say that for the cyclic process as a whole the work done must be equivalent to the difference between the heat taken in and the heat rejected.

8. The First Law of Thermodynamics. The principle of the Conservation of Energy in relation to heat and work may be expressed in the following statement, which constitutes the First Law of Thermodynamics:—*When work is done by the expenditure of heat a definite quantity of heat goes out of existence for every unit of work done; and, conversely, when heat is produced by the expenditure of work the same definite quantity of heat comes into existence for every unit of work spent.*

The word "work" is to be understood here in a comprehensive sense: it includes electrical work as well as mechanical work. Electrical work may, for instance, be done by expending heat in a thermoelectric circuit, which is a true heat-engine although there are no visibly moving parts.

9. Internal Energy. We have used in Art. 7 a phrase which requires some further explanation—the *internal energy* of a substance. No means exist by which the whole stock of energy that a substance contains can be measured. But we are concerned only with changes in that stock, changes which may arise from the substance taking in or giving out heat, or doing work, or having work spent upon it. If a substance takes in heat without doing work its stock of internal energy increases by an amount equal to the heat taken in. If it does work without taking in heat, it does the work at the expense of its stock of internal energy, and the stock is diminished by an amount equal to the work done. In general, when heat is being taken in and the substance is at the same time doing work, we have

Heat taken in = Work done + Increase of Internal Energy.

For any infinitesimally small step in the process, we may write

$$dQ = dW + dE,$$

where dQ is the heat taken in during the step, dW is the work done, and dE the increase of internal energy.

In a complete cycle there is, at the end, no change of the internal energy E, and consequently, for the cycle as a whole,

$$Q_1 - Q_2 = W,$$

where $Q_1 - Q_2$ is the net amount of heat received, namely the difference between the heat taken in and the heat rejected in the complete cycle, and W is the work done in the complete cycle.

In this notation we are supposing W to be expressed in units of heat, as well as Q and E. It would be more correct to speak of W as the thermal equivalent of the work done.

10. **Work done in Changes of Volume of a Fluid.** In an engine of the usual cylinder and piston type the working fluid does work by changes of volume. The amount of work done depends only on the relation of the pressure to the volume in these changes, and not on the form of the vessel or vessels in which the changes of volume take place. Let the intensity of pressure of the fluid (that is to say the pressure on unit of area) be P while the piston moves forward through a small distance δl. If the area of the piston is S the total force on it is PS and the work done is $PS\delta l$. But $S\delta l = \delta V$, the change of volume: hence the work done is $P\delta V$ for the small change of volume δV, or $\int_{V_1}^{V_2} P dV$ for a finite change of volume from a volume V_1 to a volume V_2 during which the pressure may vary.

In any complete cycle of operations the volume at the finish is the same as at the start, and the work done is $\int P dV$ taken round the cycle as a whole.

It is very useful to represent graphically the work which a fluid does in changing its volume. Let a diagram be drawn in which the relation of the pressure of any supposed working substance to its volume is shown by rectangular coordinates as in fig. 1. Beginning with the state represented by the point A, where the pressure is AM and volume OM, suppose the substance to expand to a state B, where the pressure is BN and the volume ON, and let the curve AB represent the intermediate states of pressure and volume. Then the work done by the substance in this expansion, which is $\int_{OM}^{ON} P dV$, is represented by the area $MABN$ under the curve AB.

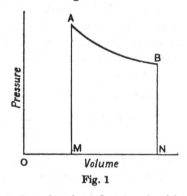

Fig. 1

Again, if the substance undergoes any complete cycle of change (fig. 2) by expanding from A through B to C and by being compressed back through D to A, work is done by it while it is expanding from A to C, equal to the area $MABCN$, and work is spent upon it while it is being compressed from C through D to A, equal

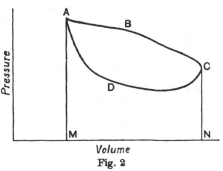

Fig. 2

to the area $NCDAM$. The net amount of work which the substance does during the cycle is equal to the algebraic sum of those areas: in other words it is equal to the area of the closed figure $ABCDA$ representing the complete cyclic operation, which area is $\int PdV$.

If on the other hand the operation were such as to trace the figure in the opposite direction, the substance being expanded from A to C through D and compressed from C to A through B, the enclosed area would be a measure of the work expended upon the substance in the cycle.

11. **Indicator Diagrams.** This pressure-volume diagram is an example, and a generalization, of the method of representing work which Watt introduced by his invention of the Indicator, an instrument for automatically drawing a diagram to represent the changes of pressure in relation to changes of volume in the action of an engine. The figure $ABCDA$ may be called the *Indicator Diagram* of the supposed action.

The indicator consists of a small cylinder containing a piston which can move in it without sensible friction but is controlled by a stiff spring. This is put in free communication with one end of the working cylinder of the engine, so that the working substance presses on the indicator piston and displaces it, against the spring, through distances that are proportional to the pressure at every instant. Connected with the indicator piston is a pencil which rises or falls with it, the connection being made, generally, through a lever that gives the movements of the indicator piston a convenient magnification. A sheet of paper on which the pencil marks its movements is caused to move through distances proportional to the motion of the engine piston, and at right angles to the path of the pencil. Thus a diagram is drawn like that of fig. 2, exhibiting a closed curve for each double stroke of the engine piston, and with coordinates which represent the changes of pressure and changes of volume. The enclosed area, when interpreted by reference to the appropriate scales of pressure and volume, measures the net amount of work done in the engine cylinder during the double stroke, so far as one side of the piston is concerned. If the engine is double-acting—that is to say, if the working substance acts successively on the two sides of the engine piston during successive strokes—a similar indicator diagram is taken for the other end of the cylinder as well. The mean effective pressure (M.E.P.) is that pressure which, multiplied by the volume range, gives the work done. It is obtained from the indicator diagram either by dividing the measured area by the horizontal length of the diagram, or by dividing the diagram into a number of vertical sections and finding the average length of the mid-ordinates.

12. **Units of Force, Pressure, and Work.** For engineering purposes, in speaking of pressure and of work, the common unit of force in British and American usage is the weight of one pound and in continental usage the weight of one kilogramme*. By the word "weight" we mean the force which gravity exerts on the mass at sea level. When scientific precision is required one must specify a locality, or rather a latitude, because gravity acts rather more strongly as we go from the equator towards the pole. The same piece of material is more strongly attracted by the earth in London than in New York to the extent of one part in 1000. If the weight of one pound of matter in mean latitude (45°) be taken as unity, its weight in any other latitude λ is

$$0.99785 \ (1 + 0.0053 \ \sin^2 \lambda):$$

The usual units of pressure are the pound per square inch and the kilogramme per square centimetre†. Another unit is the "Atmosphere," which properly means the pressure of the atmosphere with the barometer standing at 760 mm. in latitude 45°,

---

* One kilogramme is 2·20462 pounds.

† Since 1 centimetre is 0·393701 in., 1 kilogramme per sq. cm. is 14·2233 pounds per sq. in., when both are measured at the same place, so that gravity acts alike on the pound and the kilogramme. One pound per sq. in. is 0·070307 kilogramme per sq. cm.

or 759·6 mm. in London. This is equal to a pressure in London of 14·688 pounds per square inch or 1·03267 kilogrammes per square centimetre*. Continental engineers, however, often use the symbol "*At*" as a short name for one kilogramme per square centimetre: this usage will be found in Mollier's and other steam tables as well as in technical papers.

Pressures are sometimes given in inches, or in millimetres, of mercury. One inch of mercury (at 0° C.) is equivalent to 0·4912 pound per square inch; one millimetre of mercury to 1·3595 grammes per square centimetre.

The usual engineering units of work are the foot-pound and the metre-kilogramme or kilogrammetre. One kilogrammetre is 7·233 foot-pounds.

13. **Units of Heat.** For the purpose of reckoning quantities of heat we may compare them with the quantity required to warm a unit mass of water from the temperature of melting ice to the temperature at which water boils under a pressure of one atmosphere. These two points serve to determine two fixed states of temperature that are quite definite and are independent of the particular way in which temperature may be measured. The unit of heat which is obtained by taking a certain fraction of this quantity is described as the *mean thermal unit.* Thus we have a mean thermal unit which is one-hundredth part of the heat required to warm one pound of water from the melting point to the boiling point at a pressure of one atmosphere. This unit is called the pound-calory. The reason why one-hundredth part is taken is that on the Centigrade scale of temperature the interval between these fixed points is divided into 100 degrees: consequently the pound-calory is the *average* amount of heat required to warm a pound of water through one degree Centigrade, between the melting point and the boiling point as limits. The actual amount required per degree need not be the same for each degree of the scale, and in fact is not the same, for the specific heat of water is not quite constant.

Similarly, what is commonly called the British Thermal Unit (when the Fahrenheit scale is employed) would be defined as 1/180 of the quantity of heat required to warm one pound of water from

---

* 1 atmosphere is the pressure exerted by a 76 cm. column of mercury (of density 13·5951 grams per cm.³ at 0° C.); the value of gravity is 980·665 cm./sec.² (i.e. at sea level and latitude 45°); 1 atmosphere = 1·033228 kg./cm.² = 14·6959 lb./in.² = 1·013250 bar; 1 bar = 10⁶ dynes per cm.².

the melting point to the boiling point, because on the Fahrenheit scale there are 180 degrees between the two fixed points.

Again, the mean kilo-calory is one-hundredth of the amount of heat required to warm one kilogramme of water from the melting point to the boiling point, and the mean gramme-calory is one-thousandth of a kilo-calory.

It should be added that instead of the mean calory a unit is often used called the 15° calory, which means the heat required to raise unit quantity of water from 14·5° to 15·5° on the Centigrade scale.

14. Mechanical Equivalent of Heat. The experiments of Joule, begun in 1843 and continued for several years, demonstrated that when work is expended in producing heat a definite relation holds between the amount of heat produced and the amount of work spent. Causing the potential energy of a raised weight to be used up in turning a paddle which generated heat by stirring water in a vessel, and observing the rise of temperature so produced, Joule found that 772 foot-pounds of work served to heat one pound of water through one degree (Fahrenheit) on the thermometer he employed, at a particular part of the scale.

Later and more exact determinations made by Joule himself and other observers agree in showing that Joule's original figure was rather low. The general result is to fix 1400 as the number of foot-pounds (in the latitude of London) that are equivalent to the mean pound-calory as defined in Art. 13. The corresponding value of the mechanical equivalent of the "British Thermal Unit" is 777·8 foot-pounds, and that of the mean gramme-calory is 426·7 gramme-metres*.

* H. L. Callendar, who gave numerical values in his *Steam Tables* (published 1915), used the mean thermal unit for stating quantities of heat. He says there (p. 6) that his mean gramme-calory was equivalent to 4·1868 Joules, the Joule, or Watt-Second, being 10^7 ergs.

An International Steam-Table Conference which met in 1929, 1930 and 1934 has done most useful service, to be referred to later. Among its work has been to *define* what is now called the "International Steam-Table Unit" as a quantity of heat equivalent to 1/860 of an International Watt-Hour or 3600/860 International Watt-Seconds. It also accepted the official International Watt-Second to be larger than the true Joule in the ratio of 1·0003 to 1. Hence the value of the agreed International Steam-Table Unit as defined by the Conference is 1·0003 × 3600/860 or 4·1878 true Joules.

If Callendar's estimate of 4·1868 Joules for the mean gramme-calory used in his *Steam Tables* of 1915 was correct, it follows that his heat unit was less than the modern International Steam-Table Unit by only a very small quantity—about one part in 8000. The International Union of Physics (1931) accepted the Joule (10^7 ergs) as the Unit of Heat when measured in Units of Energy, and also defined the gramme-calory as the amount of heat required

The mechanical equivalent of heat enters into many of the formulas of thermodynamics. It is often called Joule's Equivalent, and is generally represented by the symbol J. The symbol A is used for the reciprocal of Joule's equivalent, or $1/J$.

15. Scales of Temperature. In the construction of an ordinary thermometer a fine tube of uniform bore is chosen, and a bulb is formed on it to contain the mercury or other liquid whose expansion is to be used as an indication of temperature. When it is filled the two fixed points are determined by placing the instrument (a) in melting ice, and (b) in the steam coming from water boiling under a pressure of one atmosphere. The position taken by the end of the column of liquid in the tube is marked for each of these two points. The distance between them is then divided into equal parts which are called degrees, 100 parts for the Centigrade scale and 180 for the Fahrenheit scale. By this construction equal steps in temperature are defined by equal amounts of expansion on the part of the selected liquid, or rather by equal amounts of difference between the expansion of the liquid itself and that of the glass in which it is contained, for it is the difference of expansion that determines the rise of the column in the tube. This common method of measuring temperature gives results that vary for different liquids and for different sorts of glass. Each of two mercury thermometers, for example, may have the fixed points correctly marked, and be of uniform bore, and yet if they are made of different sorts of glass they may give readings that differ by as much as half a degree (Centigrade) at the middle of the range between the fixed points, and may show very serious discrepancies—sometimes amounting to as much as five degrees or more—when they are applied to measure higher temperatures such as that of steam on its way to an engine. This illustrates the fact that the measurement of temperature by an ordinary thermometer gives an arbitrary scale, which cannot even be relied on to be the same in different instruments.

Measurements of temperature are much less capricious if we

to raise the temperature of 1 gramme of air-free water from 14·5° to 15·5° of the International Scale of Temperature under the pressure of one normal atmosphere. This calory is equivalent to 4·1873 Joules (accepting the ratio of the International Electrical Watt to the absolute Watt as 1·0003) and therefore has the same value as the International Steam-Table calory. The uncertainty of the mean calory is about 1 in 1000 and of the 15° calory about 1 in 4000.

[A Table of Units and Conversion Factors was agreed by the 1934 Steam-Table Conference.]

select for the expanding substance any one of the so-called per-manent gases, such as air, or nitrogen, or hydrogen, taking care of course to keep the pressure of the gas constant while it is employed to measure temperature by its changes of volume. Such an instru-ment is called a constant-pressure gas thermometer. It would be inconvenient for ordinary use; but it serves to supply a scale with which the readings of an ordinary thermometer can be compared. Thus the readings of any mercury thermometer can be corrected to bring them into agreement with the scale of a gas thermometer if that scale be adopted as the standard scale in stating temperatures.

Experiments on the expansion of various gases by heat have shown that all gases which are far from the conditions that would cause liquefaction expand very nearly alike. Thus if we compare an air thermometer with a nitrogen or a hydrogen thermometer we get practically the same scale except at extremely low temperatures such as those at which the gas is approaching the liquid state. Gases expand by almost exactly the same amount between the two fixed points, namely by 100/273 of the volume they have at the temperature of melting ice; and at intermediate points, or at points beyond the range, their agreement with one another is almost perfect. Hence the scale of the gas thermometer is much to be preferred to that of any mercury thermometer as a means of stating temperature. But there is another and even stronger reason for this preference. We shall see later that it is possible to imagine a scale of temperature, based on general thermodynamic principles, which does not depend on the properties of any particular sub-stance: that scale is called the *thermodynamic scale* of temperature, and much use is made of it in thermodynamic reasoning. The scale of a gas thermometer is practically identical with the thermo-dynamic scale. Taking the hydrogen thermometer, in which the agreement is closest, Callendar has shown* that midway between the fixed points the scale correction (that is, the difference between the numbers which state the same temperature on the hydrogen scale and the thermodynamic scale) is only 0·0018 of a degree, and that the temperature has to go up to about 1000° or down below −150° before the correction becomes as much as 0·1 of a degree. These figures are for hydrogen expanding under a constant pressure of one atmosphere. The differences between the scale of the

* H. L. Callendar, "On the thermodynamical correction of the Gas Thermo-meter," *Proc. Phys. Soc.* vol. xviii, or *Phil. Mag.* January, 1903. (Recent values of the corrections are even smaller.)

gas thermometer and the thermodynamic scale are even less if a constant-volume type of gas thermometer be used, in which increments
of temperature are measured by the increments of pressure that are
required to keep the volume of the gas constant while it is heated.

16. **Reckoning of Temperature from the "Absolute Zero."**
Experiment shows that the amount by which air or hydrogen or
any other so-called "permanent" gas expands between the two
fixed points—that is to say in passing from the temperature of
melting ice to that of boiling water (at a pressure of one atmosphere)—is about 100/273 of the volume at the lower fixed point,
care being taken that the pressure does not change. Hence if we
adopt the scale of the gas thermometer as our scale of temperature,
and use Centigrade divisions, this result may be expressed by
saying that when 273 cubic inches of gas at 0° C. are heated under
constant pressure to 1° the volume alters to 274 cubic inches. When
the gas is heated to 2° C. its volume becomes 275 cubic inches,
and so on. Similarly if the gas be cooled from 0° C. to −1° C. its
volume changes from the original 273 cubic inches to 272, and so on.
Putting this in a tabular form, let the volume be

	273 at 0° C.
It will become	272 at −1° C.
	⋮ ⋮
and finally would be	0 at −273° C.,

if the same law could be held to apply down to the lowest temperatures. Any actual gas would change its physical state before so low
a temperature were reached, becoming first liquid and then solid,
and the volume to which it would contract would consequently be
not zero but the volume of the substance in the solid state.

The above result may be concisely expressed by saying that if
temperature be reckoned not from the ordinary zero but from a
zero which is about 273 Centigrade degrees below it (more exactly
273·1), the volume of a gas, heated under constant pressure, is
proportional to the temperature reckoned from that zero. The
zero in question is spoken of as the Absolute Zero of temperature.
Denoting any temperature on the ordinary scale by t and the
corresponding temperature reckoned from the absolute zero by T,
we have (using Centigrade degrees)

$$T = t + 273 \cdot 1.$$

The absolute zero has been defined here by reference to the expansion of a gas. But it will be seen later that the thermodynamic scale starts from a zero which is absolute in the sense that no lower temperature can possibly exist, and that the zero of that scale coincides with the zero of the gas scale as defined above*.

17. **Properties of Gases: Charles' Law and Boyle's Law.** The experimental fact that all "permanent" gases expand by very nearly the same fraction of their volume for a given increase of temperature, the pressure being kept constant, is known as Charles' Law. Another fundamental property of gases, discovered by the experiments of Boyle, is that when the volume of a gas is altered by altering the pressure, the temperature being kept constant, the volume varies inversely as the pressure.

Thus if V be the volume of a given quantity of any gas, and P the pressure, then so long as the temperature remains unchanged, V varies inversely as P, or $PV=$ constant. This is Boyle's Law. It is very nearly though not exactly true in gases such as air or oxygen or nitrogen or hydrogen: the deviations from it are very slight in any gas that is in conditions far removed from those which produce liquefaction.

18. **Notion of a "Perfect" Gas.** In dealing with the properties of gases and with the thermodynamics of heat-engines it is convenient to imagine a gas which exactly conforms to laws that are only very nearly true of real gases. Such a gas is called a "perfect" gas. The properties of real gases are most easily treated as small deviations from those of imaginary "perfect" gases obeying simple laws. Among real gases helium probably comes nearest to the ideal of a perfect gas, but no real gas is in this sense strictly perfect.

In a gas which is perfect in the sense of conforming exactly to Boyle's Law we should find PV strictly constant, so long as the temperature is constant. If we define the temperature scale by reference to the expansion of the gas we should also have V varying as the temperature T (reckoned from the absolute zero) under any

---

* The position of the absolute zero is uncertain to about one-tenth of a degree. Callendar made it $-273\cdot1°$ C. and that figure is adopted here. Temperatures reckoned from the absolute zero are often distinguished by the letter K. (from Lord Kelvin): thus $273\cdot1°$ K. is the zero of the Centigrade scale. [$-273\cdot16°$ C. is the figure which the International Steam-Table Conference has agreed to accept provisionally. In principle the absolute zero is unattainable, but temperatures have been reached within $0\cdot01°$ K. of it.]

constant pressure. Combining these two statements we should have

$$PV = RT \qquad \ldots\ldots\ldots\ldots\ldots\ldots\ldots(1),$$

where R is a constant.

We may write, for any gas assumed to be perfect,

$$R = P_0 V_0 / 273 \cdot 1,$$

where P_0 and V_0 are the pressure and volume at 0° C. When the volume is reckoned per unit quantity of the gas we have a definite constant value of R for each gas, depending on the units employed and on the specific density of the gas in question.

It should be noticed that when a gas satisfying this equation is heated under constant pressure and consequently expands, R is a measure of the amount of work done by the gas in this expansion for each degree through which the temperature rises. Let the original temperature of the gas be T_1 and its volume V_1 and let it be heated under constant pressure P till the temperature is T_2 and the volume V_2. Then we have $RT_1 = PV_1$ and $RT_2 = PV_2$, from which $R(T_2 - T_1) = P(V_2 - V_1)$, which is the work done by the gas in expanding from V_1 to V_2. Let the interval of temperature be 1°, then R is equal to the work done.

Thus R is numerically expressed in units of work per unit of mass and per degree: in foot-pounds per pound or in kilogrammetres per kilogramme. If we use the Centigrade degree in both cases the ratio of the number which expresses R in foot-pounds per pound to the number which expresses it in kilogrammetres per kilogramme, or in gramme-metres per gramme, is 3·28084, namely the number of feet in a metre.

According to measurements by Regnault a cubic metre of dry air, at a temperature of 0° C. and pressure of one atmosphere as defined in Art. 12, contains 1·2928 kilogrammes*. We should accordingly have for dry air, if it were "perfect,"

$$R = 1 \cdot 03267 \times 100^2 / 1 \cdot 2928 \times 273 \cdot 1 = 29 \cdot 25,$$

in kilogrammetres per kilogramme, at the latitude of London. The factor 100^2 is required to convert the pressure into kilogrammes per square metre. The same number expresses R in gramme-metres per gramme. The corresponding value of R in foot-pounds per pound is 96·0.

In this calculation air is treated as if it conformed exactly to Boyle's Law. For the present it is to be understood that the

---

* A value which agrees quite well with recent determinations.

symbol T stands for temperature measured on the scale of a gas thermometer, from a zero which is 273·1° below the melting point of ice.

In the equation $PV = RT$, R is constant for the particular gas under consideration, but it has different values for different gases, being proportional to the volume which unit quantity of each gas occupies under standard conditions of pressure and temperature: in other words it is inversely proportional to the density. Hence the product of R by the density is the same for all "perfect" gases, that is to say for all gases which satisfy the equation. A very important property of gases, as will be seen in a later chapter, is that their density is proportional to their molecular weight. For example oxygen, whose molecular weight is nearly 16 times that of hydrogen, is nearly 16 times as dense, and so on. Hence if we multiply R by the molecular weight of the gas we get a number which would be the same for all perfect gases and is very nearly the same for real gases. The number so obtained is called the Universal Gas Constant. Once it is known the value of R for any gas is readily found by dividing the universal gas constant by the molecular weight. Using English units of length and weight, the gas constant is 2780 foot-pounds. Oxygen (O_2) has a molecular weight of 32. Accordingly for oxygen we have $R = 86·9$. Similarly for dry air, the average molecular weight of which is 28·95, R is 96·0, and for hydrogen ($H_2 = 2·016$) it is 1379. All these numbers are in foot-pounds: they assume the quantity of gas to be one pound, the volume to be in cubic feet and the pressure in pounds per square foot*. We shall return to the subject of the gas constant and the properties of particular gases in Chapter VI; meanwhile we are concerned with what happens when a given quantity of any gas is made to undergo changes of pressure, volume or temperature.

19. **Internal Energy of a Gas: Joule's Law.** *The Internal Energy of a given quantity of a gas depends only on the temperature.*

This is an inference from the fact established by experiments of Joule, that *when a gas expands without doing external work and without taking in or giving out heat, and therefore without changing its stock of internal energy, its temperature does not change.*

Joule's Law is to be regarded as strictly true only of imaginary perfect gases: in any actual gas there is a slight departure from it, which is very small indeed in a nearly perfect gas such as hydrogen.

---

* The universal gas constant is 0·08206 litre atmos./deg. mole. $= 8·315 . 10^7$ erg/deg. mole. $= 1·986$ cal./deg. mole.

The law was originally established by means of the following experiment.

Joule connected a vessel containing compressed gas with another vessel which was empty, by means of a pipe with a closed stop-cock. Both vessels were immersed in a bath of water and were allowed to assume a uniform temperature. Then the stop-cock was opened, and the gas distributed itself between the two vessels, expanding without doing external work. After this the temperature of the water in the bath was found to have undergone no appreciable change. The temperature of the gas appeared unaltered, and no heat had been taken in or given out by it, and no work had been done by it.

Since the gas had neither gained nor lost heat, and had done no work, its internal energy was the same at the end as at the beginning of the experiment. The pressure and volume had changed, but the temperature had not. The conclusion follows that the internal energy of a given quantity of gas depends only on its temperature, and not upon its pressure or volume; in other words, a change of pressure and volume not associated with a change of temperature does not alter the internal energy. Hence in any change of temperature the change of internal energy is independent of the relation of pressure to volume during the operation: it depends only on the amount by which the temperature has been changed.

The apparatus used by Joule in this experiment is shown in fig. 3. The vessel A was filled with air compressed to more than 20 atmospheres, and B was exhausted. Both vessels were immersed in a bath of water. The water in the bath was stirred and the temperature noted before the stop-cock C was opened. After the gas had come to rest in the two vessels the water was again stirred, and was found to have the same temperature as before, so far as tests made by a very sensitive thermo-meter could detect.

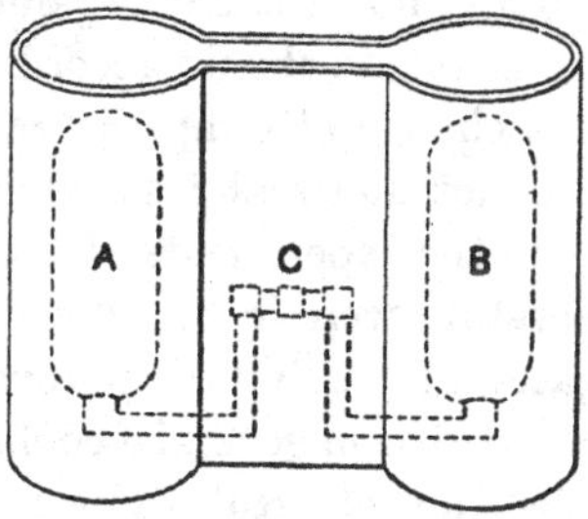

In another form of the apparatus Joule separated the bath into three portions, one portion round each of the vessels and one round the con-necting pipe. When the stop-cock was opened the water surrounding A was cooled, but this was compensated by a rise of temperature in the water

Fig. 3

surrounding B and C. The gas in A became colder in the act of expanding, but heat was given up in B and C as its eddying motion settled down, and when all was still there was neither gain nor loss of heat on the whole, so far as could be detected in this form of experiment.

It is now, however, known that a very slight change of temperature does in fact take place when a real gas expands without doing work. In later experiments by Joule and Thomson (Lord Kelvin) a more delicate method was adopted of detecting whether there is any change of internal energy when the pressure and volume change under conditions such that external work is not done. The gas was forced to pass through a porous plug by maintaining a constant high pressure on one side of the plug and a constant low pressure on the other. Care was taken to prevent any heat being gained or lost by conduction from outside. In this operation work was done upon the gas in forcing it up to the plug, and work was done by it when it passed the plug, by displacing gas under the lower pressure on the side beyond the plug. If no change of temperature took place, and if the gas conformed to Boyle's Law, these two quantities of work would be exactly equal, and consequently no external work would be done on the whole. For let P_1 be the pressure and V_1 the volume before passing the plug, and P_2 the pressure and V_2 the volume after passing the plug, the volumes being in both cases stated per pound of the gas. Then the work done upon the gas (per pound) as it approaches the plug is P_1V_1, and the work done by it as it leaves the plug is P_2V_2. If the temperature is the same on both sides these quantities are equal in a gas for which PV is constant at any one temperature. Thus a gas which is "perfect" in the sense that it conforms strictly both to Boyle's Law and to Joule's would in its passage of the plug have expanded without (on the whole) doing any work, and therefore without changing its internal energy, no heat being gained or lost. In such a gas no change of temperature should accordingly be found, as it passes the plug, and if a change of temperature is observed in any real gas it is due to the fact that real gases are not strictly "perfect."

In the experiments of Joule and Thomson* small changes of temperature were in fact detected and measured in air and other real gases, on passing the porous plug. This Joule-Thomson effect, as it is called, is in general a cooling. Observations of the Joule-Thomson effect are of great value in determining exactly the properties of

---

* See Kelvin's *Mathematical and Physical Papers*, vol. I, p. 333.

gases and vapours which are not perfect; and (as we shall see later) certain practical methods of liquefying gases under extreme cold depend upon the existence of this effect.

In the imaginary perfect gas, however, the Joule-Thomson effect is entirely absent. There is no change of temperature in passing the plug, and there is also no change of internal energy, for no work is done and (by assumption) no heat is taken in or given out.

It is important to notice that we assume the imaginary perfect gas to satisfy two conditions: it obeys Boyle's Law exactly and also Joule's Law exactly. These characteristics are independent of one another: it would be possible to have a gas satisfy one and not the other, but a gas is said to be perfect in the thermodynamic sense only when it satisfies both, and in that case certain other properties follow which will now be pointed out.

20. **Specific Heats of a Gas.** The Specific Heat of any substance means the amount of heat required per degree to raise the temperature of unit quantity of the substance, under any assumed mode of heating. Thus when a substance is heated through a small interval of temperature dT the heat taken in (per lb.) is CdT, where C is the specific heat for the particular conditions and mode of heating. In dealing with gases or other fluids two important modes of heating must be distinguished: we may heat them under conditions of constant pressure or of constant volume. We shall use the symbol C_p to represent specific heat at constant pressure, and C_v to represent specific heat at constant volume.

Consider first the operation of heating unit quantity of a perfect gas at constant volume, from temperature T_1 up to temperature T_2. The heat taken in is

$$C_v (T_2 - T_1)$$

No external work is done, for the volume (by assumption) does not change, and consequently all this heat goes to increase the stock of internal energy contained in the gas. But by Joule's Law the internal energy depends only on the temperature. Therefore if we heat the same quantity of the same gas in any other manner from T_1 to T_2, the same change of internal energy must take place.

Suppose then another manner of heating, namely at constant pressure. In that case the heat taken in is

$$C_p (T_2 - T_1).$$

During this process external work is done, because the gas expands, and its amount is

$$P (V_2 - V_1),$$

where V_1 and V_2 represent the volumes at the beginning and end of the operation respectively, and P is the pressure, which by assumption is constant. Since $PV_2 = RT_2$ and $PV_1 = RT_1$, we may write the expression for the external work in the form

$$R (T_2 - T_1).$$

This is in work units: in heat units it is

$$AR (T_2 - T_1),$$

where A is the reciprocal of Joule's equivalent (Art. 14).

The difference between the heat taken in and the work done, namely

$$(C_p - AR) (T_2 - T_1),$$

is simply an addition to the stock of internal energy. But as was pointed out above, the change of internal energy must be the same in both modes of heating, and therefore

$$C_v = C_p - AR \quad \dots\dots\dots\dots\dots\dots(2).$$

This important relation between the two specific heats in a perfect gas follows from the Laws of Boyle and of Joule.

We have here taken C_v and C_p as applying throughout a finite range of temperature from T_1 to T_2. But this range may be made infinitesimally small without affecting the argument* and in that case C_v and C_p become the specific heats at a definite temperature. The conclusion holds that for any condition of the gas

$$C_p - C_v = AR,$$

and this is true whether the specific heats are or are not independent of the temperature.

---

* Suppose the heating to be through a very small interval of temperature dT. In heating at constant volume, the heat taken in is $C_v dT$, and all of it goes to increase the internal energy by an amount dE. Hence

$$C_v dT = dE.$$

In heating at constant pressure through the same interval of temperature the heat taken in dQ does work dW and also adds to the internal energy by the amount dE. dQ is $C_p dT$; and dW is PdV, which is equal to RdT. Hence

$$C_p dT = ARdT + dE = ARdT + C_v dT.$$

From which $$C_p - C_v = AR.$$

21. Constancy of the Specific Heats in a Perfect Gas. From the above result it follows that if either of the two specific heats is constant the other must also be constant. To be constant the specific heat has to be independent both of the pressure and of the temperature.

First as to independence of pressure: we have seen (Art. 19) that the internal energy of a perfect gas depends only on the temperature and is independent of the pressure. If we heat a unit quantity of a perfect gas through 1° the change of internal energy is measured (Art. 20) by C_v, no matter what is the pressure. Hence C_v is independent of the pressure; and since, by equation (2), C_p is equal to $C_v + AR$, it follows that C_p also must be independent of the pressure.

But a gas may conform to the Laws of Boyle and Joule without having C_p and C_v constant at all temperatures, and in order completely to define our ideal perfect gas we make the further assumption that its specific heat does not change with the temperature. We shall see later that in many real gases there is a distinct increase of specific heat at high temperatures, but it is useful for the purpose of thermodynamical reasoning to imagine a gas in which the specific heat is strictly constant. Accordingly we assume this constancy as a third characteristic of the ideal perfect gas, additional to the two already specified in Arts. 18 and 19. It does not in any way conflict with these: each of the three characteristics is independent of the others.

The imaginary perfect gas which satisfies these three assumptions is no more than a convenient scaffolding by help of which we may most easily build up the theory of thermodynamics. In what follows we shall make temporary use of the ideal gas so defined as the supposed working substance of an ideal heat-engine, and thereby arrive at conclusions which, as will be shown, are of quite general validity.

22. Reversible Actions. We have now to consider particular modes in which a working substance may be expanded or compressed and may take in or give out heat, and at the outset it is important to distinguish between actions that are *reversible* and those that are irreversible.

An expansion or compression is reversible if it is carried out in such a manner that the operation can be reversed, with the result that the substance will pass back through all the stages through which it has passed during the expansion or compression and be in the same condition in all respects at each corresponding stage in both processes.

This implies that the substance must expand smoothly, without setting up any motions within itself of a kind such that their kinetic energy is frittered down into heat through internal friction. The whirls and eddies which occur as a fluid enters or expands in the cylinder of an engine are irreversible, and in ideal reversible expansion we must suppose them absent. Reversible expansion implies that there are no losses of mechanical effect from any sort of internal friction. It excludes throttling, such as occurs when a substance expands through a valve or other constricted opening into a region of lower pressure where the kinetic energy of the stream and eddies is dissipated. In such cases the motion of the stream and eddies cannot be reversed. To get the substance back to the region of higher pressure would require an expenditure of more work than was done by it during its expansion, and if we were to force it back we should find it had gained heat through the subsidence of the internal eddying motions, though no heat had come in from outside.

The kind of expansion which takes place in Joule's experiment (Art. 19) is an extreme instance of irreversible expansion.

A transfer of heat to or from any substance is reversible only if the substance is at the same temperature as the body from which it is taking heat or to which it is giving heat. Suppose, for instance, that a substance is taking in heat from a hot source and is expanding as it does so. The expansion may be reversible in itself, that is to say it may involve no internal friction, but unless the temperature of the substance be the same as that of the source, the operation as a whole—considered in its relation to the source—cannot be reversed. So considered it is reversible only when the further condition is fulfilled that compression of the substance will reverse the transfer of heat, giving back to the source the heat that was taken from it. Any thermal contact between bodies at different temperatures involves an irreversible transfer of heat.

Neither the expansions and compressions nor the transfers of heat that occur in a real engine are ever strictly reversible, some of them indeed are far from being reversible. But the study of an ideal engine, in which all the operations are reversible, is of fundamental importance in the science of thermodynamics, and it furnishes a basis for the critical analysis of actions in a real engine.

23. Adiabatic Expansion. There are two specially important kinds of reversible expansion, (1) Adiabatic and (2) Isothermal.

Adiabatic expansion or compression means expansion or compression, carried out reversibly, in which no heat is allowed to enter or leave the substance. A curve drawn to show the relation of pressure to volume during the process is called an adiabatic line. Adiabatic action would be realized if we had a substance expanding, or being compressed, without change of chemical state, and without any eddying motions, in a cylinder which (along with the piston) was totally impervious to heat.

From this definition it follows that the work which a substance does while it is expanding adiabatically is all done at the expense of its stock of internal energy; and the work which is spent upon a substance when it is being compressed adiabatically all goes to increase its stock of internal energy.

In actual heat-engines the action is never strictly adiabatic, for there are always some exchanges of heat between the working substance and the surface of the cylinder and piston. Very rapid compression or expansion may come near to being adiabatic by giving little time for any transfer of heat to occur.

After what has been said already about reversibility, it is scarcely necessary to add that expansion through a throttle-valve is not adiabatic, though it may be (and generally is) done without letting heat enter or leave the substance.

In the adiabatic expansion of any substance work is done, and since no heat is taken in or given out, there must be a decrease of internal energy equivalent to the amount of the work done by the substance.

Taking the general equation (Art. 9)

$$dQ = A\,dW + dE,$$

which applies to any small change of state on the part of any substance, we have $dQ = 0$ when the action is adiabatic, and hence for an adiabatic expansion

$$A\,dW = -dE.$$

Here dW is the work done, A is the factor required to convert an expression for work into heat units (Art. 14), and dE is the change of internal energy.

24. Isothermal Expansion. Isothermal expansion or compression means expansion or compression carried out reversibly (as regards internal action) and without change of temperature.

A curve drawn to show the relation of pressure to volume during isothermal expansion or compression is called an isothermal line.

When a substance is expanding isothermally it takes in heat to maintain its temperature constant; it therefore must be in contact with a source of heat. When it is being compressed isothermally it gives out heat, and must be in contact with a receiver which can take heat from it.

25. Adiabatic Expansion of a Perfect Gas. Consider next the behaviour of a perfect gas during adiabatic expansion or compression. We have seen that in a small adiabatic expansion of any substance (Art. 23)

$$dE = -AdW = -APdV.$$

In a perfect gas $dE = C_v dT$ (Art. 20). Hence in the adiabatic expansion of a perfect gas

$$APdV = -C_v dT.$$

But $P = RT/V$ (Art. 18). Hence

$$ARTdV/V + C_v dT = 0,$$

or, dividing by T, $ARdV/V + C_v dT/T = 0,$

which gives on integration

$$AR \log_e V + C_v \log_e T = \text{constant} \quad \ldots\ldots\ldots\ldots(3).$$

Writing $C_p - C_v$ for AR (Art. 20), and dividing by C_v, which is constant (Art. 21),

$$(C_p/C_v - 1) \log_e V + \log_e T = \text{constant}.$$

We shall write γ for the ratio of the two specific heats, namely C_p/C_v.

Thus we have

$$\gamma \log_e V - \log_e V + \log_e T = \text{constant} \quad \ldots\ldots\ldots\ldots(4).$$

Further, since PV/T is constant,

$$\log_e P + \log_e V - \log_e T = \text{constant}.$$

Adding these two equations

$$\log_e P + \gamma \log_e V = \text{constant} \quad \ldots\ldots\ldots\ldots(5),$$

which gives $PV^\gamma = \text{constant} \quad \ldots\ldots\ldots\ldots\ldots(6)$

as the equation of any adiabatic line in the pressure-volume diagram, for the adiabatic expansion of a perfect gas*.

26. Change of Temperature in the Adiabatic Expansion of a Perfect Gas. When a gas is expanding adiabatically its stock of internal energy is, as we have seen, being reduced, and hence its temperature falls, the change of internal energy being proportional to the change of temperature (Art. 20). Conversely, in adiabatic compression the temperature rises. The amount by which the temperature is changed (in a perfect gas) may be found by combining the equations

$$P_1 V_1{}^\gamma = P_2 V_2{}^\gamma \text{ and } P_1 V./P_2 V_2 = T_1/T_2.$$

Multiplying them together we have

$$\frac{T_1}{T_2} = \frac{P_1 V_1 P_2 V_2{}^\gamma}{P_2 V_2 P_1 V_1{}^\gamma},$$

whence
$$\frac{T_1}{T_2} = \left(\frac{V_2}{V_1}\right)^{\gamma-1}, \text{ or } T_2 = T_1 \left(\frac{V_1}{V_2}\right)^{\gamma-1}$$

This result of course applies to compression as well as to expansion along an adiabatic line. It may be got directly from equation (4), which can be written $\log_e T + (\gamma - 1) \log_e V = \text{constant}$; whence

$$T V^{\gamma-1} = \text{constant} \quad\quad\dots\dots\dots\dots\dots\dots(7).$$

Combining equations (6) and (7) and eliminating V, we obtain the further relation $T/P^{\frac{\gamma-1}{\gamma}} = \text{constant}$.

27. Work done in the Adiabatic Expansion of a Perfect Gas. In any kind of expansion of any fluid the work done in expanding from volume V_1 to volume V_2 is

$$W = \int_{V_1}^{V_2} P dV.$$

If the nature of the expansion be such that PV^n is constant, n being any index, then P at any point when the volume is V is $P_1 V_1{}^n/V^n$, P_1 and V_1 being the pressure and volume in the initial state. In this case, for expansion from V_1 to V_2,

$$W = P_1 V_1{}^n \int_{V_1}^{V_2} dV/V^n,$$

* Log$_e$, the "hyperbolic" or "Napierian" or "natural" logarithm of any number, is 2·3026 times the common logarithm of the number.

which gives on integration

$$W = P_1 V_1^n (V_2^{1-n} - V_1^{1-n})/(1-n),$$

or
$$W = \frac{P_1 V_1 - P_2 V_2}{n-1} \quad \dots\dots\dots\dots\dots\dots\dots(8).$$

So far we have made no assumption as to the nature of the fluid.

Apply this result to a gas expanding adiabatically, for which the index n is equal to γ (by Eq. (6), Art. 25). We then have

$$W = \frac{P_1 V_1 - P_2 V_2}{\gamma-1} = \frac{R(T_1 - T_2)}{\gamma-1} \quad \dots\dots\dots\dots(9),$$

since
$$P_1 V_1 = RT_1 \text{ and } P_2 V_2 = RT_2.$$

Further, it follows from Art. 23 that this expression (multiplied by A) is the decrease of internal energy produced by the expansion. It is also the increase of internal energy in adiabatic compression from V_2 to V_1.

28. Isothermal Expansion of a Perfect Gas. In a gas which satisfies the equation $PV = RT$, PV is constant during isothermal expansion or compression, and any isothermal line on the pressure-volume diagram is a rectangular hyperbola, the pressure varying inversely as the volume.

To find the work done in the isothermal expansion of a gas from V_1 to V_2 we have

$$W = \int_{V_1}^{V_2} P \, dV$$

and
$$P = P_1 V_1 / V,$$

from which
$$W = P_1 V_1 \int_{V_1}^{V_2} \frac{dV}{V}.$$

Integrating,
$$W = P_1 V_1 (\log_e V_2 - \log_e V_1),$$

or
$$W = P_1 V_1 \log_e \frac{V_2}{V_1}.$$

Instead of writing $P_1 V_1$ we may write PV, since the product of P and V is constant throughout the process, and again, since $PV = RT$,

$$W = RT \log_e \frac{V_2}{V_1} \quad \dots\dots\dots\dots\dots(10),$$

where T is the temperature at which the process takes place. This expression serves to give either the work that is done by a gas in isothermal expansion, or the work that is spent upon it in isothermal compression.

During the isothermal expansion or compression of a perfect gas there is no change of internal energy, since there is no change of temperature and the internal energy depends only on the temperature (Art. 19). Hence during isothermal expansion a perfect gas must take in an amount of heat equivalent to the work it does, namely $ART \log_e V_2/V_1$, and during isothermal compression from V_2 to V_1 it must give out that amount of heat.

29. Summary of Results for a Perfect Gas. It may be convenient at this point to collect the results that have been found for actions occurring in perfect gases.

It is assumed that the gas satisfies Boyle's Law (Art. 17) and Joule's Law (Art. 19) and that the specific heat (at constant pressure) is independent of the temperature. Further, the temperature is measured on the scale furnished by the expansion of the gas itself. Under these conditions we have the following results:

$$PV = RT,$$

where R is a constant depending on the specific density of the gas;

$$C_p - C_v = AR,$$

where C_p is the specific heat at constant pressure, C_v the specific heat at constant volume and A is the reciprocal of Joule's equivalent. C_p and C_v are both constant.

In adiabatic expansion:

$$PV^\gamma = \text{constant, or } P_1/P_2 = (V_2/V_1)^\gamma,$$

where γ is C_p/C_v.

$$TV^{\gamma-1} = \text{constant, or } T_1/T_2 = (V_2/V_1)^{\gamma-1}.$$

$$T/P^{\frac{\gamma-1}{\gamma}} = \text{constant, or } T_1/T_2 = (P_1/P_2)^{\frac{\gamma-1}{\gamma}}.$$

Heat taken in $= 0$.

$$\text{Work done} = \frac{R(T_1 - T_2)}{\gamma - 1} = \frac{P_1 V_1 - P_2 V_2}{\gamma - 1}.$$

$$\text{Decrease of Internal Energy} = \frac{AR(T_1 - T_2)}{\gamma - 1}.$$

In isothermal expansion:

$$PV = \text{constant, since } T = \text{constant.}$$

$$\text{Heat taken in} = ART \log_e \frac{V_2}{V_1}.$$

$$\text{Work done} = RT \log_e \frac{V_2}{V_1}.$$

Change of Internal Energy $= 0$.

30. Fundamental Questions of Heat-Engine Efficiency. We are now in a position to deal with the most fundamental questions of heat-engine efficiency, which may be stated in the following terms:

(1) Having given a source from which heat may be taken in at a high temperature, and a sink or receiver to which heat may be rejected at a lower temperature, how may heat taken from the source be utilized to the best advantage for the purpose of producing mechanical effect? In other words, how may the greatest amount of work be done by each unit of heat taken from the hot source?

(2) What fraction of the heat taken from the hot source is it theoretically possible to convert into work? In other words, what is the limiting *efficiency* of conversion?

31. The Second Law of Thermodynamics. So far as the First Law of Thermodynamics (Art. 8) goes, it is not obvious that there is anything to prevent all the heat which the source can supply from being converted into work. But it will presently be seen that a limit is imposed as a consequence of the following principle, which is known as the *Second Law of Thermodynamics*:

It is impossible for a self-acting machine, unaided by any external agency, to convey heat from one body to another at a higher temperature.

The Second Law says, in effect, that heat will not pass up automatically from a colder to a hotter body. We can force it to pass up, as in the action of a refrigerating machine, but only by applying an "external agency" to drive the machine. A heat-engine acts by letting heat pass down from a hotter to a colder body, not of course by direct conduction from one to the other, for that is a mode of transfer in which the heat would do no work, but by making the working substance alternately take in heat from the hot body and reject heat to the cold body, and thereby undergo expansions and contractions in which its pressure is on the whole greater during expansion than during contraction, with the result that a part of the heat that is passing down through the engine is converted into work. In consequence of the Second Law it is only a certain fraction of the whole heat supplied by the hot body that can be converted into work by any such process.

32. Reversible Heat-Engine. Carnot's Cycle of Operations.

To the first of the above two questions (Art. 30) a correct answer was given by Sadi Carnot in a remarkable essay, published in 1824, entitled *Réflexions sur la puissance motrice du feu et sur les machines propres à développer cette puissance.* In this essay Carnot may be said to have laid the foundation of thermodynamics. He pointed out that the greatest possible amount of work was to be obtained by letting the heat pass from the source to the receiver through an engine *working in a strictly reversible manner* not only as regards its own internal actions but also as regards the transfer of heat to it from the source and from it to the receiver. The engine conceived by Carnot is an engine every one of whose operations is reversible in the sense explained in Art. 22. He further showed how an engine might (theoretically) satisfy this condition by having its working substance go through these four reversible operations:

(1) Isothermal expansion at the temperature of the hot source (T_1). During this operation heat is taken in reversibly from the hot source.

(2) Adiabatic expansion, during which the temperature of the working substance falls from T_1 to T_2 (the temperature of the receiver).

(3) Isothermal compression at the temperature of the receiver. During this operation heat is rejected reversibly to the receiver.

(4) Adiabatic compression, by which the temperature of the working substance is raised from T_2 to T_1. This completes the cycle by bringing the substance back to the condition in which it was assumed to be at the beginning of the first operation.

In the cycle as a whole work is done by the substance: the average pressure in (1) and (2) being greater than in (3) and (4).

This cycle of operations, which is known as *Carnot's Cycle*, is entirely reversible. The working substance might be forced to go through it in the reversed direction, taking in heat from the cold body and giving out heat to the hot body. The transfers of heat would be exactly reversed, and at every stage the pressure and volume and temperature of the substance would be the same as when working direct. The work spent upon it would be equal to the work got from it in the direct action. Carnot's ideal engine accordingly affords a strictly reversible means of letting heat down from the hot source to the cold receiver.

The argument by which Carnot proved that no heat-engine can utilize heat more completely than a reversible heat-engine utilizes it, in letting heat down from a given source to a given receiver, is substantially as follows.

33. **Carnot's Principle.** To prove that no other heat-engine, working between the same source and receiver of heat, can do the same amount of mechanical work as a reversible engine by taking in a smaller quantity of heat.

Suppose there are two heat-engines R and S, one of which (R) is reversible, working between the same hot body or source of heat and cold body or receiver of heat, and each producing the same amount of mechanical work. Let Q be the quantity of heat which R takes in from the hot body. Now if R be reversed it will by the expenditure on it of the same amount of work give to the hot body the amount of heat it formerly took from it, namely Q. For this purpose set the engine S to drive R reversed. The work which S produces is just sufficient to drive R, and the two machines (S driving R) form together a self-acting machine unaided by any external agency. One of the two, namely S, takes heat from the hot body and the other, R, which is reversible, gives back to the hot body the amount of heat Q. Now if S could do its work by taking less heat than Q from the hot body, *the hot body would on the whole gain heat.* No work is being done on the system from outside, nor is any heat supplied from other sources, so whatever heat the hot body gains must come from the cold body. Therefore if S could do as much work as the reversible engine R, with a smaller supply of heat, we should be able to arrange a purely self-acting machine through which heat would continuously pass up from a cold body to a hot body. This would be a violation of the Second Law of Thermodynamics*.

* The Second Law of Thermodynamics was not formulated till 1850: Carnot's elegant argument however led to the same conclusion as deduced in the above paragraph. The two engines operated in the way described provide heat to the source and therefore work available for other purposes, which means that they together form a machine capable of perpetual motion, the possibility of which Carnot was prepared to deny on the basis of experience.

The conclusion is that S cannot do the same amount of work with a smaller supply of heat than a reversible engine; or, in modern language, that no other engine can be more *efficient* than a reversible engine, when they both work between the same two temperatures in source and receiver.

Further, let both engines be reversible. Then the same argument shows that each cannot be more efficient than the other. Hence all reversible engines taking in and rejecting heat at the same two temperatures are equally efficient.

34. **Reversibility the Criterion of Perfection in a Heat-Engine.** These results imply that, in the thermodynamic sense, reversibility is the criterion of what may be called perfection in a heat-engine. A reversible heat-engine is perfect in the sense that it cannot be improved on as regards efficiency: no other engine taking in and rejecting heat at the same two temperatures can obtain from the heat taken in a greater proportion of mechanical effect. Moreover, if this criterion be satisfied, it is, as regards efficiency, a matter of complete indifference what is the nature of the working substance or what, in other respects, is the mode of the engine's action.

It is, therefore, a complete answer to the first question in Art. 30 to say that the greatest amount of work that is theoretically possible will be done by each unit of heat if the heat is supplied to an engine which works in such a way that every one of its operations is reversible.

35. **Efficiency of a Reversible Heat-Engine.** The second question in Art. 30 could not be answered by Carnot because in his time the doctrine of the Conservation of Energy was unknown, and it was not recognized that part of the heat disappears, as heat, in passing through the engine. Carnot realized that work is done by an engine through the agency of heat, but he did not know that it is done by the conversion of heat. It is remarkable that he was nevertheless able to conceive the idea of a reversible engine and to see that it is the most effective possible means of getting work done through the agency of heat. His argument as to this is perfectly valid though it makes no use of the First Law of Thermodynamics. It is moreover extraordinarily general. There is no assumption in it as to the properties of any substance, nor as to the nature of heat, nor as to the way in which temperature is to be measured. All that he assumes about the temperatures of the source and the receiver is that one is hotter than the other. The

argument stands by itself, and the whole passage in which it is reproduced here (Art. 33) does not involve a reference to any of the results stated in earlier Articles.

But for the purpose of answering the second question of Art. 30 we shall in the first place deal with one particular reversible heat-engine, namely a reversible engine which has a perfect gas for working substance, and shall calculate its efficiency with the help of the results previously obtained for perfect gases. It will be easy to go on from that to find a general answer to the question, What is the limiting efficiency of any heat-engine?

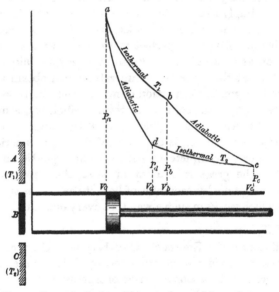

Fig. 4. Carnot's Cycle, with a gas for working substance.

36. Carnot's Cycle with a Perfect Gas for Working Substance. Consider then an ideal engine in which a substance may go through the operations of Carnot's Cycle (fig. 4). Imagine a cylinder and piston composed of perfectly non-conducting material, except as regards the bottom of the cylinder, which is a conductor. Imagine also a hot body or indefinitely capacious source of heat A, kept always at a temperature T_1, also a perfectly non-conducting cover B, and a cold body or indefinitely capacious receiver of heat C, kept always at some temperature T_2 which is lower than T_1. It is supposed that A or B or C can be applied at will to the bottom of the cylinder. Let the cylinder contain 1 lb. of a perfect gas, at temperature T_1, volume V_a, and pressure P_a to begin with. The suffixes refer to the points on the indicator diagram, fig. 4. There are four successive operations:

(1) Apply A, and allow the piston to advance slowly through any convenient distance. The gas expands isothermally at T_1, taking in heat from the hot source A and doing work. The pressure changes to P_b and the volume to V_b. The line ab on the indicator diagram represents this operation.

(2) Remove A and apply B. Allow the piston to go on advancing. The gas expands adiabatically, doing work at the expense of its internal energy, and the temperature falls. Let this go on until the temperature is T_2. The pressure is then P_c, and the volume V_c. This operation is represented by the line bc.

(3) Remove B and apply C. Force the piston back slowly. The gas is compressed isothermally at T_2, since the smallest increase of temperature above T_2 causes heat to pass into C. Work is spent upon the gas, and heat is rejected to the cold receiver C. Let this be continued until a certain point d is reached, such that the fourth operation will complete the cycle.

(4) Remove C and apply B. Continue the compression, which is now adiabatic. The pressure and temperature rise, and, if the point d has been properly chosen, when the pressure is restored to its original value P_a, the temperature will also have risen to its original value T_1. [In other words, the third operation cd must be stopped when a point d is reached such that an adiabatic line drawn through d will pass through a.] This completes the cycle.

To find the proper place at which to stop the third operation, we have (by Art. 26), for the cooling during the adiabatic expansion in stage (2),

$$(V_c/V_b)^{\gamma-1} = T_b/T_c = T_1/T_2,$$

and also, for the heating during the adiabatic compression in stage (4),

$$(V_d/V_a)^{\gamma-1} = T_a/T_d = T_1/T_2.$$

Hence $V_c/V_b = V_d/V_a,$

and therefore also $V_c/V_d = V_b/V_a.$

That is to say, the ratio of isothermal compression in the third stage of the cycle is to be made equal to the ratio of isothermal expansion in the first stage, in order that an adiabatic line through d shall complete the cycle. For brevity we shall denote either of these last ratios (of isothermal expansion and compression) by r.

The following are the transfers of heat to and from the working gas, in the four successive stages of the cycle; quantities of heat are here expressed in work units:

(1) Heat taken in from $A = RT_1 \log_e r$ (by Art. 28).

(2) No heat taken in or rejected.

(3) Heat rejected to $C = RT_2 \log_e r$ (by Art. 28).

(4) No heat taken in or rejected.

Hence, the net amount of external work done by the gas, being the excess of the heat taken in above the heat rejected in a complete cycle, is

$$R (T_1 - T_2) \log_e r;$$

this is the area enclosed by the four curves in the figure.

The *Efficiency* in this cycle, namely the fraction

$$\frac{\text{Heat converted into work}}{\text{Heat taken in}},$$

is accordingly

$$\frac{R (T_1 - T_2) \log_e r}{RT_1 \log_e r} = \frac{T_1 - T_2}{T_1}.$$

Another way of stating the result is to say that if we write Q_1 for the heat taken in from the hot source, and Q_2 for the heat rejected to the cold receiver, then from (1) and (3) above

$$Q_1/T_1 = Q_2/T_2.$$

In these expressions the temperatures T_1 and T_2 are understood to be measured on the scale of a perfect gas thermometer, and from the absolute zero.

37. Reversal of this Cycle. This cycle, being a Carnot cycle, is reversible. To realize the fact more fully we may consider in detail what will happen if we make the imaginary engine work backwards, forcing it to trace out the same indicator diagram in the opposite order. For this purpose we must expend work upon it from some other source of work. Starting as before from the point a (fig. 4) and with the gas at T_1, we shall require the following four operations:

(1) Apply B and allow the piston to advance. The gas expands adiabatically, the curve traced is ad, and when d is reached the temperature has fallen to T_2.

(2) Remove B and apply C. Allow the piston to go on advancing. The gas expands isothermally at T_2, taking in heat from C, and the curve dc is traced.

(3) Remove C and apply B. Compress the gas. The process is adiabatic. The curve traced is cb, and when b is reached the temperature has risen to T_1.

(4) Remove B and apply A. Continue the compression, which is now isothermal at T_1. Heat is now rejected to A, and the cycle is completed by the curve ba.

In this process the engine is not on the whole doing work; on the contrary, a quantity of work is spent upon it equal to the area of the diagram, or $R(T_1 - T_2) \log_e r$, and this work is converted into heat. Heat is taken in from C in the first operation, to the amount $RT_2 \log_e r$. Heat is rejected to A in the fourth operation, to the amount $RT_1 \log_e r$. In the first and third operations there is no transfer of heat. The machine is acting as a heat-pump.

The action is now in every respect the reverse of what it was before. The substance is in the same condition at corresponding stages in the two processes. The same work is now spent upon the engine as was formerly done by it. The same amount of heat is now given to the hot body A as was formerly taken from it. The same amount of heat is now taken from the cold body C as was formerly given to it. This will be seen by the following scheme:

Carnot's Cycle with a perfect gas. Direct.

Work done by the gas $= R(T_1 - T_2) \log_e r$;
Heat taken from $A = RT_1 \log_e r$;
Heat rejected to $C = RT_2 \log_e r$.

Carnot's Cycle with a perfect gas. Reversed.

Work spent upon the gas $= R(T_1 - T_2) \log_e r$;
Heat rejected to $A = RT_1 \log_e r$;
Heat taken from $C = RT_2 \log_e r$.

In the second case the heat rejected to the hot body is equal to the sum of the heat taken in from the cold body and the work spent on the substance. This of course follows from the principle of the Conservation of Energy.

38. **Efficiency of Any Reversible Engine.** The imaginary engine, then, of Art. 36 is reversible. Its efficiency, as we have seen, is

$$\frac{T_1 - T_2}{T_1},$$

where T_1 is the temperature of the source from which it takes heat and T_2 is the temperature of the receiver to which it rejects heat. But we saw, by Art. 33, that all reversible heat-engines taking in and rejecting heat at the same two temperatures are equally efficient. Hence the expression

$$\frac{T_1 - T_2}{T_1}$$

measures the efficiency of *any* reversible heat-engine, and therefore (by Art. 33) also expresses the largest fraction of the heat supplied that can possibly be converted into work by any engine whatever operating between these limits.

In other words, if we have a supply of heat at a temperature T_1, and a means of getting rid of heat at a temperature T_2, then there is no possibility of converting more than that fraction of the heat into work. This is the measure of *perfect efficiency*: it is the theoretical limit beyond which the efficiency of a heat-engine cannot go. No engine can conceivably surpass this standard, and as a matter of fact any real engine falls short of it, because no real engine is strictly reversible.

39. **Summary of the Argument.** Briefly recapitulated the steps of the argument by which we have reached this immensely important result are as follows. Following Carnot, we considered how any heat-engine works by taking in heat from a hot source and rejecting heat to a cold receiver, and established (by means of the *reductio ad absurdum* of a hypothesis which would conflict with the Second Law of Thermodynamics) the conclusion that no engine could do this more efficiently than a reversible engine does, that is to say, an engine which goes through a reversible cycle of operations. This led to the inference that all reversible engines working between the same temperatures of source and receiver were equally efficient, and consequently that an expression for the efficiency of any one of them would apply to all, and would mean the highest efficiency that is theoretically possible. Still following Carnot, we imagined a cycle which would be reversible, consisting of four stages, namely (1) isothermal expansion during which heat is taken in from the source, (2) adiabatic expansion during which the temperature of the substance falls from the temperature of the source to that of the receiver, (3) isothermal compression during which heat is rejected to the receiver, (4) adiabatic compression during which the temperature of the substance rises again to that

of the source. Up to this point there had been no assumption as
to the use of any particular working substance. We next enquired
what would happen in this cycle if a perfect gas were used as
working substance. Taking for the scale of temperature a scale based
on the expansion of a perfect gas*, and expressing on this scale
the temperatures of source and receiver as T_1 and T_2 respectively,
we found that a reversible engine, using a perfect gas for working
substance, has an efficiency of

$$\frac{T_1 - T_2}{T_1}.$$

Hence it was concluded that this expression measures the efficiency
of any reversible engine working between these limits, and that this
is the highest efficiency theoretically obtainable in any heat-engine.

This general conclusion may also be stated, with equal generality
(for any reversible engine), in the form

$$Q_1/T_1 = Q_2/T_2,$$

where Q_1 is the heat taken in by the engine from the source at T_1,
and Q_2 is the heat rejected by it to the receiver at T_2.

The efficiency of any heat-engine may be written

$$\frac{Q_1 - Q_2}{Q_1} \text{ or } 1 - \frac{Q_2}{Q_1},$$

whether the engine be reversible or not. In a reversible engine,
or, as we may now call it, a thermodynamically perfect engine,
this becomes

$$1 - T_2/T_1.$$

In an engine which falls short of reversibility a smaller fraction
of the heat supply is converted into work and the heat rejected is
relatively larger; Q_2/T_2 is greater than Q_1/T_1.

40. Absolute Zero of Temperature. The zero from which
T_1 and T_2 are measured is the zero of the gas thermometer, which
was defined (Art. 16) as the temperature at which the volume of the
gas would vanish if the same law of expansion continued to apply.
But we can now give it another meaning. Taking the expression
for the efficiency of a reversible heat-engine

$$1 - T_2/T_1,$$

* That is to say, a scale in which the temperature is proportional to the
volume of the gas, when the pressure is kept constant.

we see that if the cold receiver were at the temperature of the absolute zero (so that $T_2 = 0$) the efficiency would be equal to 1: in other words, all the heat supplied to the engine would be converted into work. It is clearly impossible to imagine a receiver colder than that, for it would make the efficiency greater than 1 and thereby violate the First Law of Thermodynamics by making the amount of work done greater than the heat supplied. Hence the zero which we found on the gas scale is also an absolute thermodynamic zero, a temperature so low that it is inconceivable on thermodynamic grounds that there can be any lower temperature. The term "absolute zero" has consequently acquired a new meaning: without reference to the properties of any substance we see that it represents a limit below which temperature cannot go. This justifies the use of the word "absolute" as applied to a zero of temperature.

41. Conditions of Maximum Efficiency. From the above result it will be obvious that the availability of heat for transformation into work depends essentially on the range of temperature through which the heat is let down, from that of the hot source to that of the cold body into which heat is rejected; it is only in virtue of a difference of temperature between bodies that conversion of any part of their heat into work becomes possible. No mechanical effect could be produced from heat, however great the amount of heat present, if all bodies were at a dead level of temperature. Again, it is impossible to convert the whole of any supply of heat into work, because it is impossible to have a body at the absolute zero of temperature as the sink into which heat is rejected.

If T_1 and T_2 are given as the highest and lowest temperatures of the range through which a heat-engine is to work, it is clear that the maximum of efficiency can be reached only when the engine takes in all its heat at T_1 and rejects at T_2 all that is rejected. With respect to every portion of heat supplied to the engine the greatest ideal efficiency is

$$\frac{\text{Temperature of reception} - \text{temperature of rejection}}{\text{Temperature of reception}}.$$

Any heat taken in at a temperature below T_1, or rejected at a temperature above T_2, will be less capable of conversion into work than if it had been taken in at T_1 and rejected at T_2, and hence, with a given pair of limiting temperatures, it is essential to maximum efficiency that no heat be taken in by the engine except at the top of the range, and no heat rejected except at the bottom of the range. Further, as we have seen in Art. 33, when the temperatures at which heat is received and rejected are assigned, an engine attains the maximum of efficiency if it be reversible.

It may be useful to repeat here that in the transformation into work of heat supplied from a given source, the condition of reversibility is satisfied in the whole operation from source to receiver if (1) no part of the working substance is brought into contact during the operation with any body at a sensibly different temperature, and (2) there is no dissipation of energy through internal friction. The first condition excludes any unutilized drop in temperature; the second excludes eddying motions and such like sources of waste, which arise in consequence of expansion through throttle-valves or constricted orifices, or in consequence of any cause that sets up dissipative motions within the substance. In a piston and cylinder engine we have to think of the substance as expanding by the gradual displacement of the piston, doing work upon it, and not wasting energy to any sensible extent by setting portions of itself into motion. There are to be no local variations of pressure within the cylinder, such as might occur in a fast-running engine through the inertia of the expanding fluid.

When we proceed to deal in a later chapter with steam jets in relation to steam turbines, we shall see that it is possible to have (in theory) a reversible action, though the work done by the substance in expanding is employed to give kinetic energy to the substance itself as a whole by forming a jet, because in that case the energy of the jet is recoverable when proper care is taken to control the formation of the jet. But the eddying motions spoken of here are of a different class: their energy is irrecoverable and for that reason they violate the condition of reversibility.

It may also be worth while to repeat here that no real heat-engine can work between the source and the receiver in a strictly reversible manner. It cannot wholly escape eddying motions: it cannot wholly escape transfers of heat between the working substance and bodies at other temperatures. In particular, since the working substance must in practice take in heat at a reasonable rate from the hot source, the source is usually much hotter than the substance while heat is being taken in. This is, in practice, the most serious breach of reversibility in the transformation of heat by a steam-engine. It means that between the temperature of the source and the highest temperature reached by the working substance in its cycle of operations, there is a wasteful drop, a drop that is not utilized thermodynamically. If it were practicable in the steam-engine to avoid the drop between the temperature of the furnace gases and that of the water in the boiler a greatly increased efficiency of conversion would be attainable.

If we leave this drop out of account, and take for the upper limit T_1, not the temperature of the furnace gases but the temperature in the boiler, and if we also take for T_2 the temperature in the condenser, the fraction

$$\frac{T_1 - T_2}{T_1}$$

will measure the greatest fraction of the heat supplied to the boiler that can be converted into work, under ideally favourable (in other words, strictly reversible) conditions between the boiler and the condenser. The performance of any real engine falls short of this because it includes irreversible features, the chief of which are throttling actions in the steam-passages and exchanges of heat between the steam and the metal of the cylinder and piston. But although this limit of efficiency cannot be actually reached, it affords a valuable criterion with which to compare the performance of any real engine, and establishes an ideal for engine designers to aim at.

It is important to realize that a substance may expand reversibly although it is taking in heat from a source hotter than itself: in other words, there may be an irreversible drop of heat between the source and the substance, but no irreversible action within the substance. Thus the fluid in a boiler is at a definite temperature lower than that of the furnace while it is taking in heat from the furnace; there is accordingly an irreversible drop in this transfer of heat: but the formation and expansion of the steam may go on in a reversible manner. We can imagine all the *internal* actions of the working substance to be reversible, although as regards transfers of heat from the source or to the receiver there is not reversibility. In that event the engine will still work as efficiently as possible *between its own limits of temperature*, namely the limits at which the substance takes in and rejects heat, though it is no longer the most efficient possible contrivance for utilizing the full range of temperature from source to receiver.

Thus if we interpret T_1 and T_2 as the limits of temperature of the working substance itself—without any reference to a source or a receiver—T_1 being the temperature of the substance while it is taking in heat, and T_2 the temperature of the substance while it is rejecting heat, and if the internal actions of the substance are reversible, then $(T_1 - T_2)/T_1$ still measures the efficiency of the engine. This fraction still expresses the greatest efficiency that is theoretically possible in any heat-engine working between the limits T_1 and T_2.

When we speak of a substance as taking in heat at a stated temperature, or rejecting heat at a stated temperature, it is to be understood that the temperature of the substance itself is meant, though that may not be the temperature of the source or receiver; and when we speak of a substance as expanding or being compressed in a reversible manner we do not imply that it may not be taking in heat from a source hotter than itself or rejecting heat to a receiver colder than itself. A cycle of operations may be internally reversible, that is to say, reversible so far as actions within the working substance are concerned, although it happens to be associated with an irreversible transfer of heat to the working substance from the source or from the working substance to the receiver*.

42. Thermodynamic Scale of Temperature. Reference was made in Art. 15 to the fact (first pointed out by Lord Kelvin†) that thermodynamic principles allow a scale of temperature to be defined which is independent of the properties of any particular substance, real or imaginary. Up to the present we have based the scale on the properties of a perfect gas, taking a scale in which the degrees (or equal intervals of temperature) correspond to equal amounts of expansion on the part of a perfect gas kept at constant pressure. Using this scale we have seen that a reversible engine which works between the limits T_1 and T_2, and takes in any quantity of heat Q_1 at T_1, rejects at T_2 a quantity Q_2 equal to $Q_1 T_2/T_1$, and has an efficiency equal to $(T_1 - T_2)/T_1$.

Now imagine that the heat Q_2, which is rejected by this engine, forms the supply of a second reversible engine taking in heat at T_2 and rejecting heat at a lower temperature T_3, such that the interval of temperature through which it works $(T_2 - T_3)$ is the same as the interval through which the first engine works $(T_1 - T_2)$. Call each of these intervals ΔT. Let the heat Q_3 rejected by this second

---

* We may imagine a source at T_1 and receiver at T_2 to be substituted for the actual source and receiver, if these have a wider range of temperature, without affecting the action of the working substance.

† *Mathematical and Physical Papers*, vol. I, p. 100; also pp. 283-286.

engine pass on to form the supply of a third reversible engine, working through an equal interval ΔT and rejecting heat Q_4 to a fourth reversible engine, and so on. We imagine a series of engines, every one of which is reversible, each passing on its rejected heat to form the supply of the next engine in the series, and each working through the same number of degrees on the perfect gas thermometer, ΔT. The efficiencies of the successive engines are

$$\Delta T/T_1, \quad \Delta T/T_2, \quad \Delta T/T_3, \text{ etc.}$$

The amounts of heat supplied to them are

$$Q_1, \quad Q_2 = Q_1 T_2/T_1, \quad Q_3 = Q_2 T_3/T_2 = Q_1 T_3/T_1, \text{ etc.}$$

Multiply in each case the heat taken in by the efficiency to find the amount of work done by each engine in the series, and we find that the amount of work done is the same for all the engines, namely

$$\frac{Q_1 \Delta T}{T_1}.$$

Accordingly, we might define the interval of temperature for each engine, without reference to a perfect gas or to any other thermometric substance, as that interval which makes every engine in the series do the same amount of work; and if we did so we should get a scale of temperature which is identical with the scale of the perfect gas thermometer.

The above method of obtaining a thermodynamic scale of temperature may be put thus: Starting from any arbitrary condition of temperature at which we may imagine heat to be supplied, let a series of intervals be taken such that equal amounts of work will be done by every one of a series of reversible engines, each working with one of these intervals for its range, and each handing on to the engine below it the heat which it rejects, so that the heat rejected by the first forms the supply of the second, and so on. Then call these intervals of temperature equal. What the above proof shows is that the intervals thus defined to be equal are also equal when measured on the scale of the perfect gas thermometer: in other words, the thermodynamic scale and the perfect gas scale coincide at every point. Any temperature T reckoned from zero on the scale of a perfect gas thermometer is also an absolute temperature on the thermodynamic scale.

The conception, then, of a chain of reversible heat-engines, each working through a small definite range, furnishes for the statement of temperature a scale which is really absolute in the sense of being independent of all assumptions about expansion or other behaviour of any substance. As the heat goes down from engine to engine in the chain, part of it is converted into work at each step, and the remainder passes on to form the heat-supply of the next engine. We have only to think of the steps as being such that the amount of heat converted into work is the same for each step, and that the remainder passes from engine to engine till all is converted. Thus if we have n engines in the chain, and if the whole quantity of heat supplied to the first engine is Q_1, then the steps are such that each engine converts the quantity Q_1/n of heat into work. When n steps are completed there is no heat left: all is converted into work. This means that the absolute zero of temperature has been reached: we may in fact define the absolute zero as the temperature which is reached in this manner. It is imagined to be reached by coming down through a finite number of steps of temperature, each step representing a finite fall in temperature. We define the absolute or thermodynamic scale by saying that these steps are to be taken as equal to one another. From this it will be seen that the conception of an absolute zero, and of an absolute thermodynamic scale with uniform intervals, does not depend on any notion about perfect gases or about the properties of any particular substance. We reach the absolute zero when, on going down through the chain of perfect engines, we come to a point at which the last fraction of the heat has been converted into work. That fixes the absolute zero. And we call the steps by which we have come equal steps of temperature, the steps being determined by the consideration that each engine in succession is to do the same amount of work out of the residue of heat received from the engine immediately before it in the series. That fixes the scale. Moreover the steps can be so taken, that the scale they give will agree at two fixed points with the ordinary thermometric scale, and will contain between those fixed points the same number of steps as the ordinary scale contains degrees. Thus suppose the initial temperature, at the top of the chain, is that of the boiling point of water, and that we have 373 engines in the chain. Then we find that it takes 100 steps to come down to the temperature of melting ice, and 273 more steps* to complete the conversion of the remaining heat into work. This means that the uniform step of temperature on the thermodynamic scale is equal to the average of the intervals called degrees on any centigrade thermometer, when that average is taken between the

---

* More exactly 273 and a fraction (Art. 16).

freezing point and the boiling point (0° and 100°), and that the absolute zero is at a point 273 of such steps* below the freezing point. But the thermodynamic scale would agree from point to point with the indications of the thermometer throughout the whole of the scale only if the thermometer could use a perfect gas as its expanding substance. Even with hydrogen, which is very nearly a perfect gas, there are slight divergences which were mentioned in Art. 15.

43. Reversible Engine receiving Heat at Various Temperatures. In Carnot's cycle it was assumed that there was only one source and one receiver of heat. All the heat that was taken in was taken in at T_1; all the heat that was rejected was rejected at T_2. But an engine may take in heat in stages, at more temperatures than one, and may also reject heat in stages. With regard to every quantity of heat so taken in, the result still applies that the greatest fraction of it that can be converted into work is represented by the difference between its temperatures of reception and rejection, divided by the absolute temperature of reception. And this is the fraction that will be converted into work provided the processes within the engine are reversible.

Thus if Q_1 represents that part of the whole supply of heat which is taken in at T_1 and Q_2 represents what is taken in at some other temperature T_2, Q_3 at T_3, and so on, and if T_0 be the temperature at which the engine rejects heat, the whole work done, if the engine is reversible, is

$$W = \frac{Q_1\,(T_1 - T_0)}{T_1} + \frac{Q_2\,(T_2 - T_0)}{T_2} + \frac{Q_3\,(T_3 - T_0)}{T_3} + \text{etc.}$$

We here take, for simplicity of statement, a single temperature of rejection T_0.

A mechanically analogous machine would be a great water-wheel, working by gravity, and receiving water into its buckets from reservoirs at various levels, some of which are lower than the top of the wheel. Let M_1, M_2 and so on be the weights of water received at heights l_1, l_2, etc. above any datum level, and let l_0 be the height above the same datum level at which the water leaves the wheel. If the wheel is perfectly efficient (and here again the test of perfect efficiency is reversibility) the work done is

$$M_1\,(l_1 - l_0) + M_2\,(l_2 - l_0) + M_3\,(l_3 - l_0) + \text{etc.}$$

Comparing the two cases we see that the quantity Q_1/T_1 is the

* More exactly 273 and a fraction (Art. 16).

analogue in the heat-engine of M_1 in the water-wheel, Q_2/T_2 is the analogue of M_2, and so on. The amount of work which can be got out of a given quantity of heat by letting it down to an assigned level of temperature is not simply proportional to the product of the quantity of heat by the fall of temperature, but to the product of Q/T by the fall of temperature. On the strength of this analogy Zeuner has called the quantity Q/T the "heat weight" of a quantity of heat Q obtainable at a temperature T.

Another way of putting the matter has a wider application. Let the engine as before take in quantities of heat represented by Q_1, Q_2, Q_3, etc. at T_1, T_2, T_3, and let it reject heat at T', T'', T''', etc., the quantities rejected being respectively Q', Q'', Q''', etc. Then by the principle that in a reversible cycle the heat rejected is to the heat taken in as the absolute temperature of rejection is to the absolute temperature of reception, we have

$$\frac{Q'}{T'}+\frac{Q''}{T''}+\frac{Q'''}{T'''}+ \ldots = \frac{Q_1}{T_1}+\frac{Q_2}{T_2}+\frac{Q_3}{T_3}+ \ldots,$$

from which

$$\Sigma \frac{Q}{T}=0,$$

when the summation is effected all round the reversible cycle. In this summation heat taken in is reckoned as positive and heat rejected as negative. If the cycle is not reversible, the heat rejected will be relatively greater, and therefore, for a non-reversible cycle, $\Sigma (Q/T)$ will be a negative quantity.

Some of the processes may be such that changes of temperature are going on continuously while heat is being taken in or given out, and if so we cannot divide the reception or rejection of heat into a limited number of steps, as has been done above. But the equation may be adapted to the most general case by writing it

$$\int \frac{dQ}{T}=0,$$

integration being performed round the whole cycle.

This holds for any internally reversible cycle. It means that when a substance has passed through any series of reversible changes which cause it to return to its initial state, the quantities of heat which it has taken in and given out are so related to the temperature of the substance at each stage as to make this integral vanish for the cycle as a whole. If the cycle is not reversible $\int dQ/T$ is a negative quantity, because the amount of heat rejected is relatively larger than when the cycle is reversible.

44. Entropy. We have now to introduce an important thermo-dynamic quantity which serves many useful purposes. The *Entropy* of a substance is a function of its state which is most conveniently defined by reference to the heat taken in or given out when the state of the substance undergoes change in a reversible manner. In any such change, the heat taken in or given out, divided by the absolute temperature of the substance, measures the change of entropy. Thus if a substance which is either expanding reversibly or not expanding at all takes in heat δQ when its temperature is T, its entropy increases by the amount $\delta Q/T$. We shall see that the entropy of any substance in a definite state is a definite quantity, which has the same value when the substance comes back again to the same state after undergoing any changes. To give the entropy a numerical value we must start from some arbitrary point where, for convenience of reckoning, the entropy is taken as zero. We are concerned only with *changes* of entropy, and consequently it does not matter, except for convenience, what zero state is chosen for the purpose of calculating the entropy.

Starting then from any suitable zero, each element δQ of the heat taken in has to be divided by T, which is the absolute temperature of the substance when δQ is being taken in. The sum

$$\Sigma \frac{\delta Q}{T}$$

measures the entropy of the substance, on the assumption that no irreversible change of state has occurred during the process. We shall denote the entropy of any substance by ϕ. If the temperature is changing continuously while heat is being taken in, the change of entropy from any state a to any other state b is

$$\phi_b - \phi_a = \int_a^b \frac{dQ}{T},$$

provided there is no irreversible action within the substance during its change of state.

This definition of the entropy of a substance as a quantity which is to be measured by reckoning $\int_a^b \frac{dQ}{T}$ while the substance passes by a reversible process from any state a to any other state b, is consistent with the fact that the entropy is a definite function of the state of the substance, which means that it has only one possible value so long as the substance is in the same state. To prove this we must show that the same value is obtained for the entropy no matter what reversible operation be followed in passing from one state to the other: in other words, that $\int_a^b \dfrac{dQ}{T}$ is the same for all reversible operations by which a substance might pass from state a to state b. Consider any two reversible ways of passing from state a to state b. If we suppose one of them to be reversed the two together will form a complete cycle for which (by Art. 43) $\int \dfrac{dQ}{T} = 0$. Hence $\int_a^b \dfrac{dQ}{T}$ for one of them must be the same as for the other. It is therefore a matter of indifference, in the reckoning of entropy, by what "path" or sequence of changes the substance passes from a to b provided it be a reversible path: starting from any zero state the reckoning of the entropy in a given state will always give the same value, which shows that the entropy is simply a function of the actual state and does not depend on previous conditions.

It is chiefly because the entropy of a substance is a definite function of the state, like the temperature, or the pressure, or the volume, or the internal energy, that the notion of entropy is important in engineering theory. The entropy of a substance is usually reckoned per unit of mass, and numerical values of it reckoned in this manner are given in tables of the properties of steam and of the other substances which are used in heat-engines and refrigerating machines.

But we may also reckon the entropy of a body as a whole when the state of the body is fully known, or the change of entropy which a body undergoes as a whole when it takes in or gives out heat. And we may also reckon the total entropy of a system of bodies by adding together the entropies of the several bodies that make up the system.

45. Conservation of Entropy in Carnot's Cycle. As a simple illustration of the uses to which the idea of entropy may be put, consider the changes of entropy which a substance undergoes when it is taken through Carnot's cycle (Art. 32). All four operations are reversible. In the first, which is isothermal expansion at T_1, the entropy of the substance increases by the amount Q_1/T_1, where Q_1 is the amount of heat taken in from the hot source. In the second operation no heat is taken in or given out and there is no change of entropy. In the third operation a quantity of heat Q_2 is rejected at T_2: the entropy of the substance accordingly

falls by the amount Q_2/T_2. In the fourth operation there is again no transfer of heat and no change of entropy. It is only in the first and third operations that changes of entropy occur. Moreover they are equal, for $Q_1/T_1 = Q_2/T_2$, which shows that the substance has the same entropy as at first, when it has returned to the original state.

During the first operation, while it was taking in heat, its entropy rose from the initial value, which we may call ϕ_a, to a value ϕ_b such that

$$\phi_b = \phi_a + Q_1/T_1.$$

During the third operation, while the substance was rejecting heat, its entropy fell again from ϕ_b to ϕ_a, and

$$\phi_a = \phi_b - Q_2/T_2.$$

Taking the cycle as a whole, the thermal equivalent of the work done by the substance is $Q_1 - Q_2$, and is accordingly equal to

$$(T_1 - T_2)\,(\phi_b - \phi_a).$$

Further, the source of heat has lost an amount of entropy equal to Q_1/T_1, and the receiver has gained an equal amount of entropy, namely Q_2/T_2. We may therefore regard the reversible engine of Carnot as a device which transfers entropy from the hot source to the cold receiver without altering the amount of the entropy so transferred. The amount of heat alters in the process of transfer, for an amount of heat $Q_1 - Q_2$ disappears, which is the thermal equivalent of the work done; but the amount of entropy in the system as a whole does not change.

If, on the other hand, we had to do with an engine which is not reversible, working between the same source and receiver, Q_2 would be relatively larger, since less of the heat taken in is converted into work. Hence Q_2/T_2 would be greater than Q_1/T_1 and the amount of entropy would therefore increase in the transfer.

46. **Entropy-Temperature Diagram for Carnot's Cycle.** It is instructive to represent the changes of entropy in a Carnot cycle by means of a diagram the two coordinates of which are the entropy of the working substance and its temperature (fig. 5). The first operation (isothermal expansion) is represented by ab, a straight line drawn at the level of temperature T_1: during this operation the entropy of the substance rises from ϕ_a to ϕ_b. This is followed by adiabatic expansion bc during which the temperature falls but the entropy does not change. Then isothermal compression cd at

temperature T_2, during which the entropy returns to the initial value. Finally adiabatic compression *da* completes the cycle.

The area of the closed figure *abcd* measures (in heat units) the work done during the cycle. The area *mabn* measured to the base line, which is the absolute zero of temperature, is the heat taken in from the source. The area *mdcn* is the heat rejected to the receiver. These figures are rectangles.

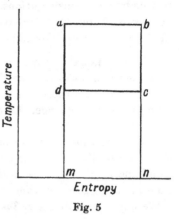

Fig. 5

All this is true whatever be the working substance. Neither in Art. 45 nor here is any assumption made as to that. The diagram (fig. 5) applies to any engine going through the reversible cycle of Carnot whether it use a gas (as in Art. 36) or any other substance.

47. Entropy-Temperature Diagrams for a Series of Reversible Engines. We may apply this method of representation to exhibit the action of the imaginary chain of reversible engines which was used in Art. 42 to establish a thermodynamic scale of temperature.

Starting from any temperature T_1 let a reversible engine take in heat at that temperature, and go through the Carnot cycle of operations represented by the rectangle *abcd* (fig. 6). For this purpose it takes in heat equivalent to *mabn* and rejects heat equivalent to *mdcn*. Let its rejected heat pass on to the next engine of the series, which goes through the Carnot cycle *dcef*, and let the interval of temperature *df* be so chosen as to make the work done by

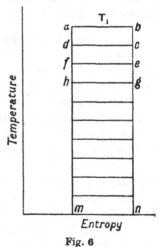

Fig. 6

the second engine equal to the work done by the first. From the geometry of the figure it is obvious that this requires *df* to be equal to *ad*, so that the area *abcd* may be equal to the area *dcef*. Similarly in order that the work done by the third engine should be the same,

we must have $fh = df = ad$, and so on. Thus these intervals constitute
equal steps in a scale of temperature which is based entirely on
thermodynamic considerations, the condition determining the steps
being simply this, that the same amount of work shall be done by
the heat as it passes down through each step.

48. No change of Entropy in Adiabatic Processes. It follows
from the definition of entropy given in Art. 44 that when a substance
is expanded or compressed in an adiabatic manner (Art. 23) its
entropy does not change. An adiabatic line is consequently a line
of constant entropy, or, as it is sometimes called, an *isentropic* line.
Just as isothermal lines can be distinguished by numbers T_1, T_2,
etc. denoting the particular temperature for which each is drawn,
so adiabatic lines can be distinguished by numbers ϕ_1, ϕ_2, etc.
denoting the particular value of the entropy for each.

We might accordingly *define* the entropy of a substance as that
characteristic of the substance which does not change in adiabatic
expansion or compression, and this definition would be consistent
with the method of reckoning entropy described in Art. 44. It is
only in a reversible process that the change of entropy of a substance is to be determined by reference to the heat it takes in or
gives out. The definition of an adiabatic process (Art. 23) excludes
any process that is not reversible.

49. Change of Entropy in an Irreversible Operation. It is
important in this connection to realize that a substance may increase its entropy without having any heat communicated to it
from outside. When a substance expands in an *irreversible* manner,
as by passing through a throttle-valve from a region of high pressure
to a region of lower pressure, it gains entropy. Work is then done
by the substance on itself, in giving energy of motion to each
portion as it passes through the valve, and this energy of motion
is frittered down into heat as the motion subsides through internal
friction. The effect is like that produced by the communication
of some heat, though none is taken in from outside the substance.
Expansion through a throttle-valve may be regarded as consisting
of two stages. The first stage is a more or less adiabatic expansion
during which the substance does work in setting itself in motion:
the second stage is the loss of this motion and the consequent
generation within the substance itself of an equivalent amount of
heat. There is accordingly a gain of entropy, which occurs because
the process as a whole is not reversible.

We cannot directly apply the definition of entropy given in Art. 44 to determine the amount by which the entropy of a substance changes in an irreversible operation such as throttling. But when the final state is known it is in general easy to calculate the entropy corresponding to that state, by considering the amount by which the entropy would have changed if the substance had come to that state by a reversible operation for which $\int dQ/T$ measures the change.

When a substance has passed through any complete cycle of operations its entropy is the same at the end as at the beginning, for the original state has been restored in all respects. This is true of an irreversible cycle as well as of a reversible cycle. But for an irreversible cycle $\int dQ/T$ does not vanish. It has a negative value (Art. 43) and it does not measure change of entropy, for it is only in an internally reversible action that the change of entropy is dQ/T.

50. **Sum of the Entropies in a System.** It is instructive to enquire how the sum of the entropies of all parts of a thermodynamic system is affected when we include not only the working substance but also the source of heat and the sink or receiver to which heat is rejected. Consider a cyclic action in which the working substance takes in a quantity of heat Q_1 from a source at T_1 and rejects a quantity Q_2 to a sink at T_2. When the cycle is completed the source has lost entropy to the amount Q_1/T_1: the working substance has returned to the initial state, and therefore has neither gained nor lost entropy: the sink has gained entropy to the amount Q_2/T_2. If the cycle is a reversible one $Q_1/T_1 = Q_2/T_2$, and therefore the system taken as a whole, consisting of source, substance and sink, has suffered no change in the sum of the entropies of its parts. But if the cycle is not reversible the action is less efficient, Q_2 bears a larger proportion to Q_1 and Q_2/T_2 is greater than Q_1/T_1. Hence in an irreversible action the sum of the entropies of the system as a whole becomes increased. This conclusion has a very wide application: it is true of any system of bodies in which thermal actions may occur. It may be expressed in general terms by saying that when a system undergoes any change, the sum of the entropies of the bodies which take part in the action remains unaltered if the action is reversible, but becomes increased if the action is not reversible. No real action is strictly reversible, and hence any real action occurring within a system of

bodies has the effect of increasing the sum of the entropies of the bodies which make up the system. This is a statement, in terms of entropy, of the principle that in all actual transformations of energy there is what Lord Kelvin called a universal tendency towards the dissipation of energy*. Any system, left to itself, tends to change in such a manner as to increase the aggregate entropy, which is calculated by summing up the entropies of all the parts. The sum of the entropies in any system, considered as a whole, tends towards a maximum, which would be reached if all the energy of the system were to take the form of uniformly diffused heat; and if this state were reached no further transformations would be possible. Any action within the system, by increasing the aggregate entropy, brings the system a step nearer to this state, and to that extent diminishes the availability of the energy in the system for further transformations.

This is true of any limited system. Applied to the universe as a whole, the doctrine suggests that it is in the condition of a clock once wound up and now running down. As Clausius, to whom the name entropy is due, has remarked, "the energy of the universe is constant: the entropy of the universe tends towards a maximum."

An extreme case of thermodynamic waste occurs in the direct conduction of a quantity of heat Q from a hot part of the system, at T_1, to a colder part at T_2, no work being done in the process. The hot part loses entropy by the amount Q/T_1: the cold part gains entropy by the amount Q/T_2, and as the latter is greater there is an increase in the aggregate quantity of entropy in the system as a whole.

51. Entropy-Temperature Diagrams. We shall now consider, in a more general manner, diagrams in which the action of a substance is exhibited by showing the changes of its entropy in relation to its temperature. Such a diagram forms an interesting and often useful alternative to the pressure-volume or indicator diagram. One example, namely the entropy-temperature diagram for a Carnot cycle, has already been sketched in fig. 5.

Let $d\phi$ be the small change of entropy which a substance undergoes when it takes in the small quantity of heat dQ at any temperature T, it being assumed that in the process the substance

---

* *Mathematical and Physical Papers*, vol. I, p. 511.

undergoes only a reversible change of state. Then, by the definition of entropy (Art. 44),

$$d\phi = \frac{dQ}{T},$$

whence $\qquad\qquad Td\phi = dQ,$

and $\qquad\qquad \int Td\phi = \int dQ,$

the integration being performed between any assigned limits. Now if a curve be drawn with T and ϕ for coordinates, $\int Td\phi$ is the area under the curve. This by the above equation is equal to $\int dQ$, which is the whole amount of heat taken in while the substance passes through the states which that portion of the curve represents. Let ab, fig. 7, be any portion of the curve of ϕ and T. The area of the cross-hatched strip, whose breadth is $\delta\phi$ and height T, is $T\delta\phi$, which is equal to δQ, the heat taken in during the small change $\delta\phi$. The whole area $mabn$ or

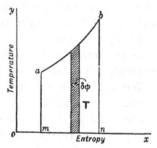

$\int Td\phi$ between the limits a and b is the whole heat taken in while the substance changes in a reversible manner from the state represented by a to the state represented by b. Similarly, in changing reversibly from state b to state a by the line ba the substance rejects an amount of heat which is measured by the area $nbam$. The base line ox corresponds to the absolute zero of temperature.

Fig. 7. Entropy-Temperature Curve.

When an entropy-temperature curve is drawn for any complete cycle of changes it forms a closed figure, since the substance returns to its initial state. To find the area of the figure we have to integrate throughout the complete cycle, and provided there has been no irreversible action within the substance,

$$\int Td\phi = Q_1 - Q_2,$$

Q_1 being the heat taken in and Q_2 the heat rejected. But the difference between these is the heat converted into work, hence

$$\int Td\phi = W,$$

when the integration extends round a complete cycle and W is expressed in thermal units. Thus an entropy-temperature diagram, so long as it represents changes of state all of which are reversible, but not otherwise, has the important property in common with a pressure-volume diagram that the enclosed area measures the work done in a complete cycle.

But the entropy-temperature diagram has an advantage not possessed by the pressure-volume diagram, in that it exhibits not only the work done, but also the heat taken in and the heat rejected, by means of areas under the curves. An illustration of this has already been given in speaking of the Carnot cycle (Art. 46), and others will be found in Chapter III.

52. Perfect Engine using Regenerator. Besides the cycle of Carnot there is (theoretically) one other way in which an engine can work between a source and receiver so as to make the whole action reversible, and thereby transform into work the greatest possible proportion of the heat that is supplied. Suppose there is, as part of the engine, a body (called a "regenerator") into which the working substance can temporarily deposit heat, while the substance falls in temperature from the upper limit T_1 to the lower limit T_2, and suppose further that this is done in such a manner that the transfer of heat from the substance to the regenerator is reversible. This condition implies that there is to be no sensible difference in temperature between the working substance and the material of the regenerator at any place where they are in thermal contact. Then when we wish the substance to pass back from T_2 to T_1 we may reverse this transfer, and so recover the heat which was deposited in the regenerator. This alternate storing and restoring of heat serves instead of adiabatic expansion and compression to make the temperature of the working substance pass from T_1 to T_2 and from T_2 to T_1 respectively. It enables the temperature of the substance to fall to T_2 before heat is rejected to the receiver, and to rise to T_1 before heat is taken in from the source.

This idea is due to Robert Stirling, who in 1827 designed an engine to give it effect. For the present purpose it will suffice to describe the regenerator as a passage (such as a group of tubes) through which the working fluid can travel in either direction, whose walls have a very large capacity for heat, so that the amount alternately given to or taken from them by the working fluid causes no more than an insensible rise or fall in their temperature. The temperature of the walls at one end of the passage is T_1, and this falls continuously down to T_2 at the other end. When the working fluid at temperature T_1 enters the hot end and passes through, it comes out at the cold end at temperature T_2, having stored in the walls of the regenerator a quantity of heat which it will pick up again when passing through in the opposite direction. During the return journey of the working fluid through the regenerator from the cold to the hot end its temperature rises from T_2 to T_1 by picking up the heat which was deposited when the working fluid passed through from the hot end to the cold. The process is strictly reversible, or rather would be so if the regenerator had an unlimited capacity for heat, if no conduction of heat took place along its walls from the hotter parts towards the cold end, and if there were no loss by conduction or radiation from its external surface. A regenerator satisfying these conditions is of course an ideal impossible to realize in practice.

53. **Stirling's Regenerative Air-Engine.** Using air as the working substance, and employing his regenerator, Stirling made an engine which, allowing for practical imperfections, is the earliest example of a reversible engine. The cycle of operations in Stirling's engine was substantially this (in describing it we treat air as a perfect gas):

(1) Air, which had been heated to T_1 by passing through the regenerator, was allowed to expand isothermally through a ratio r, taking in heat from a furnace and raising a piston. Heat taken in (per lb. of air) $= RT_1 \log_e r$ (by Art. 28).

(2) The air was caused to pass through the regenerator from the hot to the cold end, depositing heat and having its temperature lowered to T_2, without change of volume. Heat stored in regenerator $= C_v (T_1 - T_2)$. The pressure of course fell in proportion to the fall in temperature.

(3) The air was then compressed isothermally at T_2, through the same ratio r to its original volume, in contact with a receiver of heat. Heat rejected $= RT_2 \log_e r$.

(4) The air was again passed through the regenerator from the cold to the hot end, taking up heat and having its temperature raised to T_1. Heat restored by the regenerator $= C_v (T_1 - T_2)$. This completed the cycle.

The efficiency is

$$\frac{RT_1 \log_e r - RT_2 \log_e r}{RT_1 \log_e r} = \frac{T_1 - T_2}{T_1}.$$

The indicator diagram of this action is shown in fig. 8. Stirling's engine is important, not as a present-day heat-engine (though it has been revived in small forms after a long interval of disuse), but because it is typical of the only mode, other than Carnot's plan of adiabatic expansion and adiabatic compression, by which the action of a heat-engine can be made reversible.

A modified form of regenerative engine was devised later by Ericsson, who kept the pressure instead of the volume constant while the working substance passed through the regenerator, and so got an indicator diagram made up of two isothermal lines and two lines of constant pressure.

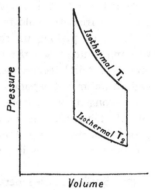

Fig. 8. Ideal Indicator diagram of Air-Engine with Regenerator (Stirling).

The entropy-temperature diagram of a regenerative engine is of the type shown in fig. 9.

The isothermal operation of taking in heat at T_1 is represented by ab; bc is the cooling of the substance from T_1 to T_2 in its passage through the regenerator, where it deposits heat: cd is the isothermal rejection of heat at T_2; and da is the restoration of heat by the regenerator while the substance passes through it in the opposite direction, by which the temperature of the substance is raised from T_2 to T_1. Assuming the action of the regenerator to be ideally perfect, bc and ad are precisely similar curves whatever be their form. The area of the figure is then equal to the area of the rectangle which would represent the ordinary Carnot cycle (fig. 5). The equal areas $pbcq$ and $ndam$ measure the heat stored and restored by the regenerator.

When the working substance is air and the regenerative changes take place either under constant volume, as in Stirling's engine, or under constant pressure, as in Ericsson's, the specific heat C being treated as constant, ad and bc are logarithmic curves with the equation

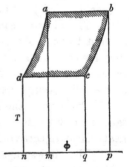

Fig. 9. Entropy-temperature diagram of perfect engine using a Regenerator.

$$\phi = \int \frac{CdT}{T} = C \log_e T + \text{constant},$$

C being C_v in Stirling's process and C_p in Ericsson's.

54. **Joule's Air-Engine.** A type of air-engine was proposed by Joule which, for several reasons, possesses much theoretical interest. Imagine a chamber C (fig. 10) full of air (temperature T_2), which is kept cold by circulating water or otherwise; another chamber A heated by a furnace and full of hot air in a state of compression (temperature T_1); a compressing cylinder M by which air may be pumped from C into A, and a working cylinder N in which air from A may be allowed to expand before passing back into the cold chamber C. We shall suppose the chambers A and C to be large, in comparison with the volume of air that passes in each stroke, so that the pressure in each of them may be taken as sensibly constant. The pump M takes in air from C, compresses

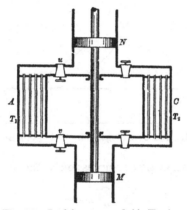

Fig. 10. Joule's proposed Air-Engine.

it adiabatically until its pressure becomes equal to the pressure in A, and then, the valve v being opened, delivers it into A. The indicator diagram for this action on the part of the pump is the diagram $fdae$ in fig. 11. While this is going on, the same quantity of hot air from A is admitted to the cylinder N, the valve u is then closed, and the air is allowed to expand adiabatically in N until its pressure falls to the pressure in the cold chamber C. During the back stroke of N this air is discharged into C. The operation of N is shown by the indicator diagram $ebcf$ in fig. 11. The area $fdae$ measures the work spent in driving the pump; the area $ebcf$ is the work done by the air in the working cylinder N. The difference, namely, the area $abcd$, is the net amount of work obtained by carrying the given quantity of air through a complete cycle. Heat is taken in when the air has its temperature raised

on entering the hot chamber A. Since this happens at a pressure
which is sensibly constant, the heat taken in

$$Q_A = C_p(T_b - T_a),$$

where $T_b = T_1$, the temperature of A, and T_a is the temperature
reached by adiabatic compression in the pump. Similarly, the
heat rejected

$$Q_C = C_p(T_c - T_d),$$

where $T_d = T_2$, the temperature of C, and T_c is the temperature
reached by adiabatic expansion in N. Since the expansion and
compression both take place between the same terminal pressures,

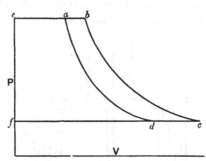

Fig. 11. Indicator diagram in Joule's Air-Engine.

the ratio of expansion and compression is the same. Calling it r,
we have

$$\frac{T_a}{T_d} = \frac{T_b}{T_c} = r^{\gamma-1}$$

(Art. 26), and hence also

$$\frac{T_b}{T_a} = \frac{T_c}{T_d}, \text{ and } \frac{T_b - T_a}{T_a} = \frac{T_c - T_d}{T_d}.$$

Hence

$$\frac{Q_A}{Q_C} = \frac{T_a}{T_d} = \frac{T_b}{T_c},$$

and the efficiency

$$\frac{Q_A - Q_C}{Q_A} = \frac{T_a - T_d}{T_a} = \frac{T_b - T_c}{T_b}.$$

This is less than the efficiency of a perfect engine working between
the same limits of temperature $\left(\dfrac{T_1 - T_2}{T_1}\right)$ because the heat is not
taken in and rejected at the extreme temperatures.

The atmosphere may take the place of the chamber C: that is
to say, instead of having a cold chamber, with circulating water to

absorb the rejected heat, the engine may draw a fresh supply at each stroke from the atmosphere, and discharge into the atmosphere the air which has been expanded adiabatically in N.

The entropy-temperature diagram for this cycle is drawn in fig. 12, where the letters refer to the same stages as in fig. 11. After adiabatic compression da, the air is heated in the hot chamber A, and the curve ab for this process has the equation

$$\phi - \phi_a = \int_{T_a}^{T} \frac{C_p \, dT}{T} = C_p \, (\log_e T - \log_e T_a).$$

Then adiabatic expansion gives the line bc, and cd is another logarithmic curve for the rejection of heat to C by cooling under constant pressure. The ratio $\dfrac{T_a}{T_b}$, which is represented by $\dfrac{ea}{eb}$ in fig. 11 and by $\dfrac{ma}{nb}$ in fig. 12, shows the proportion which the volume of the pump M must bear to the volume of the working cylinder N. The need of a large pump would be a serious drawback in practice, for

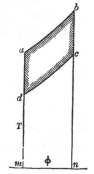

Fig. 12. Entropy-temperature diagram in Joule's Air-Engine.

it would not only make the engine bulky but would cause a relatively large part of the net indicated work to be expended in overcoming friction within the engine itself.

In the original conception of this engine by Joule it was intended that the heat should reach the working air through the walls of the hot chamber, from an external source. But instead of this we may have combustion of fuel going on within the hot chamber itself, the combustion being kept up by the supply of fresh air which comes in through the compressing pump, and, of course, by supplying fuel either in a solid form from time to time through a hopper, or in a gaseous or liquid form. In other words, the engine may operate as an *internal-combustion* engine. Internal-combustion engines, essentially of the Joule type, employing solid fuel have been used on a small scale, but by far the most important development of the type is to be found in engines which work by the explosion or burning of a mixture of air with combustible gas or the vapour of a combustible liquid. The thermodynamics of internal-combustion engines will be considered in a later chapter.

We shall also see later (Chapter IV) that a practicable refrigerating machine, using air for working substance, is obtained by making Joule's Air-Engine work as a heat-pump.

CHAPTER II

PROPERTIES OF FLUIDS

55. **States of Aggregation or Phases.** In the previous chapter the only substances whose properties were discussed were imaginary ones, namely perfect gases. We have now to treat of real substances, such as steam, carbonic acid, or ammonia, which serve as working substances in heat-engines or refrigerating machines, and to examine their action and properties in the light of thermodynamic principles.

Any such substance may exist in three states of aggregation, solid, liquid and gaseous. These states are now generally called phases. Some substances, such as sulphur or iron, have more than one solid phase, depending on changes in atomic arrangement. We are mainly concerned with the liquid and gaseous phases, in either of which the substance is spoken of as a fluid. The working fluid in an engine is often a mixture of the same substance in the two phases of liquid and vapour; but in some stages of the action it may consist entirely of liquid, in others entirely of vapour. The vapour of a substance may be either saturated or superheated. A vapour mixed with its liquid, and in equilibrium with it, must be saturated. Any attempt to heat the mixture would result in more of the liquid turning into saturated vapour. But when a vapour has been removed from its liquid it may be heated to any extent, thereby becoming superheated. Thus when steam is formed in a boiler it is saturated as it leaves the water, but it may be superheated on its way to the engine by passing through hot pipes which cause its temperature to rise above that of the boiler.

A gas such as hydrogen or oxygen or nitrogen is a superheated vapour which can be reduced to the saturated condition by greatly lowering its temperature.

At any one pressure the saturated vapour of a substance can have but one temperature: the superheated vapour at the same pressure may have any temperature higher than that.

In the change of phase from solid to liquid, and again in the change from liquid to vapour, heat is taken in, though the substance does not rise in temperature while the change is going on. The heat so taken in was said in the phraseology of old writers to become latent, and the name Latent Heat is still applied to it. Thus the heat taken in by unit mass of a substance in passing, without change of pressure, from the solid to the liquid phase is called the latent heat of the liquid, and the heat taken in by unit mass in passing, without change of pressure, from liquid to vapour is called the latent heat of the vapour. The latent heat of water, for example, is 80 thermal units, which means that unit mass of ice takes in 80 units of heat while it melts. If we assume the pressure to be one atmosphere, this happens at the temperature which is taken as the lower fixed point (0° C.) in graduating a thermometer.

The temperature at which a solid melts is only slightly affected by the pressure and this is also true of the latent heat of melting. Thus the latent heat of water is practically the same at all pressures ordinarily met with.

At a pressure of one atmosphere water boils at the temperature which is taken for the upper fixed point of the thermometer (namely 100° C.), and the amount of heat taken in per unit of mass is 539·1 units, which measures the latent heat of the vapour. We shall see immediately that the temperature at which the change from liquid to vapour occurs, and also the amount of heat taken in during the change, depends greatly on the pressure. At higher pressures the temperature of boiling is higher and the amount of latent heat is less.

Changes of phase from solid to liquid or from liquid to vapour, in pure substances, are reversible. Under the same conditions as to pressure, for example, water boils and steam is condensed at the same temperature, and the same quantity of heat is taken in during the one process as is given out during the other.

In describing the properties of fluids it will save repetition if we speak particularly of water, taking it as typical of the rest. It is itself of special interest to the engineer, being the working substance of the steam-engine, and the numerical values by which its properties are expressed are better known than those that relate to other fluids. But the definitions and thermodynamical principles which will be stated must be understood as applying to fluids in general.

We have now to consider in more detail some of the points that have been briefly summarized in this Article.

56. **Formation of a Vapour under Constant Pressure.** The properties of steam, or of any other vapour, are most conveniently stated by referring in the first instance to what happens when it is formed *under constant pressure*. This is substantially the process which occurs in the boiler of a steam-engine when the engine is at work. To fix the ideas we may suppose that the vessel in which

steam is to be formed is a long upright cylinder fitted with a frictionless piston which may be loaded so that it exerts a constant pressure on the fluid below. Let there be, to begin with, at the foot of the cylinder a quantity of the liquid (which for convenience of statement we shall take as one unit of mass), and let the piston rest on it with a pressure P. Let heat now be applied to the bottom of the cylinder. As heat enters it produces the following effects in three stages:

(1) The temperature of the liquid rises until a certain temperature T_s is reached, at which vapour begins to be formed. The value of T_s depends on the particular pressure P which the piston exerts. Until the temperature T_s is reached there is nothing but liquid below the piston.

(2) Vapour is formed, more heat being taken in. The piston, which is supposed to continue to exert the same constant pressure, rises. No further increase of temperature occurs during this stage, which continues until all the liquid is converted into vapour. During this stage the vapour which is formed is *saturated*. The volume which the piston encloses at the end of this stage—the volume, namely, of unit mass of saturated vapour at pressure P and consequently at temperature T_s—will be denoted by V_s.

(3) If more heat be allowed to enter after all the liquid has been converted into vapour, the volume will increase and the temperature will rise. The vapour is then *superheated*: its temperature is above the temperature of saturation.

57. **Saturated and Superheated Vapour.** The difference between saturated and superheated vapour may be expressed by saying that if one mixes liquid with vapour at the same temperature some of the liquid will turn to vapour if the vapour is superheated, but none if the vapour is already saturated. Steam in contact with water, and in thermal equilibrium with it, is necessarily saturated. When saturated the properties of a vapour differ considerably, as a rule, from those of a perfect gas, but when superheated they approach those of a perfect gas more and more closely the farther the process of superheating is carried, that is to say, the more the temperature is raised above T_s, the temperature of saturation corresponding to the given pressure P.

58. **Relation of Pressure to Temperature in Saturated Steam.** The temperature T_s at which vapour is formed under the conditions described in Art. 56, which is called the temperature of saturation,

depends on the value of P. The relation of pressure to temperature along with other properties of saturated steam was examined by Regnault, in experiments ranging from temperatures below the zero of the centigrade scale, where the vapour whose pressure was measured was that given off by ice, up to 220° C.

Since then many measurements of the properties of steam have been made by other observers, who besides checking and correcting Regnault's figures have much extended the range of the experiments. Much of this work is recent; it has dealt with steam under conditions about which exact information became necessary to engineers as the use developed of high pressure and high superheat. Steam tables which had been calculated by Callendar and others for a limited range of pressures became inadequate under the altered conditions of practice, and much research was undertaken to supplement the earlier data. The technical demand for fuller knowledge and for agreement as to numerical results led to an International Conference which met in London in 1929, in Berlin in 1930 and in New York in 1934. After reviewing the available measurements, the Conference first agreed on a "skeleton" steam table which stated certain figures as representing the most probable values of the chief properties, along with small "tolerances" indicating the limits within which these values might still be regarded as uncertain*. These tolerances have been considerably reduced as a result of the most recent conference.

The following figures for the relation of pressure to temperature in saturated steam are obtained from the data agreed at the 1934

Saturation pressure

Tempera-ture ° C.	Pounds per sq. in.	Kg. per sq. cm.	Tolerance kg. per sq. cm. (plus or minus)
0	0·08858	0·006228	0·000006
50	1·7890	0·12578	0·00006
100	14·696	1·03323	0
150	69·033	4·8535	0·0032
200	225·54	15·857	0·008
250	576·90	40·560	0·013
275	862·79	60·660	0·018
300	1246·1	87·611	0·024
325	1748·8	122·95	0·03
350	2398·5	168·63	0·04

* See *World Power*, vol. xv, April 1931.

As regards experiments on the relation of the saturation pressure to the temperature, see Holborn and Henning, *Ann. Physik*, vol. xxvi, 1908, p. 833; Holborn and Baumann, *Ann. Physik*, vol. xxxi, 1910, p. 945; Keyes and Smith, *Mechanical Engineering*, vol. lii, 1930, p. 126; and especially A. Egerton and

Conference. A column is added giving the pressure in pounds per sq. in.

59. Tables of the Properties of Steam. Modern steam tables may be said to take their origin from the work of the late Professor H. L. Callendar who in 1900 pointed out that it was possible to frame an equation of state (connecting P, V and T) which would be sufficiently valid over a limited range to allow the chief properties of steam to be inferred within that range from a comparatively small number of experimental data by the help of certain thermodynamic relations. He showed that sets of figures might be compiled in this way which would not only agree with the data but would also be thermodynamically consistent with one another. The suggestion was taken up by Dr Richard Mollier of Dresden who in 1906 published the first edition of his *Neue Tabellen und Diagramme für Wasserdampf*, founded on the work of Callendar. Besides giving numerical tables Mollier developed graphic methods of exhibiting the properties of steam which have proved of great service to engineers. In 1915 Callendar published tables covering a range up to 500 pounds per sq. inch*. At that time the experimental data for high pressures were very meagre. Later researches by Callendar himself and by others showed that in the upper part of the range Callendar's tabulated figures required substantial correction.

In dealing subsequently with still higher pressures Callendar endeavoured to modify his equation of state so that it should give better agreement with experimental results in the upper regions of the scale. This departure led to the *Revised Callendar Steam Tables* (1931) which were published after his death. The equations were designed to interpret certain features in the neighbourhood of the critical point which are not accepted by other authorities.

Other calculators have used somewhat different methods and have stressed different experimental data, with the result that the various published tables show small discrepancies, especially in the region of high pressure. The figures fall for the most part within the tolerances admitted by the International Conference of 1930. Any one of the four tables named below† will serve the

G. S. Callendar, *Phil. Trans. Roy. Soc.* A, vol. CCXXXI, 1933, p. 147, where the authors compare the results of their own experiments with the most authoritative earlier work; also Osborne, Stimson, Fiock and Ginnings, *J. Bur. Standards*, vol. X, 1933, p. 155.

* *The Callendar Steam Tables*, 1915; London: Edward Arnold. These tables were reprinted in Callendar's book on *Properties of Steam*, 1920.

† (1) *Steam Tables and Mollier Diagrams*, J. H. Keenan; New York: The

engineer sufficiently for most purposes. Each of them covers, for superheated as well as saturated steam, the whole range of pressures and temperatures now employed in practice.

For the convenience of readers of this book short tables have been compiled which will be found in the Appendix. They are based on what the author considers to be the most accurate data at present available. These tables take the place of selections from the original Callendar tables which were printed in the First Edition; they cover a much wider range.

60. Specific Volume of Dry Saturated Steam. Among the quantities shown in steam tables is the specific volume of dry saturated steam, V_s, that is to say the volume per unit of mass when the steam has no water mixed with it and is also not superheated. To secure these conditions in direct measurements of the volume is not easy, but values of V_s may also be inferred from other quantities. The following table states in cubic metres per kilogramme the values of V_s that are taken from data agreed by the International Conference of 1934 along with the tolerances that were admitted. The volumes in cubic ft. per lb. are also given in the second column.

Specific volume of Dry Saturated Steam

Tempera-ture ° C.	Cub. ft. per lb.	Cub. m. per kg.	Tolerance cub. m. per kg. (plus or minus)
0	3304·8	206·31	0·21
50	192·94	12·045	0·012
100	26·802	1·6732	0·0017
150	6·2866	0·39246	0·00039
200	2·0372	0·12718	0·00013
250	0·8019	0·050061	0·00005
300	0·3464	0·021625	0·000035
350	0·1410	0·008802	0·000035

From these figures along with those of Art. 58 the relation of pressure to volume in saturated steam is readily found. An old empirical formula connecting these quantities, $PV^{\frac{16}{16}} = \text{constant}$, fits well for pressures below 200 pounds per sq. inch but fails at high pressures.

American Society of Mechanical Engineers (1930). (2) *The Revised Callendar Steam Tables*; London: Edward Arnold (1931). (3) *Tabellen und Diagramme für Wasserdampf*, O. Knoblauch, E. Raisch, H. Hausen, W. Koch; Munich: R. Oldenburg (1932). (4) *Neue Tabellen und Diagramme für Wasserdampf*, R. Mollier; Berlin, seventh ed.: J. Springer (1932). This supersedes earlier editions, including an English version published in 1927.

61. Boiling and Evaporation. The familiar case of water boiling in a kettle or other open vessel is only a special example of the formation of steam under constant pressure. There the constant pressure is that of the atmosphere, and consequently the temperature at which the water boils is about 100° C.*

Water in the open *evaporates* slowly at any temperature lower than that at which it boils. Though the pressure of the vapour so formed is lower than that of the atmosphere—and may be very much lower—the vapour is able to escape from the surface by diffusion into the air: the pressure on the surface of the water is still that of the atmosphere. As the temperature of water in the open is raised this slow evaporation from the surface becomes more rapid, but it is only when the temperature reaches the value which corresponds (for saturated steam) to the given atmospheric pressure that the water boils: the vapour is then formed in bubbles at the pressure of the atmosphere, and it escapes not by diffusion but by displacing the superincumbent air.

62. Mixture of Vapour with other Gases: Dalton's Principle. In what has been said about the relation of pressure and volume to temperature in the saturated state, it has been assumed that in the process of formation there is simply a mixture of the liquid with its vapour, no other substance being present. This is substantially true in a steam boiler or in the evaporator of a refrigerating machine. But the case is different when the vapour has to mix with another gas or gases. A principle discovered by Dalton then applies, that the pressure in any closed space containing a mixture of two or more gases at any given temperature is equal to the sum of the pressures which each of the gases would exert separately if the others were absent, that is to say if each of the gases (at the same temperature) alone occupied the whole space. These pressures, which are added together to make up the actual pressure, are called "partial pressures." Dalton's principle is very nearly true of real gases and vapours at moderate pressures, and is exactly true of the ideal perfect gases of thermodynamic theory. An important instance of its application is considered in the next Article.

63. Evaporation into a Space containing Air: Saturation of the Atmosphere with Water-Vapour. When water evaporates in a

---

* Water in the open boils at 100° C. when the atmospheric pressure has its standard value, which corresponds to a barometer reading (corrected to 0° C.) of 760 mm. at sea level in latitude 45°, or 759·6 mm. in London (see Art. 12).

closed space containing air, the process goes on until a definite amount of it has become mixed, as vapour, with the air already there. When this has happened, and a state of equilibrium is reached, the air is said to be saturated with water-vapour. The amount of water-vapour that a given volume of air will take up in this way depends upon the temperature: by Dalton's principle it is substantially the same as would be required to fill the same space with saturated steam at that temperature if the air were not present. Each constituent of the mixed gases in the closed space behaves as if it alone occupied the whole volume and it contributes a "partial pressure" just as if the other constituents were not there. This principle affords a ready means of calculating the quantity of water-vapour contained in a saturated atmosphere when the temperature and the pressure are assigned.

As an example, suppose air at 30° C. to be saturated with water-vapour. At that temperature 1 lb. of saturated steam has a volume of 527 cubic feet, and therefore one cubic foot weighs 0·0019 lb. Consequently each cubic foot of the air takes up 0·0019 lb. of water-vapour in reaching the state of saturation at that temperature. And since the corresponding pressure of water-vapour is 0·615 pound per sq. inch, the pressure in an enclosed space containing this moist air is greater by 0·615 pound per sq. inch than it would be if the water-vapour were removed and the dry air alone were left to fill the same space at the same temperature. In other words 0·615 pound per sq. inch is the "partial pressure" of the water-vapour present in the air under the assumed conditions.

When the amount of water-vapour present in air is less than enough to cause saturation the water-vapour is held in a super-heated state. If the temperature of the mixture be lowered, a point is reached at which the water-vapour in the air becomes saturated, and any further lowering of the temperature causes some of the vapour to be deposited as liquid on the walls of the containing vessel, or on any particles of dust that may be present.‑ Any solid particles will serve as nuclei for condensation. The water condensed on such nuclei forms a mist of minute drops which fall so slowly that they seem to be held in suspension. The temperature at which water begins to be deposited from moist air is called the dew-point. Condensation of some of the water contained in air will also occur on any cold surface (colder than the dew-point) with which the air comes in contact: this results from local cooling of the air close to the surface in question. Thus in a refrigerating plant with pipes that convey a liquid colder than the freezing point through the warm atmosphere of the engine-room, a coating of ice forms round the pipes. For the same reason an effective way to dry air is to make it cold and drain away the water condensed in the process: at the lowest temperature the air remains saturated, but the amount of water required to saturate it at a low temperature is very small, and when it is allowed to become warm again without taking up more water it will be far from saturation.

64. Heat required for the Formation of Steam under Constant Pressure: Heat of the Liquid and Latent Heat. Return now to the imaginary experiment of Art. 56, where steam is formed under the constant pressure of a loaded piston—nothing but water or water-vapour being present—and enquire what amount of heat has to be supplied in each stage of the operation. In the *first stage* the substance is wholly in the condition of water which is being heated from the initial temperature to T_s, the temperature at which the second stage begins. During this first stage the heat taken in per lb. is roughly equal at low pressures to one thermal unit for each degree by which the temperature of the water rises. It would be exactly equal to that if the specific heat of water were constant and equal to unity, but this is not the case. At about 30° C. the specific heat of water is a little less than unity; it passes a minimum value thereabouts and then increases, becoming substantially greater than unity at such temperatures as are found in steam boilers. Thus for instance to heat 1 lb. of water from 0° C. to 200° C. under a pressure sufficient to keep steam from forming takes 203·5 thermal units instead of 200: to heat it to 300° C. would take 321 units.

During this first stage, while the substance is still liquid, nearly all the heat that is taken in goes to increase the stock of internal energy. There is scarcely any external work done, for the volume is only slightly increased. Thus in heating water from 0° C. to 200° C. the volume changes from 0·0160 to 0·0185 cubic ft. and the external work done is only 81 foot-pounds. This is equivalent to barely 0·06 thermal unit, and is negligible in comparison with the 203·5 units of heat that are taken in.

In the *second stage*, the liquid changes into saturated steam without change of temperature. The heat that is taken in during this stage constitutes what is called the Latent Heat of the vapour. We shall denote it by L. For steam formed under a pressure of one atmosphere (saturation temperature 100° C.) the latent heat is 539·1: with lower pressures of formation it is greater, and with higher pressures it is less. At the end of the second stage the substance contains no liquid; it is then dry saturated steam: at any earlier point, when it consists partly of saturated steam and partly of water, it may be spoken of as wet steam.

The latent heat of a vapour may be defined as the amount of heat which is taken in by unit mass of the liquid while it all changes into saturated vapour under constant pressure, the liquid having been previously heated up to the temperature at which the vapour is formed.

A considerable part of the heat taken in during this process is spent in doing external work, since the substance expands against the constant pressure P. It is only the remainder of the so-called latent heat L that can be said to remain in the fluid and to constitute an addition to its stock of internal energy. The amount spent in doing external work during the second stage is

$$AP (V_s - V_w),$$

where V_s is the volume of the saturated vapour and V_w is the volume of the liquid at the same temperature and pressure, A being the factor for converting units of work into thermal units. The excess of L above this quantity measures the amount by which the internal energy increases during the second stage*.

65. **Total External Work done.** In the two stages together the whole amount of external work done is to be found by taking the whole increase of volume and multiplying it by the pressure. Then when water at t_0 is converted into dry saturated steam at t_s under a constant pressure P the whole external work done is AP (V_s at $t_s - V_w$ at t_0). Nearly the whole of this, as we have already seen, is done in the second stage.

66. **Internal Energy of a Fluid.** No matter what changes a substance may undergo, its internal energy will return to the same value when the substance returns to the same condition in all respects. In other words the internal energy is a function of the actual state of the substance and is independent of the way in which that state has been reached. Thus the internal energy of

---

* Thus when 1 gram of water at 200° C. and a pressure of 225·54 pounds per sq. in. is converted into steam, of the 463·35 calories taken in, 46·79 calories are spent in doing external work and 416·56 calories go to increase the stock of internal energy.

1 lb. of saturated steam at a particular pressure is a definite quantity which is the same whether the steam has been formed by boiling under constant pressure or in any other manner. Steam formed in a closed vessel of constant volume, for example, would have the same internal energy as steam at the same pressure but formed under conditions of constant pressure, though the amount of heat taken in during its formation would be different, for no external work is done in the process of formation in a closed vessel of constant volume. In that case the heat taken in would be equal to the increase of internal energy.

We have no means of measuring the total stock of internal energy in a substance, and can deal only with changes in the stock. But by taking some arbitrary starting point as a zero from which the internal energy E is reckoned we can give E a numerical value for any other state of the substance. That value really expresses the difference from the internal energy in the zero state. The usual convention is to write $E = 0$ when the substance is in the liquid condition at a temperature of $0°$ C., and at a pressure equal to the vapour-pressure corresponding to that temperature. We may call this, for brevity, the zero state of the substance*.

67. The "Total Heat" of a Fluid. We come now to another function which is of great use in thermodynamic calculations. It is generally called the Total Heat and is represented here† by the letter I.

The Total Heat I is defined for any state of the substance by the equation
$$I = E + APV.$$

That is to say I is equal to the sum of the internal energy and the external work which would be done if the substance could be imagined to start from no volume at all and to expand to its actual volume, under a constant pressure equal to its actual pressure. Since the pressure, volume, and internal energy are all functions of the actual state, I is also a function of the actual state: its value is

* From the tables in the Appendix, the internal energy at $0°$ C. for saturated vapour is given as 567·2 calories on the basis that $E = 0$ for water at $0°$ C. When the water is converted into steam at $0°$ C. under constant pressure, the heat taken in L is 597·26 calories and of this 30·06 calories represents the external work done: the difference measures E. The values of E increase slowly with temperature up to $240°$ C.

† Some writers call the Total Heat the "Enthalpy." Callendar uses the letter H for this function. In view of the fact that Rankine gave H another meaning the author prefers a different symbol. Mollier represents the Total Heat by i, and this usage is followed by other German writers.

independent of how the state has been reached. In steam, for example, the heat taken in during formation depends on how the steam is formed, but the Total Heat I depends only on the final condition. The total heat can be calculated for any condition of a substance, whether in the state of liquid or of saturated or super-heated vapour. It is expressed in thermal units per unit of mass and therefore has the same numerical value whether stated in lb.-calories per lb. or gramme-calories per gramme. Values of the total heat of saturated steam and also of water under saturation pressure at various temperatures are given in the tables. The total heat of saturated steam increases with rising pressure but passes a maximum at about 410 pounds per sq. inch.

It follows from the definition of I that in the zero state of any substance, at which E is reckoned to be zero, I is not zero but has a small positive value depending on the volume of the liquid and its pressure at that state. Since E is then zero I is equal to AP_0V_0, where P_0 is the pressure at the zero state, namely the vapour-pressure at $0°$ C., and V_0 is the volume of the liquid at $0°$ C. and pressure P_0. For water this quantity AP_0V_0 is quite negligible, amounting as it does to $0\cdot000146$ thermal unit. For carbonic acid it is about 1 thermal unit, for ammonia and sulphurous acid it is much less.

68. **Change of the Total Heat during Heating under Constant Pressure.** An important property of the function I is that when any substance is heated under constant pressure the change of I is equal to the heat taken in. To prove this, let Q be the heat taken in while the substance expands under constant pressure P from a state in which the volume is V_1 and the internal energy is E_1 to another state in which the volume is V_2 and the internal energy is E_2. Then the external work done is $P(V_2-V_1)$ and, by the conservation of energy,

$$Q = E_2 - E_1 + AP(V_2 - V_1),$$

which may be written

$$Q = (E_2 + APV_2) - (E_1 + APV_1)$$

or $\qquad\qquad Q = I_2 - I_1,$

where I_1 is the total heat in the first state and I_2 is the total heat in the second state.

69. **Application to Steam formed under Constant Pressure, from Water at $0°$ C.** The above proposition applies to every stage

of the imaginary experiment of Art. 56. Referring to that experiment, assume that to begin with there is under the piston 1 lb. of water at 0° C. and at the pressure at which steam is to be formed. By definition of the total heat,

$$I = E + APV,$$

E at the beginning may be taken as zero*. Hence the value of I for the water at 0° C. may be taken as APV_0, where V_0 is the volume of 1 lb. of water at 0° C. and P is the pressure at which steam is to be formed. At the end of the first stage

$$I_w = Q_1 + APV_0,$$

where I_w represents the value of I for water at the temperature at which steam is about to form. When values of I_w are known this allows Q_1, the heat taken in during the first stage, to be accurately determined. Values of I_w are included in the steam tables.

During the second stage an amount of heat equal to L is taken in at constant pressure, and the total heat changes from I_w to I_s, where I_s is the total heat of saturated steam. Hence

$$I_s = L + I_w$$
$$= L + Q_1 + APV_0.$$

The sum $L + Q_1$ is the whole heat of formation, in the experiment of Art. 56. Thus the "total heat" of steam is equal to the heat of formation under constant pressure, *plus* a small quantity which is the thermal equivalent of the work that would be done in lifting the piston far enough to admit the original volume of the water. The quantity APV_0 forms a very small part of the "total heat": it is only one thermal unit when the pressure of formation is 500 pounds per sq. inch and is much less at low pressures.

These remarks and the following tabular scheme will serve to show how the total heat of saturated steam (or other vapour) is related to the heat of formation under constant pressure. But the student should accustom himself to think of the total heat without reference to any process of formation, as a property which a substance possesses in its actual state—a property which is just as simply a function of the state as is the temperature, or the pressure,

---

* The convention of Art. 66 makes $E = 0$ for water at 0° C. and pressure P_0. Here we have water at 0° C. and pressure P, which is higher than P_0: but the higher pressure does not cause the internal energy of water at 0° C. to differ appreciably from zero.

or the volume, or the internal energy, or the entropy, which we shall have to consider presently*.

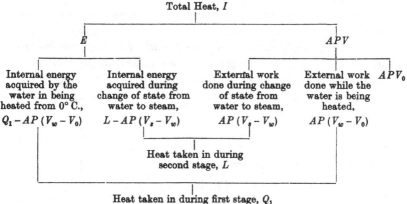

Total Heat, I

E APV

| Internal energy acquired by the water in being heated from 0° C., $Q_1 - AP(V_w - V_0)$ | Internal energy acquired during change of state from water to steam, $L - AP(V_s - V_w)$ | External work done during change of state from water to steam, $AP(V_s - V_w)$ | External work done while the water is being heated, $AP(V_w - V_0)$ | APV_0 |

Heat taken in during second stage, L

Heat taken in during first stage, Q_1

70. Total Heat of a mixture of Liquid and its Saturated Vapour. It follows from Art. 68 that while a liquid is being converted into vapour, under constant pressure, the total heat I increases in proportion to the amount of vapour that is formed. At any intermediate stage in the process, if we call q the fraction that is vaporized and $1 - q$ the fraction that is still liquid, the total heat of the mixture is

$$qI_s + (1 - q) I_w,$$

which may be written

$$I_w + qL.$$

Similarly, while a vapour is being condensed under constant pressure, I becomes less by an amount measured by the heat given out, which is proportional, at any intermediate stage, to the fraction then condensed.

71. Total Heat of Superheated Vapour. The state of a superheated vapour is completely specified when the temperature and pressure are known. Accordingly its volume, its total heat, and such other quantities as depend only on the state, are then determinate, no matter how the process of superheating may have been carried out. Thus a pound of superheated steam at a given pressure

* The function here called the total heat I, namely $E + APV$, was introduced by Willard Gibbs (*Trans. of the Connecticut Academy*, vol. III; *Collected Scientific Papers*, vol. I, p. 92), and was first called the "Total Heat" by Callendar (*Phil. Mag.* 1908, vol. v, p. 50). Its great importance in technical thermodynamics was emphasized by Mollier, who employed it in charts for exhibiting the properties of steam and other substances. The use of such charts will be described later.

and temperature has a definite volume V', a definite total heat I', and so on, whether it has been formed under constant pressure or in any other way.

In engineering practice the superheating of steam is usually done at constant pressure: the steam on leaving the boiler passes through a group of tubes forming a superheater which is kept hot by the furnace gases. While the steam is taking up heat in these tubes its pressure remains equal, or nearly equal, to that in the boiler. The

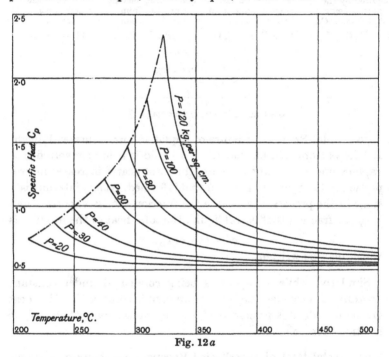

Fig. 12 a

process corresponds to the third stage in Art. 56. The amount of heat taken in during superheating depends on the specific heat of the vapour, which—unlike the specific heat of a perfect gas—is by no means constant. At high pressures it is much greater than at low pressures, and it changes as the process of superheating goes on, being notably greater in the initial stages than after the steam has acquired a fair amount of superheat.

These characteristics will be seen in fig. 12 a which is based on experiments by Knoblauch and Koch*. It shows how the specific

* *Zeit. des Ver. deutsch. Ing.* vol. 72, 1928, p. 1733. See also Knoblauch's *Steam Tables.*

heat changes when steam is superheated at various constant pressures. The lowest of the curves drawn here is for a pressure of 20 kg. per sq. cm. (284 pounds per sq. inch). At that pressure the specific heat is initially 0·74 and as superheating proceeds it falls to 0·53. At the highest pressure shown, 120 kg. per sq. cm., the specific heat is initially 2·35 and falls to 0·67 at 500° C. With further heating the specific heat for each pressure would pass a minimum and begin to rise slowly. The broken line on the left shows the temperatures of saturation from which the several processes of superheating take their start.

When we know the mean specific heat at constant pressure from saturation up to the actual temperature it is easy to work out the total heat I' of superheated vapour by adding to I_s the quantity that would be taken in during a constant-pressure superheating process. Thus

$$I' = I_s + \overline{C}_p\,(T' - T_s),$$

where T_s is the temperature of saturation for the actual pressure, T' is the actual temperature, and $\overline{C}_p$ is the mean specific heat between T_s and T'. The value of I' so found is the same whether the superheated vapour has been formed under constant pressure or in any other way.

72. **Constancy of the Total Heat in a Throttling Process.** An important property of the function I, in any substance, is that it does not change when the substance passes through a valve or other constricted opening, such as the porous plug of the Joule-Thomson experiment mentioned in Art. 19, by which it becomes

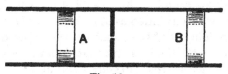

Fig. 13

throttled or "wire-drawn" so that its pressure drops. A practical instance of this kind of action occurs when steam passes through a partially closed orifice or "reducing valve." Eddies are formed in the fluid as it rushes through the constricted opening, and the energy expended in forming them is frittered down into heat as they subside.

To prove that I is constant in such an operation we shall consider what happens while a unit quantity of the substance passes through a constricted opening (as in fig. 13), and, to make the matter clear, imagine this unit quantity to be separated from the rest of the substance by two frictionless pistons, one of which (A) slides in the pipe that leads to the constriction and the other (B) slides in the pipe that leads away from it. On one side, as the substance comes up, let its pressure be P_1, volume V_1 and internal energy E_1. On the other side, after passing the constriction, let its pressure be P_2, volume V_2 and internal energy E_2. As each portion approaches the constriction, work is done upon it by the substance behind pushing in the imaginary piston A, and the amount of that work done while unit quantity is passing is P_1V_1. After each portion has passed the constriction it does work upon the substance in front by pushing out the imaginary piston B, and the amount of that work is P_2V_2 for the whole unit quantity. Any excess of the work done by the substance on piston B over the work done upon it by piston A must be supplied by a reduction in its stock of internal energy. Hence

$$AP_2V_2 - AP_1V_1 = E_1 - E_2,$$

from which

$$E_2 + AP_2V_2 = E_1 + AP_1V_1,$$

or

$$I_2 = I_1.$$

Thus the total heat does not change in consequence of the throttling. The imaginary pistons were introduced only to make the reasoning more intelligible; the argument holds good whether they are there or not. It applies to any fluid, and to any action in which there is a frictional fall of pressure.

We might accordingly describe the quantity I as *that property of a substance which does not change in a throttling process*[*].

73. Entropy of a Fluid. In reckoning the entropy of a fluid we follow the same convention as in reckoning internal energy: the entropy of the liquid at 0° C. is taken as zero. Consider, as before, a process in which the liquid is first heated under constant pressure and then vaporized at that pressure. During the heating of the liquid from an initial temperature T_0 to any temperature T (on the absolute scale) the entropy increases by the amount

$$\int_{T_0}^{T} \frac{dQ}{T} = \int_{T_0}^{T} \frac{\sigma dT}{T},$$

where σ is the specific heat at constant pressure.

[*] It is assumed that no heat is taken in or given out, and also that the velocity in the pipes is so small that no account need be taken of any difference in the kinetic energy of the stream in the pipes before and after passing the constriction, once the eddies have subsided. If the stream has acquired an appreciable amount of kinetic energy after the process, there will be a corresponding reduction in I. (See Art. 104.)

In steam tables there is a column for ϕ_w, the entropy of water at various temperatures*, the pressure in each case being the saturation pressure at that temperature. It is the amount of entropy which the water has at the end of the first stage in Art. 56, when steam is just about to be formed.

During the second stage an additional amount of heat L is taken in at constant temperature T_s, namely the temperature at which steam is formed under the given pressure. Hence the entropy increases by the amount $\dfrac{L}{T_s}$, and we have, for the entropy of saturated steam,

$$\phi_s = \phi_w + \frac{L}{T_s}.$$

Values of ϕ_s are given in the tables.

During superheating there is a further increase of entropy as the substance takes in more heat. Like the volume or the total heat the entropy of any vapour is a function of the actual state, independent of the way in which the vapour has been formed.

74. Mixed Liquid and Vapour: Wet Steam. In many of the actions that occur in steam-engines and refrigerating machines we have to do not with dry saturated vapour but with a mixture of saturated vapour and liquid. In the cylinder of a steam-engine, for example, the steam is generally *wet*; it contains a proportion of water which varies as the stroke proceeds. When any such mixture is in a state of thermal equilibrium the liquid and vapour have the same temperature, and the vapour is saturated. What is called the *dryness* of wet steam is measured by the fraction q of vapour which is present in unit mass of the mixture. When the dryness is known it is easy to determine other quantities. Thus, reckoning in every case per unit mass of the mixture, we have:

Latent Heat of wet steam $= qL = q\,(I_s - I_w)$ ....(1),

Total Heat of wet steam, $I_q = I_w + qL = I_s - (1-q)\,L$ ......(2),

Volume of wet steam, $V_q = qV_s + (1-q)\,V_w$ ....(3),

which is very nearly equal to qV_s unless the mixture is so wet as to consist mainly of water;

Entropy of wet steam, $\phi_q = \phi_w + \dfrac{qL}{T_s} = \phi_s - \dfrac{(1-q)\,L}{T_s}$ ....(4).

* See Appendix.

From (2) it follows that when the total heat I_q of wet steam is known, the dryness may be found by the equation

$$q = \frac{I_q - I_w}{I_s - I_w} \quad \dots\dots\dots\dots\dots\dots(5).$$

Combining (2) and (4), and eliminating q, we have

$$I_q = I_w + T_s(\phi_q - \phi_w) \quad \dots\dots\dots\dots\dots(6),$$

which is a convenient expression for finding the total heat of wet steam when the data are the temperature and the entropy. An alternative form is

$$I_q = I_s - T_s(\phi_s - \phi_q) \quad \dots\dots\dots\dots\dots(7).$$

In these expressions I_w is the total heat of water, and I_s that of dry saturated steam, at the temperature of the wet mixture.

All these formulas apply to a mixture of any liquid with its vapour.

75. **Specification of the State of any Fluid.** We have now spoken of the following quantities, which are functions of the state of the substance. They all depend on the actual state, not on how that state has been reached:

The temperature, T.
The pressure, P.
The volume, V.
The Internal Energy, E. ⎫ These four are reckoned per unit
The Total Heat, I. ⎬ quantity of the substance.
The Entropy, ϕ. ⎭

A substance may change its state in many different ways: it may for instance take in heat at constant volume or while expanding; it may expand or be compressed with or without taking in heat; expansion may take place through a throttle-valve or under a piston. But in any change of state whatever, the amount by which each of these quantities is altered depends only on what the initial and final states are, and not at all on the particular process by which the change of state has been effected.

There are other quantities, such as the heat taken in, or the work done, which depend on how the change of state has taken place. In dealing with them we have to distinguish between one process of change and another, even when both processes bring the substance from the same initial to the same final condition.

The working substance may be a liquid, or a mixture of liquid and vapour, or a dry-saturated or superheated vapour. The condition of a dry-saturated vapour is only a boundary condition between that of wetness and that of superheat. To specify completely the state at any instant it is enough to give either the pressure or the temperature and one of the other four quantities named in this list. Thus if P and V are given the state is fully defined: all the other quantities can then be determined, provided, of course, we have sufficient experimental knowledge of the characteristics of the substance. Or we may specify the state by giving another pair of quantities, such as T and ϕ, or P and I, or ϕ and I.

More generally, any two of these six quantities will serve as data in specifying the state, so long as the substance is in equilibrium and is homogeneous; but when it is a mixture of liquid and vapour the pressure and temperature do not suffice without some other particular such as the dryness q.

With regard to these functions it may be useful to repeat here that

T is constant in isothermal expansion;

ϕ is constant in adiabatic expansion;

I is constant in expansion through a throttle-valve or porous plug.

76. **Isothermal Expansion of a Fluid: Isothermal Lines on the Pressure-Volume Diagram.** A saturated vapour can expand isothermally only when it is wet: the process corresponds to the second stage in the experiment of Art. 56; it goes on at constant pressure and involves change of part of the liquid in the wet mixture into vapour. Similarly, isothermal compression of a wet vapour involves condensation of part of it. Isothermal lines on the pressure-volume diagram for a mixture of vapour and liquid are straight lines of uniform pressure.

It is instructive to consider the general form of the isothermal lines as the fluid passes successively through the stages of being (1) entirely liquid, (2) a mixture of vapour and liquid, (3) entirely vaporous, by having its pressure gradually reduced under conditions such that the temperature remains constant throughout the process. Imagine for instance a cylinder to contain a quantity of the liquid under pressure applied by a loaded piston, and let the cylinder stand on a body at a definite constant temperature, which will supply enough heat to it to maintain the temperature un-

changed when the pressure of the piston is gradually relaxed and
the volume consequently increases. Starting from a condition of
very high pressure, say at A_1 (fig. 14), when the contents of the
cylinder are wholly liquid, let the load on the piston be slowly
reduced so that the pressure gradually falls. The contents at first
remain liquid, until the pressure falls to the saturation value for the
given temperature, namely the pressure at which vapour begins to
form. Thus we have in the pressure-volume diagram a line $A_1 B_1$
to represent what happens while the pressure is falling during this
first stage; the contents are then still liquid. The volume of the
liquid increases, but only very slightly, in consequence of the
pressure being relaxed, and hence $A_1 B_1$ in the diagram is nearly but
not quite vertical. At B_1 vapour begins to form, and continues

forming until all the liquid be
comes vapour. This is represented
by $B_1 C_1$, a stage during which
there is no change of pressure.
At C_1 there is nothing but saturated vapour. Then, if the fall of
pressure continues, a line $C_1 D_1$
is traced, the progressive fall of
pressure being associated with a
progressive increase of volume.
The temperature, by assumption,
is kept constant throughout. At
D_1, or at any point beyond C_1, the
vapour has become superheated,
because its pressure is lower than
the pressure corresponding to
saturation, and hence its temperature is higher than the temperature corresponding to saturation at the actual pressure. Any

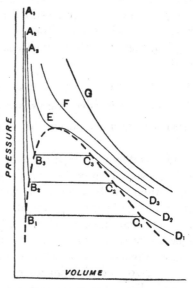

Fig. 14. Isothermal Lines.

such line $ABCD$ is an isothermal for the substance in the successive
states of liquid (A to B), liquid and vapour mixed (B to C),
saturated vapour (at C), superheated vapour (C to D). Now take a
much higher temperature. We get a similar isothermal $A_2 B_2 C_2 D_2$;
and at a still higher temperature another isothermal $A_3 B_3 C_3 D_3$,
and so on. The higher the temperature the nearer do B and C
approach each other, and if the temperature be made high enough
the horizontal portion of the isothermal line vanishes.

77. The Critical Point: Critical Temperature and Critical Pressure. A curve (shown by the broken line) drawn through $B_1B_2B_3$, etc. is continuous with one passing through $C_1C_2C_3$, and it is only within the region of which this curve is the upper boundary that any change from liquid to vapour takes place. The branch $B_1B_2B_3$, which shows the volume of the liquid, meets the branch $C_1C_2C_3$, which shows the volume of the saturated vapour, in a rounded top. The summit of this curve represents a state which is called the *Critical Point*. The temperature for an isothermal line E that would just touch the top of this curve is called the *Critical Temperature*. We might define the critical temperature in another way by saying that if the temperature of a vapour is above the critical temperature no pressure, however great, will cause it to liquefy. The pressure at the critical point is called the *Critical Pressure*; at any higher pressure the substance cannot exist as a non-homogeneous mixture, partly liquid and partly vapour.

Starting from D and increasing the pressure, the temperature being kept constant, we may trace any of the isothermals backwards. The initial state is then that of a gas (a superheated vapour). If the temperature is low enough we have a discontinuous process $DCBA$: as the pressure increases C is reached when the vapour is saturated and condensation begins: at B condensation is complete, and from B upwards towards A we are compressing liquid. At any point between C and B the substance exists in two states of aggregation; part is liquid and part is vapour. But if the temperature is above the critical temperature the isothermal is one that lies altogether outside of the boundary curve, shown by the broken line; in that case the substance does not suffer any sharp change of state as the pressure rises. It passes from the state of a gas to that of a liquid in a continuous manner, following a course such as is indicated by the lines F or G, and at no stage in the process is it other than homogeneous.

The continuity of the liquid and gaseous states in any fluid will be realized if one thinks of a process by which the fluid may pass from one to the other state without any

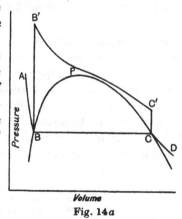

Fig. 14a

abrupt change such as occurs in boiling. Suppose for example that we start with liquid at B (fig. 14a) and heat it at constant volume to a temperature above the critical temperature. This operation is represented by the line BB'. Then let it expand isothermally along the line $B'C'$ which lies above the critical point P, and finally let it be cooled at constant volume from C' to C. This brings it to the initial temperature and pressure, but it is now saturated vapour. During each of the three steps the fluid has remained homogeneous; the passage from liquid to vapour has occurred in a manner so continuous that there is no traceable point of transition from one phase to the other.

The critical temperature for steam is about 374° C., and the corresponding pressure 3200 pounds per square inch. In the action of an ordinary steam-engine the critical pressure is not approached except in certain Benson boiler installations. But with carbonic acid, whose critical temperature is only about 31° C., the properties of the fluid in the neighbourhood of the critical point are important in connection with refrigerating machines which employ carbonic acid as working substance.

The so-called permanent gases, oxygen, nitrogen, hydrogen and so forth, are vapours which under ordinary conditions are very highly superheated. Their critical temperatures are so low that it is only by extreme cooling that they can be brought into a condition which makes liquefaction possible. The critical temperature of hydrogen is −240° C. or 33° absolute. Even helium, the most refractory of the gases, has been liquefied, but only by cooling it to a temperature within about 5 degrees of the absolute zero.

78. **Adiabatic Expansion of a Fluid.** In the adiabatic expansion of any fluid its temperature, pressure, energy, and total heat fall, but the entropy remains constant since the operation is reversible and no heat is communicated to or taken from the substance. A case which is important in practice is that of a wet vapour, such as the mixture of steam and water which expands in the cylinder of an engine. When a saturated vapour expands adiabatically it usually becomes wet; and if it is initially wet (unless very wet*) it becomes wetter. The fact that its entropy

---

* When the mixture is very wet to begin with, it becomes drier during adiabatic expansion, because so much of the portion which was initially liquid vaporizes under the reduced pressure that this more than makes up for condensation in the portion which was initially vapour (see Art. 100).

remains unaltered allows the change of condition to be investigated as follows, if we assume that the liquid and vapour in the mixture are in thermal equilibrium throughout the process.

For greater generality we shall suppose the vapour to be wet to begin with. Let the initial temperature be T_1 and the initial dryness q_1. In this state (Art. 74) the entropy is

$$\phi = \phi_{w_1} + \frac{q_1 L_1}{T_1},$$

L_1 being the latent heat of the vapour and ϕ_{w_1} the entropy of the liquid, both at the temperature T_1. These quantities are found in the tables. Let the substance expand adiabatically to any lower temperature T_2, at which the latent heat is L_2 and the entropy of the liquid is ϕ_{w_2}: we have to find the resulting value of the dryness, q_2. The entropy may now be expressed as

$$\phi_{w_2} + \frac{q_2 L_2}{T_2},$$

and since there has been no change of entropy this is equal to the initial value ϕ. Hence

$$q_2 = \frac{T_2}{L_2} (\phi - \phi_{w_2}).$$

This equation serves to determine the dryness after expansion, and once it is known the volume V_q is readily found as in Art. 74. Its exact value is $q_2 V_{s_2} + (1 - q_2) V_{w_2}$, which is practically equal in ordinary cases to $q_2 V_{s_2}$, V_{s_2} being the volume of saturated vapour at the temperature T_2. The pressure is the saturation pressure corresponding to T_2. Thus the calculation fixes a point in the adiabatic line of the pressure-volume diagram, for expansion from the initial conditions. A series of points may be found in the same way, corresponding to successive assumed temperatures which are reached in the course of the expansion, if it is desired to trace the line.

In the special case when the vapour is dry and saturated to begin with, the constant entropy ϕ is equal to ϕ_{s_1} and the expression for the wetness after expansion to any temperature T_2 becomes

$$q_2 = \frac{T_2}{L_2} (\phi_{s_1} - \phi_{w_2}).$$

Similarly, if the substance is entirely liquid in the initial state, the pressure being sufficient to prevent vapour from forming, adiabatic expansion will cause some of it to vaporize. Its initial entropy is ϕ_{w_1}, and since this does not change,

$$q_2 = \frac{T_2}{L_2}(\phi_{w_1} - \phi_{w_2}),$$

after expansion to a temperature T_2*.

So far, this Article has dealt with the expansion of saturated or wet vapour. When a *superheated* vapour expands adiabatically its expansion is divisible into two distinct stages. The first stage brings it down to the state of saturation; in the second stage it is a wet vapour and the foregoing methods of calculation apply.

The amount of adiabatic expansion which will bring a super-heated vapour down to saturation is determined from the fact that the entropy is constant. We have only to find at what temperature (or pressure) the entropy of saturated vapour is equal to that of the superheated vapour in the given initial state. This comparison is readily made when tables or charts giving the properties of the substance are available, as they are for steam, ammonia, carbon-dioxide, and certain other fluids. Charts of the type which will be described in the next chapter serve well for the examination of cases in which the vapour is superheated before expansion. Such cases occur frequently in steam-engine practice; most modern engines of large power are supplied with highly superheated steam, which tends to become wet as its expansion proceeds. With a suitable chart it is easy to trace the whole course of any adiabatic expansion through the region of superheat, past the point of saturation, and finally in the region of wetness.

It must not be supposed that the expansion of steam in an actual engine is adiabatic, for there is always some transfer of heat between the working fluid and the metal of the cylinder and piston. There is no perfectly non-conducting material for the surfaces. The ideal of adiabatic expansion is approximated to when the action

---

* As an example of the calculation, let steam initially dry and saturated at a temperature of 190° C. ($P = 182 \cdot 04$ lb. per sq. in.) expand adiabatically to a pressure of one atmosphere (temperature 100° C.). The entropy, which remains constant during expansion, is $1 \cdot 5540$, ϕ_{w_2} is $0 \cdot 8120$ and L_2 is $589 \cdot 06$. With these data q_2 is $0 \cdot 860$, 14 per cent. of the steam has become liquefied and the volume which was originally $2 \cdot 504$ cub. ft. per lb. is $23 \cdot 052$ cub. ft. after expansion.

With water initially at 190° C. ($P = 182 \cdot 04$ lb. per sq. in.), let adiabatic expansion occur to a pressure of one atmosphere, q_2 becomes $0 \cdot 154$: in other words $15 \cdot 4$ per cent. of the water vaporizes in consequence of the expansion. The resulting volume is $4 \cdot 127$ cub. ft. per lb. Complete recondensation would occur on compressing to $182 \cdot 04$ lb. per sq. in., when the volume would become $0 \cdot 01829$ cub. ft.

occurs so fast as not to allow time for any considerable transfer of heat.

Sudden, and therefore practically adiabatic, expansion from a high pressure may be used to produce a very low temperature. It was in this way that gases such as oxygen and nitrogen were for the first time liquefied. The gas was compressed and was cooled in the compressed state to a fairly low temperature. It was then suddenly expanded, and the further cooling which resulted from this expansion caused a portion of it to become liquid.

79. Supersaturation. In discussing adiabatic expansion we have assumed that there is at every step in the expansion a condition of equilibrium in the fluid, that is to say equilibrium between the part that is vapour and the part that is liquid. But it is known, as a result of experiment, that when a vapour is suddenly cooled by adiabatic expansion the condition of equilibrium is not reached at once. Suppose the vapour to be initially dry and saturated: on expansion a part of it must condense if equilibrium is to be established. This condensation takes an appreciable time; it is a surface phenomenon, taking place partly on the inner surfaces of the containing vessel and partly by the growth of drops throughout the volume. Consequently the sudden expansion of a vapour may produce temporarily a condition that is called *supersaturation*, in which the substance continues for a time to exist as a homogeneous vapour, although its pressure and temperature are such that the condition of equilibrium would require a part of it to be condensed. In this supersaturated state the density of the vapour is abnormally high, higher than the density of saturated vapour at the actual pressure. The temperature is also abnormally low, lower than the temperature of saturation at the actual pressure: for this reason the supersaturated vapour might be called *supercooled**. The supersaturated condition is not stable: it disappears through the condensation of a part of the vapour, and the resulting mixture of vapour and liquid has its temperature raised by the latent heat which is given out in this condensation. We shall see later, in connection with the theory of steam jets (Art. 135), that expansion involving supersaturation may occur under practical conditions†.

* "Undercooled" is an expression often used.

† An interesting example of supersaturation occurs when dust-free air saturated with water-vapour is suddenly expanded. So long as particles of dust are present a mist forms (on slight expansion) by the condensation of water on them as nuclei, but if they are removed before such an expansion the mist does not form and the vapour becomes supersaturated. If however the ratio

The supercooling of a vapour without condensation is analogous to the supercooling of a liquid without crystallization. In both cases there is a departure from the state of equilibrium, and in both cases the restoration of equilibrium involves an *irreversible* action within the substance. The normal adiabatic expansion of a vapour, dealt with in Art. 78, is reversible, but if there has been supercooling there is an irreversible development of heat within the fluid when the supercooled vapour passes into the stable state of a mixture of liquid and saturated vapour.

80. Change of Internal Energy and of Total Heat in Adiabatic Expansion. "Heat-Drop." When a fluid expands adiabatically from any condition a to any other condition b the decrease of internal energy $E_a - E_b$ is equal to the thermal equivalent of the work done in the expansion. This is because it takes in no heat and consequently the work which it does in expanding is done at the expense of its stock of internal energy.

Referring to the pressure-volume diagram (fig. 15) the work done during expansion from a to b is measured by the area *mabn*, consequently in adiabatic expansion

$$E_a - E_b = A \text{ (area } mabn).$$

Further, the decrease of total heat which the substance undergoes during the process is equal to the thermal equivalent of the area *eabf*. To prove this, we have, by the definition of the total heat (Art. 67),

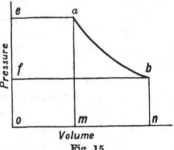

Fig. 15

$$I_a = E_a + AP_a V_a,$$
and $$I_b = E_b + AP_b V_b;$$
from which

$$I_a - I_b = E_a - E_b + A\,(P_a V_a - P_b V_b)$$
$$= A \text{ (area } mabn + \text{area } eamo - \text{area } fbno)$$
$$= A \text{ (area } eabf).$$

This is true whatever be the condition of the fluid before expansion:

of expansion is large, so that there is much supercooling, a mist forms even in the absence of dust: in that case it appears that drops of the liquid form about smaller nuclei, which may consist of ionized molecules, or of groups accidentally linked in the course of the molecular collisions that occur in any gas. (See Aitken, *Trans. R.S.E.* vol. xxx, and *Nature*, March 1, 1888 and Feb. 27, 1890: also C. T. R. Wilson, *Phil. Trans. R.S.* vol. clxxxix, 1897.)

it applies for example to superheated as well as to saturated or wet steam, or to any gas.

It may be instructive to the student to have the same proof put in a somewhat different form. From the equation which defines the total heat I in any state, namely,

$$I = E + APV,$$

we have by differentiation

$$dI = dE + Ad(PV)$$
$$= dE + APdV + AVdP.$$

But in any small change of state it follows from the conservation of energy that the increase of internal energy *plus* the work done by the fluid is equal to the heat taken in, or

$$dE + APdV = dQ,$$

where dQ is the heat taken in during the change. Hence in any small change of state

$$dI = dQ + AVdP.$$

In an adiabatic operation $dQ = 0$, and hence in that case

$$dI = AVdP.$$

Therefore if the fluid expands adiabatically from state a to state b the resulting decrease in its total heat, namely

$$I_a - I_b = A \int_b^a VdP.$$

This integral is the area $eabf$ of the pressure-volume diagram (fig. 15). It is the whole work done in a cylinder when the fluid is admitted at the pressure corresponding to state a, then expanded adiabatically to state b, and then discharged at the pressure corresponding to state b.

The decrease of total heat in expansion, $I_a - I_b$, is called the "Heat-drop." It is a quantity of much importance in the theory of heat-engines. The above equation shows that under adiabatic conditions the whole work done in the cylinder, when expressed in heat units, is measured by the heat-drop. In the next chapter this principle will be applied to infer from the heat-drop the work that can be done in steam-engines under various assumed conditions.

Given the state in which the steam is supplied and the limit of pressure down to which it may be expanded, the adiabatic heat-drop measures the greatest amount of work that is ideally obtainable in any engine, whether of the piston or the turbine type, provided there is no further supply of heat to the steam in the course of its action.

CHAPTER III

THEORY OF THE STEAM-ENGINE AND OTHER VAPOUR ENGINES

81. Carnot's Cycle with Steam or other Vapour for Working Substance. We are now in a position to study the action of a heat-engine employing water and steam, or any other liquid and its vapour, as the working substance. To simplify the first consideration of the subject as far as possible, let it be supposed that we have, as before, a long cylinder, composed of non-conducting material except at the base, and fitted with a non-conducting piston; also a source of heat A at some temperature T_1; a receiver of heat, or as we may now call it, a condenser, C, at some lower temperature T_2; and also a non-conducting cover B (as in Art. 36). Then Carnot's cycle of operations can be performed as follows. To fix the ideas, suppose that there is unit mass of water in the cylinder to begin with, at the temperature T_1.

(1) Apply A, and allow the piston to rise against the constant pressure P_1 which is the saturation pressure corresponding to the temperature T_1. The water will take in heat and be converted into steam, expanding isothermally at the temperature T_1. This part of the operation is shown by the line ab in fig. 16.

(2) Remove A and apply B. Allow the expansion to continue adiabatically (bc), with falling pressure, until the temperature falls to T_2. The pressure will then be P_2, namely, the pressure which corresponds in the steam table to T_2, which is the temperature of the cold body C.

(8) Remove B, apply C, and compress. Steam is condensed by rejecting heat to C. The action is isothermal, and the pressure remains P_2. Let this be continued until a certain point d is reached, which is to be chosen so that adiabatic compression will complete the cycle.

(4) Remove C and apply B. Continue the compression, which is now adiabatic. If the point d has been rightly chosen, this will complete the cycle by restoring the working fluid to the state of water at temperature T_1.

The indicator diagram for the cycle is drawn in fig. 16, the lines bc and da having been calculated by the method of Art. 78, for a particular example in which the initial pressure is 90 pounds per square inch ($T_1 = 433$), and the expansion is continued down to the pressure of the atmosphere, 14·7 pounds per square inch ($T_2 = 373$).

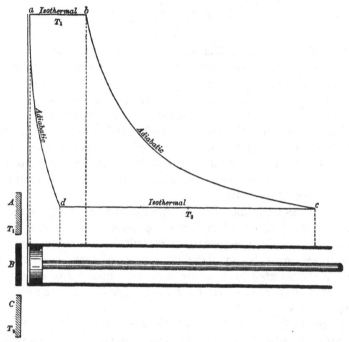

Fig. 16. Carnot's Cycle with water and steam for working substance.

Since the process is reversible, and since heat is taken in only at T_1 and rejected only at T_2, the efficiency (by Art. 38) is

$$\frac{T_1 - T_2}{T_1}.$$

The heat taken in per unit mass of the liquid is L_1, and therefore the work done is

$$\frac{L_1 (T_1 - T_2)}{T_1},$$

a result which may be used to check the calculation of the lines in the diagram by comparing it with the area which they enclose. It will be seen that the whole operation is strictly reversible in the thermodynamic sense.

Instead of supposing the working substance to consist wholly of water at a and wholly of steam at b, the operation ab might be taken to represent the partial evaporation of what was originally a mixture of steam and water. The heat taken in would then be $(q_b - q_a) L_1$, and as the cycle would still be reversible the area of the diagram would be

$$\frac{L_1 (q_b - q_a) (T_1 - T_2)}{T_1}.$$

82. Efficiency of a Perfect Steam-Engine. Limits of Temperature. If the action here described could be realized in practice, we should have a thermodynamically perfect steam-engine using saturated steam. Like any other perfect heat-engine, an ideal engine of this kind has an efficiency which depends upon the temperatures between which it works, and upon nothing else. The fraction of the heat supplied to it which such an engine would convert into work would depend simply on the two temperatures, and therefore on the pressures, at which the steam was produced and condensed respectively.

It is interesting therefore to consider what are the limits of temperature between which steam-engines may be made to work. The temperature of condensation is limited by the consideration that there must be an abundant supply of some substance to absorb the rejected heat; water is actually used for this purpose, so that T_2 has for its lower limit the temperature of the available water-supply.

To the higher temperature a limit is set by the difficulties that attend the use of high-pressure steam. Recent practice employs much higher pressures than were formerly attempted, and in exceptional instances plants have been installed with pressures of the order of 2000 pounds per square inch. This means that the upper limit of temperature, so far as saturated steam is concerned, is something like 350° C. By superheating the steam the range of temperature has been extended to 450° and even 500° C.*, but the chief part of the heat-supply is still taken in at the temperature of saturation. Superheating has great practical advantages in keeping the steam more nearly dry during its expansion, but it does not much increase the theoretical efficiency above the Carnot value corresponding to the boiler temperature. The Carnot efficiency working between 600° K. and 300° K. would be 50 per cent.

* There is difficulty in finding materials to withstand the conditions for sufficient time: a gradual "creep" occurs owing to the plasticity of the metal.

Under the most favourable conditions, a steam-engine comes far short of taking full advantage of the high temperature at which heat is produced in the combustion of coal. From the thermodynamic point of view its worst fault is the irreversible drop of temperature between the combustion-chamber of the furnace and the boiler. The combustion of the fuel supplies heat at a high temperature: but a great part of the convertibility of that heat into work is at once sacrificed by the fall in temperature which takes place before the conversion into work begins.

83. Entropy-Temperature Diagram for a Perfect Steam-Engine. The imaginary steam-engine of Art. 81 has the same very simple entropy-temperature diagram as any other engine which follows Carnot's cycle. The four operations are represented by the four sides of a rectangle (fig. 17). The first operation changes water (at the upper limit of temperature) into saturated steam at the same temperature; the entropy accordingly changes from ϕ_w to ϕ_s. This is shown by the constant-temperature line ab in fig. 17. In the second operation—which is adiabatic expansion—the entropy does not change, and the temperature falls to the lower limit, at which heat is to be rejected: this is represented by the line of constant entropy bc. In the third operation, cd, the temperature remains constant and the entropy is restored to its original value, heat being rejected to the cold body. In the fourth operation—which is adiabatic compression—the entropy does not change, and the temperature rises to the upper limit; the substance has returned to its initial state in all respects. In order to be comparable with other diagrams which will follow, fig. 17 is sketched for a particular example in which P_1 is 180 pounds per sq. inch, and P_2 is 1 pound per sq. inch ($t_1 = 189 \cdot 48°$ C., and $t_2 = 38 \cdot 75°$ C.).

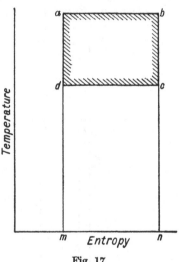

Fig. 17

Expressed in terms of entropy, the heat taken in (during ab) is $T_1 (\phi_s - \phi_w)$. This is represented by the area under ab measured

down to the absolute zero of temperature, namely the area *mabn*.
The heat rejected (during *cd*) is $T_2(\phi_s - \phi_w)$ and is represented by
the area *ncdm*. The thermal equivalent of the work done in the
cycle is accordingly $(T_1 - T_2)(\phi_s - \phi_w)$, and is represented by the
area *abcd*, enclosed by the lines which represent the four reversible
operations. The efficiency is

$$\frac{(T_1 - T_2)(\phi_s - \phi_w)}{T_1(\phi_s - \phi_w)} = \frac{T_1 - T_2}{T_1}.$$

In the example for which the diagram is drawn, with the data
stated above, the numerical value of this is 0·326.

84. **Use of "Boundary Curves" in the Entropy-Temperature
Diagram.** In fig. 18 the diagram of fig. 17 is drawn over again,
with the addition of a curve through *a* which represents the values
at various temperatures of ϕ_w, the entropy of water when steam

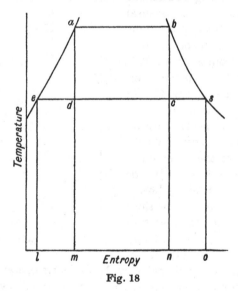

Fig. 18

is just about to form, and a curve through *b* which represents at
various temperatures the value of ϕ_s, the entropy of dry saturated
steam. These curves are called Boundary Curves. They are readily
drawn from the data in the steam tables. Any point on the
boundary curve through *a* would relate to the entropy of water;
between the two curves any point in the diagram relates to a
mixture of water and steam; to the right of the boundary curve

through b any point would relate to steam in the superheated state. We are not at present concerned with the outlying regions but only with the space between the two curves, within which the points c and d fall. Let the line cd be produced both ways to meet the boundary curves in e and s. Then the ratio of cs to es represents the fraction of the steam which becomes condensed during the adiabatic expansion bc from the condition of saturation at b.

To prove this we may first consider the meaning of any horizontal (isothermal) line such as se on the entropy-temperature diagram between the two boundary curves. It represents complete condensation of 1 lb. of dry saturated steam, under constant temperature and pressure. During its conversion from the condition of dry saturated steam (at s) to water (at e) the steam gives out a quantity of heat which is measured by the area under the line, namely the area $osel$. Any intermediate point in the line represents a mixture of water and steam; thus c represents a mixture which, though it has actually been produced by adiabatic expansion from b, might have been produced by partial condensation from s under constant pressure, a process which would be represented by sc, or by partial evaporation under the same constant pressure from e, a process which would be represented by ec. Now if the mixture at c were completely condensed under constant pressure to e, the heat given out would be measured by the area $ncel$. This heat is given out by the condensation of that part of the mixture which consisted of steam. Hence the fraction which existed at c as steam, or in other words the *dryness* of the mixture at c, is measured by the ratio of the areas $ncel$ to $osel$, which is equal to the ratio of the lengths ec to es. Hence also the ratio cs to es measures the wetness of the mixture at c.

An entropy-temperature diagram on which the boundary curves are drawn therefore gives a convenient means of determining the wetness of steam at any stage in the process of adiabatic expansion. It is only necessary to draw a vertical line through the point representing the initial condition. That line represents the adiabatic process, and the segments into which it divides a horizontal line drawn from one boundary curve to the other at any level of temperature represent the proportions of water and steam in the resulting mixture. This is true not only of the final stage, when adiabatic expansion is complete, but of any intermediate stage; for the argument given above obviously applies to a horizontal line drawn at any temperature between the two boundary curves.

Similarly the point d, which represents the wet mixture at the beginning of adiabatic compression da, shows by the ratio of segments ds to de what is the proportion of water to steam at which the third stage of the cycle has to be arrested, in order that adiabatic compression may bring the mixture wholly to the state of water when the cycle is completed by the operation da.

The student should compare this graphic method of studying the wetness resulting from adiabatic expansion with the calculations given in Art. 78. He will observe that both have the same basis. At any temperature T the length es of the isothermal line drawn from the water boundary curve at e to the steam boundary curve at s is L/T, and the intercept ec up to any intermediate point c on that line is qL/T, where q is the dryness of the mixture at the point c. The same principle of course holds for the entropy-temperature diagram of any other fluid.

85. Modified Cycle omitting Adiabatic Compression. Consider next a modification of the Carnot cycle of Art. 81. Let the first and second operations occur as they do there, but let the third operation be continued until the steam is wholly condensed. The substance then consists of water at T_2, and the cycle is completed by heating it, in the condition of water, from T_2 to T_1. In the simple engine of Art. 81, where all the operations occur in a single

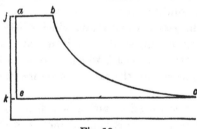

Fig. 19

vessel, this could be done by increasing the pressure exerted by the piston from P_2 to P_1, after condensation is complete, then removing the cold body C and applying the hot body A. The water is therefore heated at P_1 and no steam is formed till the temperature reaches T_1.

The pressure-volume diagram (or indicator diagram) of a cycle modified in this manner is shown by $abce$ in fig. 19. The sketch is not drawn to scale. As before, ab is the operation of forming steam, from water, at T_1 and P_1; bc is adiabatic expansion from T_1

and P_1 to T_2 and P_2. Then *ce* is complete condensation at T_2 and P_2. The fourth operation *ea* now involves two stages, first raising the pressure of the condensed water from P_2 to P_1 and then heating it from T_2 to T_1. During both of these stages the changes of volume are negligible in comparison with those that take place in the other operations.

The entropy-temperature diagram for this modified cycle is shown by *abce* in fig. 20, where the same letters as in fig. 19 are used for corresponding operations. As in the Carnot cycle, *ab* represents the

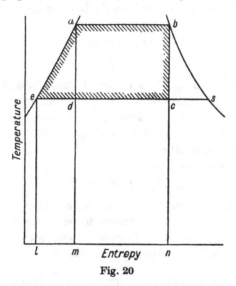

Fig. 20

conversion of a pound of water at T_1 into dry-saturated steam at T_1, and *bc* represents its adiabatic expansion to T_2, resulting in a wet mixture at *c*, the dryness of which is measured by the ratio *ec/es*. Then *ce* represents the complete condensation at T_2 of the steam in this wet mixture, and *ea*, which practically coincides with the boundary curve, represents the re-heating of the condensed water from T_2 to T_1, after its pressure has been raised to P_1 so that no steam is formed during this operation*.

The working substance behaves reversibly throughout all these operations, and therefore the work done in the cycle is represented

---

* The line *ea* in both diagrams, figs. 19 and 20, really stands for a broken line *ea'a*, where *ea'* represents the raising of pressure from P_2 to P_1 at constant temperature T_2, and *a'a* represents the heating from T_2 to T_1 at constant pressure P_1. In fig. 19 *a'* practically coincides with *a*; in fig. 20 *a'* practically coincides with *e*.

by the area *abce* in the entropy-temperature diagram of fig. 20. The diagram further exhibits the heat taken in and the heat rejected. The whole heat taken in is measured by the area *leabn*, and of this the area *leam* measures the heat taken in during the last operation, while the water is being re-heated, and the area *mabn* measures the heat taken during the first operation, while the water is turning into steam. The area *ncel* measures the heat rejected, namely during the condensing process *ce*.

To express algebraically the work done in the cycle, refer to the indicator diagram, fig. 19, and let the lines *ba* and *ce* be produced to meet the line of no volume in *j* and *k*. Then, by Art. 80, the area *jbck* is an amount of work equivalent to the difference of total heats

$$I_b - I_c,$$

namely the "heat-drop" of a pound of steam in expanding adiabatically from the condition at *b* to the condition at *c*. The small area *jaek* is $(P_1 - P_2) V_{w_2}$, where V_{w_2} is the volume of a pound of water at T_2, which we may take to be practically constant for the purposes of this calculation.

Hence the expression

$$I_b - I_c - A (P_1 - P_2) V_{w_2}$$

is the thermal equivalent of the work done in the cycle. If figs. 19 and 20 were both carefully drawn to scale for any particular example, a measurement of the enclosed area *abce* in either figure would give a result in agreement with this calculation.

86. Engine with Separate Organs. The importance of the modified cycle described in Art. 85 lies in the fact of its being the ideal performance of an engine with a separate condenser into which the working fluid passes after expansion, and from which it is pumped, in the cold state, back to the boiler. The engine of fig. 16 had one organ only—a cylinder which also served as boiler and as condenser. We come nearer to the conditions that hold in practice if we think of an engine with separate organs, shown diagrammatically in fig. 21, namely a boiler *A* kept at T_1, a non-conducting cylinder and piston *B*, and a surface condenser *C* kept at T_2. To these must be added a feed-pump *D* which returns the condensed water to the boiler. Provision is made by which the cylinder can be put into connection with the boiler or condenser at will.

With this engine the cycle of fig. 19 can be performed. An indicator diagram for the cylinder *B* is sketched in fig. 22. Steam is admitted from the boiler, giving the line jb. At b "cut-off" occurs, that is to say the valve which admits steam from the boiler to the cylinder is closed. The steam in the cylinder is then expanded adiabatically to the pressure of the condenser, giving the line bc. At c the "exhaust" valve is opened which connects the cylinder

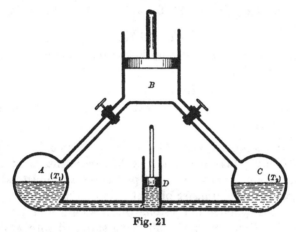

Fig. 21

with the condenser. The piston then returns, discharging the steam to the condenser and giving the line ck. The area $jbck$ represents the work done in the cylinder B. The condensed water

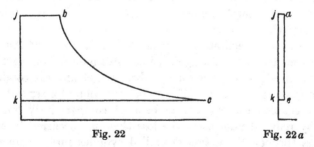

Fig. 22   Fig. 22 a

is then returned to the boiler by the feed-pump, and the indicator diagram showing the work expended upon the pump during this operation is sketched in fig. 22 a. It is the rectangle $keaj$; where ke represents the up-stroke in which the pump fills with water at the pressure P_2, and aj represents the down-stroke in which it discharges water to the boiler against the pressure P_1. If we superpose the diagram of the pump on that of the cylinder we get their difference, namely $abce$ (fig. 19), to represent the net amount of work done by the fluid in the cycle. It is the excess of the work done by the fluid in the cylinder over that spent upon it in the pump.

Taking the two parts separately, the adiabatic heat-drop,

$$I_b - I_c,$$

is the thermal equivalent of the work done by the fluid in the cylinder, and

$$(P_1 - P_2) V_{w_2}$$

is the work spent upon the fluid in the feed-pump. Accordingly the difference, namely

$$I_b - I_c - A (P_1 - P_2) V_{w_2},$$

is, as before, the thermal equivalent of the work obtained in the cycle as a whole.

87. The Rankine Cycle. This cycle is commonly called the Rankine Cycle*. Like the Carnot cycle it represents an ideal that is not practically attainable, for it postulates a complete absence of any loss through transfer of heat between the steam and the surfaces of the cylinder and piston. But it affords a very valuable criterion of performance by furnishing a standard with which the efficiency of any real engine may be compared, a standard which is less exacting than the cycle of Carnot, but fairer for comparison, inasmuch as the fourth stage of the Carnot cycle is necessarily omitted when the steam is removed from the cylinder before condensation. A separate condenser is indispensable in any real engine that pretends to efficiency.

The use of a separate condenser was in fact one of the great improvements which distinguished the steam-engine of Watt from the earlier engine of Newcomen, where the steam was condensed in the working cylinder itself. The introduction of a separate condenser enabled the cylinder to be kept comparatively hot, and thereby reduced immensely the loss that had occurred in earlier engines through the action of chilled cylinder surfaces upon the entering steam. But a separate condenser, greatly though it adds to efficiency in practice, excludes the compression stage of the Carnot cycle, and consequently makes the Rankine cycle a more appropriate ideal with which the performance of an ordinary engine should be compared.

The efficiency of the Rankine cycle is less than that of a Carnot cycle with the same limits of temperature. This is because, in the

* Also sometimes associated with Clausius' name.

Rankine cycle, the heat is not all taken in at the top of the range. In the Rankine cycle, as in Carnot's, all the internal actions of the working substance are, by assumption, reversible, and consequently each element of the whole heat-supply produces the greatest possible mechanical effect when regard is had to the temperature at which that element is taken in. But part of the heat is taken in at temperatures lower than T_1, namely while the working substance is having its temperature raised from T_2 to T_1 in the fourth operation. Hence the average efficiency is lower than if all had been taken in at T_1, as it would be in the cycle of Carnot.

Each pound of steam does a larger amount of work in the Rankine cycle than it does in the Carnot cycle. This will be apparent when the areas are compared which represent the work in the corresponding diagrams: the area *abce* with the area *abcd* in fig. 20. But the quantity of heat that has to be supplied for each pound in the Rankine cycle is also greater, and in a greater ratio: it is measured by the area *leabn*, as against *mabn*. Hence the efficiency is less in the Rankine cycle. One may put the same thing in a different way by saying that, in the Rankine cycle, of the whole heat-supply the part *leam* does only the comparatively small amount of work *ead* and the remainder of the heat-supply, namely *mabn*, does the same amount of work as it would do in a Carnot cycle.

88. Efficiency of a Rankine Cycle. Taking in the first instance a Rankine cycle in which the steam supplied to the cylinder is dry and saturated, the whole amount of heat taken in is the quantity required to convert water at P_1 and T_2 into saturated steam at P_1. This quantity is $I_{s_1}-\{I_{w_2}+A(P_1-P_2)V_{w_2}\}$, for the total heat of the water at P_1 and T_2 is greater than I_{w_2} by the quantity $A(P_1-P_2)V_{w_2}$.

The work done is (by Art. 85) equal to the heat-drop *minus* the work spent in the feed-pump, or $I_{s_1}-I_c-A(P_1-P_2)V_{w_2}$, where I_c is the total heat of the wet mixture after adiabatic expansion.

The efficiency in the cycle as a whole is therefore

$$\frac{I_{s_1}-I_c-A(P_1-P_2)V_{w_2}}{I_{s_1}-I_{w_2}-A(P_1-P_2)V_{w_2}}.$$

The feed-pump term $A(P_1-P_2)V_{w_2}$ is relatively so small that it is often omitted in calculations relating to ideal efficiency, just as it is omitted in stating the results of tests of the performance of real engines. In such tests it is customary to speak of the work done per lb. of steam, without making any deduction for the work that has to be spent per lb. in returning the feed-water to the boiler. But in the complete analysis of a Rankine cycle the feed-pump term should, in strictness, be taken into account; it is only then that the area of the entropy-temperature diagram gives a true measure of the work done.

If however we are concerned only with the work done in the cylinder of the ideal engine, then the heat-drop alone has to be reckoned. It is the exact measure of that work. The ratio of the heat-drop to the heat supplied shows what proportion of the supply is converted into work in the cylinder, under the ideal conditions of adiabatic action.

89. Calculation of the Heat-Drop. It is therefore essential to be able to calculate the heat-drop in ideal engines under any assigned initial and final conditions. For this purpose we have to find I_0, the total heat of wet steam after adiabatic expansion. One way of doing so would be first to calculate the dryness q and then apply equation (2) of Art. 74, $I_q = I_w + qL$. But equations (6) and

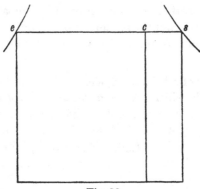

Fig. 23

(7) of that Article give a more convenient method, which is available here because we know the entropy of the mixture. These expressions may be directly obtained by considering what amount of heat the wet mixture would have to part with if it were to be wholly condensed, and what amount of heat it would have to take up if it were to be wholly vaporized, under the constant pressure corresponding to the temperature of saturation T in either case. To bring a mixture at c (fig. 23) into the condition of water at e would require the removal of a quantity of heat equal to the area

under ec, measured down to the base-line which is at the absolute zero of temperature. This quantity is $T (\phi - \phi_w)$, where ϕ is the entropy at c and ϕ_w is the entropy of water (at e). On the other hand, to bring it to the condition of saturated steam would require the addition of a quantity of heat equal to the area under cs, namely $T (\phi_s - \phi)$. Hence the total heat of the mixture at c is

$$I_c = I_w + T (\phi - \phi_w),$$

or

$$I_c = I_s - T (\phi_s - \phi),$$

as in Art. 74.

The entropy ϕ of the wet mixture is the constant entropy under which adiabatic expansion has taken place: it is to be calculated from the initial conditions. This method of finding the total heat after adiabatic expansion makes no assumption as to what the state of the steam was before expansion: it is equally valid whether the steam was dry, wet, or superheated to begin with. What is assumed is that after expansion the steam is wet, and that will in general be true even if there be a large amount of initial superheat. It is also assumed (Art. 78) that the vapour and liquid in the wet mixture are in equilibrium.

In the Rankine cycle of Art. 87 it was assumed that the steam was dry and saturated at the beginning of the adiabatic expansion. Consequently its initial total heat was I_{s_1}, and ϕ throughout expansion was equal to ϕ_{s_1}. Under these conditions the total heat after adiabatic expansion is

$$I_c = I_{s_2} - T_2 (\phi_{s_2} - \phi_{s_1});$$

and the heat-drop is

$$I_{s_1} - I_c = I_{s_1} - I_{s_2} + T_2 (\phi_{s_2} - \phi_{s_1}).$$

To take a numerical example, let the steam be supplied in a dry-saturated state at a pressure P_1 of 180 pounds per sq. inch, and let it expand adiabatically to a pressure P_2 of 1 pound per sq. inch, at which $T_1 = 462 \cdot 58°$, $T_2 = 311 \cdot 85°$, $\phi_{s_1} = 1 \cdot 5551$, $\phi_{s_2} = 1 \cdot 9779$, $I_{s_1} = 665 \cdot 2$ and $I_{s_2} = 614 \cdot 5$.

Hence the total heat after adiabatic expansion to the assumed pressure of condensation is

$$I_c = 614 \cdot 5 - 311 \cdot 85 (1 \cdot 9779 - 1 \cdot 5551) = 482 \cdot 6.$$

And the heat-drop

$$I_{s_1} - I_c = 665 \cdot 2 - 482 \cdot 6 = 182 \cdot 6.$$

If we consider the Rankine cycle as a whole, the feed-pump term $A(P_1 - P_2) V_w$ is

$$\frac{144}{1401}(180-1)\,0\cdot0161 = 0\cdot3.$$

Deducing this relatively unimportant feed-pump term from the heat-drop we have 182·3 pound-calories as the thermal equivalent of the net amount of work done in the Rankine cycle.

The heat supplied is

$$I_{s_1} - I_{w_2} - A\,(P_1 - P_2)\,V_{w_2} = 665\cdot2 - 38\cdot7 - 0\cdot3 = 625\cdot2.$$

Hence the efficiency of this Rankine cycle is

$$\frac{182\cdot6}{625\cdot2} = 0\cdot292.$$

A Carnot cycle with the same limits of temperature would have the efficiency 0·326. The difference between this and 0·292 shows the loss which results in the Rankine cycle from not supplying all the heat to the best possible thermodynamic advantage, namely at the top of the temperature range.

90. The Function G. In some steam tables numerical values are given of a function G, defined by the equation

$$G = T\phi - I,$$

which applies to steam in any state, wet, dry-saturated, super-heated, or supercooled. By the help of this function the process of calculating the heat-drop may be slightly shortened. G has the important property that it is constant during a process of evaporation or condensation at constant pressure. For in any step of such a process $\delta I = T\delta\phi$ and T is constant; consequently $\delta G = 0$. Hence the value of G for a wet mixture at temperature T and entropy ϕ, such as the mixture at c (fig. 20) resulting from adiabatic expansion, is the same as G_s, the (tabulated) value of G for dry-saturated steam at the same pressure. Therefore to find I_c, the total heat of the wet mixture, we have

$$I_c = T\phi - G_c = T\phi - G_s.$$

The heat-drop is then determined as before, by subtracting I_c from the total heat before expansion*.

* G (with its sign reversed) is one of three functions to which Willard Gibbs gave the name of "Thermodynamic Potentials": see his *Scientific Papers*, vol. I. He represented them by the symbols ψ, χ, and ζ. Of these, ψ is $E - T\phi$. This function was called by Helmholtz the "Free Energy"; its chief uses are in chemical thermodynamics but it also has some engineering applications. The function χ is the total heat I, namely $E + APV$, and is, as we have seen,

Taking the numerical example as in Art. 89, T is 311·85°, ϕ is 1·5551 and G_s (for saturated steam at a pressure of 1 pound per sq. inch) is 2·59 by the tables. This gives

$$I_c = 311·85 \times 1·5551 - 2·59 = 482·4.$$

So that the heat-drop is 665·2 − 482·4 = 182·8.

Or we may obtain the heat-drop even more directly thus, when tabulated values of G are available. The relation

$$I = T\phi - G$$

holds for any state of the substance. Hence between any two points b and c on the same adiabatic line the heat-drop

$$I_b - I_c = (T_b - T_c)\,\phi - (G_b - G_c).$$

In the present example G_b is the value of G for saturated steam at $P = 180$, which (by the tables) is 54·24. G_c is equal to the value for saturated steam at $P = 1$, which is 2·59. The difference of temperature $T_b - T_c$ is 150·77 degrees. Hence the heat-drop is

$$150·77 \times 1·5551 - (54·24 - 2·59) = 182·8,$$

which agrees with the result found above by less direct methods. The use of the function G in this connection is only a matter of convenience. The procedure in Art. 89 gives the heat-drop readily enough, though not quite so shortly, without the help of G.

91. **Extension of the Rankine Cycle to Steam supplied in any State.** In the Rankine cycle described in Arts. 86–88 the steam was supplied to the cylinder in the dry-saturated state. But the term Rankine cycle is equally applicable whatever be the condition of the working substance on admission, whether wet, dry-saturated, or superheated. As regards the action in the cylinder, all that is assumed is that the substance is admitted at a constant pressure P_1, is expanded adiabatically to a pressure P_2 and is discharged at that pressure, and that in the process there is no transfer of heat to or from the metal, or any other irreversible action. In these conditions the heat-drop in adiabatic expansion from P_1 to P_2 is the thermal equivalent of the area $jbck$ in fig. 22 (compare also Art. 80) and therefore measures the work done in the cylinder, no matter what the condition of the substance on admission may be.

of particular importance in the thermodynamics of engineering. The function ζ is $E - T\phi + APV$ or $I - T\phi$; hence $G = -\zeta$. This function is useful in treating of the equilibrium of different states or phases of the same substance. One example of such equilibrium occurs in wet steam, which is a mixture of two phases. These functions will be referred to again in Chapter VIII.

This applies to wet steam or superheated steam just as much as to dry-saturated steam.

92. Rankine Cycle with Steam initially Wet. A complete

Rankine cycle for steam that is wet on admission to the cylinder is shown on the entropy-temperature diagram by the figure $ab'c'e$ (fig. 24). The point b' is so placed that the ratio ab' to ab is equal to q_1, the assumed dryness on admission. The line $b'c'$ represents

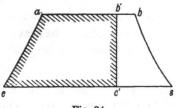

Fig. 24

adiabatic expansion from P_1 to P_2, $c'e$ represents condensation at P_2, and ea represents as before the heating of the condensed water.

The total heat before adiabatic expansion is $I_{w_1}+q_1L_1$ or $I_{s_1}-(1-q_1)L_1$, and the heat supplied is the excess of this quantity above

$$I_{w_2}+A(P_1-P_2)V_{w_2}.$$

The entropy ϕ during adiabatic expansion is

$$\phi_{w_1}+q_1L_1/T_1 \text{ or } \phi_{s_1}-(1-q_1)L_1/T_1.$$

The total heat after adiabatic expansion is

$$I_{s_1}-T_2(\phi_{s_2}-\phi) \text{ or } T_2\phi-G_2.$$

The heat-drop is got by subtracting this from the total heat before adiabatic expansion. Or the heat-drop may be found, as soon as ϕ is calculated, by using the expression

$$(T_1-T_2)\phi-(G_1-G_2).$$

The efficiency which, as before, is practically equal to the heat-drop divided by the heat supplied, is slightly less than when the steam is saturated before expansion; the reason being that the proportion of heat supplied at the upper limit of temperature is now rather less, because part of the water remains unconverted into steam.

As a numerical example let q_1 be 0·9 and let the limits of pressure be the same as in the example of Art. 89. Then the total heat per lb. of the mixture before expansion, which is $I_{s_1}-0.1L_1$, is

$$665.2-0.1\times473.0=617.9.$$

The heat supplied is $617.9-39.0=578.9$. The entropy is

$$\phi_{s_1}-qL_1/T_1=1.5551-\frac{0.1\times473.0}{462.58}=1.4528.$$

The total heat after expansion, $I_{s_2} - T_2 (\phi_{s_2} - \phi)$, is 450·7, or from $T_2 \phi - G_2$, 450·5; the heat-drop is therefore $617·9 - 450·6 = 167·3$, and the same figure is obtained for it by the direct formula

$$(T_1 - T_2) \, \phi - (G_1 - G_2),$$

thus $(462·58 - 311·85) \, 1·4528 - (54·24 - 2·59) = 167·3$. Allowing for the feed-pump term, the efficiency in the complete Rankine cycle is $\dfrac{167·0}{578·9} = 0·288$, as against 0·292 when there was no initial wetness.

In practice the steam supplied to an engine would be wet only if there were condensation in the steam-pipe, such as would occur if it were long or insufficiently covered with non-conducting material, or if the boiler "primed." Priming is a defective boiler action which causes unevaporated water to pass into the steam-pipe along with the vapour. A moderate amount of wetness reduces the ideal efficiency only very slightly*. But its practical effect in reducing the efficiency of an actual engine is much greater, because the presence of water in steam increases the exchanges of heat between it and the metal of the cylinder, and consequently makes the real action depart more widely from the adiabatic conditions which are assumed in the ideal operations of the Rankine cycle.

93. **Rankine Cycle using Superheated Steam.** On the other hand if the steam be superheated before it enters the engine, the exchanges of heat between it and the metal are reduced; the action becomes more nearly adiabatic, and the performance of the real engine approaches more closely the ideal of the Rankine cycle. This is the chief reason why superheating improves the efficiency of a real engine of the cylinder and piston type. In steam turbines it is specially beneficial for by keeping the steam comparatively dry throughout expansion it reduces internal friction and tends to prevent pitting of the blades by the impact of water particles. It is universally employed in the large turbine plants of power stations, and in some of them provision is made for resuperheating the partly expanded steam at one or more stages, because the greatest amount of initial superheating that can be safely given to steam at pressures of the order of 1000 pounds per sq. inch is insufficient to keep the steam sufficiently dry in the later stages of its expansion.

In the entropy-temperature diagram (fig. 25) the line bb' represents the process of superheating steam that was dry-saturated

---

* In the absence of any superheating, steam from a boiler would be more or less wet, "priming" is therefore an inherent characteristic of a boiler.

at b. During this process its entropy and its temperature both increase, and when the pressure and temperature at any stage in the superheating are known the corresponding entropy is found from the tables relating to superheated steam. If we assume that the pressure during superheating is constant, and equal to the boiler pressure, the line bb' is an extension, into the region of superheat, of the constant-pressure line ab. During the process

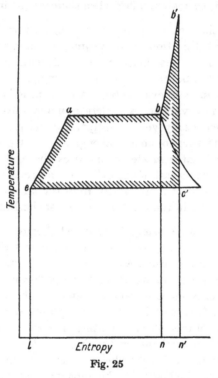

Fig. 25

of superheating the steam takes in a supplementary quantity of heat equal to the area under the curve bb', measured down to the base-line, namely $nbb'n'$. This quantity of heat may also be found from the tables, being equal to the excess of the total heat $I_{b'}$, over that of saturated steam at the same pressure. In the subsequent process of adiabatic expansion $b'c'$ the steam loses superheat, and if the process is carried so far that the adiabatic line through b' crosses the boundary curve, it becomes saturated and then wet, and the final condition is that of a wet mixture at c'. The total heat of this wet mixture is found by the method already described.

The work done in the Rankine cycle as a whole is the area $eabb'c'$, and the heat taken in is the area $leabb'n'$. Both these quantities are readily calculated without the help of the diagram. To find the work done in the cycle we calculate the heat-drop during adiabatic expansion, namely, $I_{b'} - I_{c'}$, and subtract from that the small term which is the thermal equivalent of the work done in the feed-pump, namely, $A\,(P_1 - P_2)\,V_{w_2}$. The whole heat supplied is

$$I_{b'} - I_{w_2} - A\,(P_1 - P_2)\,V_{w_2}.$$

Of this the part supplied during superheating is only $I_{b'} - I_b$, or the area under bb' measured down to the base-line nn'.

The change of temperature and volume that occur during the adiabatic expansion of superheated steam may be readily found when tables are available that show, for each pressure, the volume and the entropy in relation to the temperature. Since the entropy remains constant during adiabatic expansion we have only to look up, for any pressure lower than the initial pressure, the tabulated values of temperature and volume that correspond to an entropy equal to the initial entropy. This applies down to the temperature of saturation: after that the expanding steam becomes wet and its volume at any stage is found from the diagram or by calculating the dryness as in Art. 78.

As a numerical example we may again take $P_1 = 180$ and $P_2 = 1$, and assume that superheating raises the temperature of the steam to $400°$ C. The steam tables show that the total heat of the super-heated steam is 778·0 and its entropy is 1·7573. The heat-supply is $778 \cdot 0 - 39 \cdot 0 = 739 \cdot 0$. After adiabatic expansion the steam is wet, and its total heat, which is $I_{s_2} - T_2\,(\phi_{s_2} - \phi)$, is 545·7. The adiabatic heat-drop is therefore 232·3 and the efficiency of the cycle

$$\frac{232 \cdot 0}{739 \cdot 0} = 0 \cdot 314.$$

This is rather better than the figure for saturated steam because a portion of the heat, but not a very large portion, is now supplied at a higher temperature. Since the entropy remains constant and at 14·7 pounds per sq. inch the entropy of saturated steam is 1·7573; the steam becomes wet when expansion is continued any further. If it is continued till the pressure becomes 1 pound per sq. inch, since

$$q_2 = \frac{T_2}{L_2}(\phi_{s_1} - \phi_{w_2}) = \frac{311 \cdot 85}{575 \cdot 29}(1 \cdot 7573 - 0 \cdot 1323) = 0 \cdot 881,$$

the steam will then be 12 per cent. wet.

Fig. 25 a gives an example of reheating. It uses data taken from modern practice, but is sketched here for the ideal conditions of adiabatic action. The steam is initially superheated to 400° C. from a saturation temperature of 300° (line bb′) at the corresponding pressure of 1246 pounds per sq. inch. Then adiabatic expansion to a

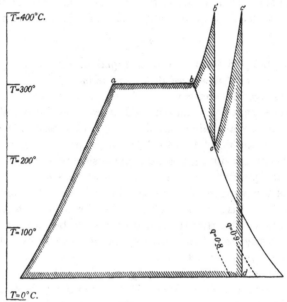

Fig. 25 a

saturated state (line b′c) brings it to a temperature of about 210°. From this it is reheated to 400° (line cc′), and is then expanded (line c′d) to the condenser temperature of 25° C. Even the second superheating does not suffice to keep it completely dry in the last stages of expansion: the diagram shows that under the assumed adiabatic conditions it has become wet at d to the extent of fully 15 per cent.*

94. Reversibility in the Rankine Cycle. Whatever the initial state may be, whether dry-saturated, wet, or superheated, the *internal* action of the working substance in the Rankine cycle is reversible. The only irreversible event in the cycle occurs when the cold condensate is returned to the boiler; the working substance is thereby caused to take in a part of the whole heat-supply when

* For a discussion of reheating in steam turbine practice see papers by H. L. Guy, *Proc. Inst. Mech. Eng.* 1927 and 1929.

its own temperature is lower than that of the hot source. But once the heat is taken in it is all used to the best advantage, in actions that are strictly reversible, and the work which the steam does during expansion to the assigned lower limit is the greatest conceivable amount that can be done.

That work is measured by the adiabatic heat-drop. Whatever be the nature of a real engine—whether of the piston or the turbine type—the adiabatic heat-drop serves as a standard with which to compare the work that is actually got out of the steam from admission to exhaust. The amount of work which steam will do in passing through any real engine cannot exceed—and necessarily falls short of—the ideal amount as expressed by the adiabatic heat-drop. Designers of steam turbines make much use of the adiabatic heat-drop in estimating the expected performance, after making a deduction which is determined by experience with similar machines*.

In any engine the ratio of the actual amount of work done per lb. of steam to the amount that would be done in the ideal Rankine engine under corresponding conditions of supply and exhaust, is called the *Efficiency-Ratio*†. To secure high efficiency in the conversion of heat into work there are obviously two separate conditions to be aimed at: (1) that there shall be a large heat-drop relatively to the heat of formation of the steam: in other words a high value for the ideal efficiency; (2) that there shall be a large Efficiency-Ratio. The second condition depends on practical features of design with which we are not at present concerned. But as regards the ideal efficiency it is important to notice that great advantage can be obtained by making full use of the highest vacuum which can practically be obtained.

By working out the efficiency of the Rankine cycle with different boiler pressures, but with the same pressure of exhaust throughout, it will be seen that when the boiler pressure is fairly high no great advantage is obtained by raising it. Neither the heat-drop nor the heat-supply is much affected. Very large increase in pressure must be made to gain substantial advantage.

* To facilitate such estimates tables are published giving the heat-drop under a wide range of initial conditions as to pressure and superheat, and final conditions as to pressure of condensation. A set of such tables, entitled *The Revised Heat Drop Tables*, 1931, founded on the revised Callendar Tables of that date, has been calculated by Dr Herbert Moss and published by Edward Arnold and Co. for the British Electrical and Allied Industries Research Association.

† See Report of a Committee on the Thermal Efficiency of Steam-Engines, *Min. Proc. Inst. Civ. Eng.* vol. cxxxiv, 1898.

On the other hand it is of great advantage to have what engineers call a "high vacuum"—that is to say to make the pressure of condensation as low as possible. If a high vacuum can be maintained and effectively utilized we obtain from the steam the work which it is capable of doing under conditions of low pressure but of very large volume in the last stages of the expansion. In a given initial steam-pressure both the heat-drop and the possible efficiency are largely increased by reducing the pressure in the condenser.

To secure in a real engine the full benefit of a high vacuum the steam must continue to do useful work in expanding down to the pressure at which condensation is to take place. In engines of the cylinder and piston type this is impracticable for two reasons: the volume of the steam becomes excessive, and the mechanical friction of the piston against the cylinder becomes relatively so great as to absorb all the work done in the final stages. But with the steam turbine these considerations do not apply; there is then nothing to prevent the steam from continuing to do useful work as it expands right down to the pressure of the condenser, and special pains are accordingly taken to maintain a good vacuum in the condenser of a steam turbine. It is largely for this reason that good steam turbines achieve in practice a much greater efficiency than even the best engines of the cylinder and piston type.

95. **Effect of Incomplete Expansion.** When steam is released from the cylinder at a pressure substantially higher than the pressure in the condenser its expansion is said to be incomplete. The effect is to lose available work represented by the toe that is cut off the pressure-volume diagram, as in fig. 26, and to make a corresponding reduction in the efficiency. Release takes place at c and the pressure falls to f while the piston is stationary.

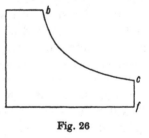

Fig. 26

To exhibit incomplete expansion on the entropy-temperature diagram, imagine that instead of letting part of the steam escape from the cylinder by opening the exhaust-valve, we produce the same effect within the cylinder itself (as might be done in the engine of Art. 81) by applying a receiver of heat which will bring the pressure down to the lower limit P_2 by causing part of the contents to condense before the piston begins its return stroke. The piston being stationary, the volume of the working substance does not alter during this process. If we imagine the receiver of heat to have a temperature which falls progressively from that of the steam at c to the final temperature (T_2) at f, this removal of heat takes place reversibly. The work done by the steam is not affected by substituting this reversible process for the action of the condenser, because the pressure in the cylinder is in no way altered by the substitution, but we are now able to draw a curve that will represent the process on the entropy-temperature diagram.

This is done in fig. 27, where the curve cf represents the condensation of part of the steam at constant volume, while the piston is at rest before beginning its return stroke. The constant volume in this process is to be reckoned per lb. of steam: it is the volume of the cylinder divided by the quantity of fluid in it: in other words

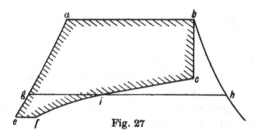

Fig. 27

it is the volume per lb. of the wet steam at c. Call that volume V_c. Then at any level of temperature such as gih, a point i on the constant-volume curve which represents the process is found by taking

$$\frac{gi}{gh} = \frac{V_c}{V_h},$$

where V_h is the volume of 1 lb. of saturated steam at that temperature. The area of the figure within the shaded lines represents the thermal equivalent of the work done in the complete cycle. The corner cut off by the curve cf shows what is lost by incomplete expansion as compared with the work done in a Rankine cycle. In the example sketched in fig. 27 the initial pressure (at b) is 180 pounds per square inch, and the steam is released after adiabatic expansion to 15 pounds per square inch.

If adiabatic expansion were entirely absent, and steam were admitted at P_1 during the whole of the forward stroke of the piston, and discharged at P_2 during the backward stroke, the entropy-temperature diagram would take the form shown in fig. 28, where *bif* is a constant-volume line representing the fall of pressure from P_1 to P_2. This corresponds, in the ideal cycle, to the action of primitive steam-engines such as Newcomen's, before Watt introduced the practice of cutting off the supply of steam at an early stage in the stroke and allowing the remainder of the stroke to be performed by expansion under falling pressure. Points in the curve *bif* are found in the manner shown above.

In this case the work done in the cylinder, per lb. of steam, is $(P_1 - P_2) V_{s_1}$. The net amount of work done, allowing for the feed-

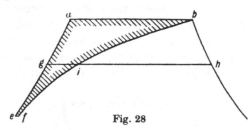

Fig. 28

pump, is $(P_1 - P_2)(V_{s_1} - V_{w_2})$, and the thermal equivalent of this quantity should be equal to the area within the shaded lines of the entropy-temperature diagram. The example drawn in fig. 28 assumes P_1 to be 180 pounds per sq. inch, and P_2 to be 1 pound per sq. inch. It gives a theoretical efficiency of barely 0·075.

The efficiency of actual primitive engines working without expansion was much less than this, not only because the pressure was less, but because at every stroke a large part of the steam entering the cylinder became at once condensed upon the walls, and consequently the volume of steam taken from the boiler was greater than the volume swept through by the piston.

96. Modified Cycle with Regenerative Feed-heating. We have seen that the feature which makes the Rankine cycle less efficient than the cycle of Carnot is the return of the condensate to the boiler as feed-water in a cold state. In modern steam turbines a device has been developed which to a great extent removes this source of loss. Some steam is "bled" from the turbine at each of a series of stages during the expansion and is brought into thermal contact with the condensate at a corresponding series of points in the course of its passage back to the boiler. Thus the feed is progressively heated by taking up heat from a part of the working steam, and matters are so arranged that at each of the heating points there is no great difference between its temperature and that of the bled steam. When the points are numerous and the differences of temperature are small this process of feed-heating approximates to a reversible operation, and the cycle is thereby caused to approximate to the cycle of Carnot. In the ideal case this approximation would be complete: we should then have all the internal actions reversible, with no heat taken in from the hot source except at the upper limit T_1, and no heat rejected to the cold receiver except at the lower limit T_2. The efficiency would consequently then be $(T_1 - T_2)/T_1$ as in the cycle of Carnot.

In large turbines steam is bled at several points and is returned to the boiler along with the condensate it has served to warm. Even with a few points of bleeding the practical result is a substantial increase of efficiency.

In the ideal action of this regenerative cycle the work done by each pound of steam supplied to the turbine is less than it would be in the Rankine cycle, because all the steam does not suffer the full heat-drop. But the heat which each pound takes in from the source is also less, and in a greater proportion: hence the increased efficiency.

97. Binary Vapour Engines. Instead of employing a single working substance in a heat-engine to work through the whole temperature range, two or more substances may be employed in thermodynamic series, the first rejecting heat to the second at an intermediate temperature, and the second then continuing the work of expansion in a second cylinder. This idea first took shape in the steam and ether engine of Dr Tremblay about 1850, where the object was to extend the lower part of the working range by having the incompletely expanded steam give up its remaining heat to a more volatile fluid such as ether, which would exert a fairly high pressure at low temperatures and might continue the process of expansion without excessive increase of volume. The idea did not find favour: the advantages of binary working were insufficient to balance the practical drawbacks.

In recent years the idea has been revived experimentally and with a somewhat different object, namely to extend the upper part of the temperature range. For this purpose a less volatile liquid than water is required, one that may be vaporized at a higher temperature and without excessive pressure, and mercury has been used, with steam as the second working substance. Mercury will boil at a temperature of 500° C. with a pressure not much exceeding

100 pounds per square inch*. An alternative to mercury is diphenyl-oxide, the critical temperature of which is about 580° C. and its critical pressure about 400 pounds per square inch†.

98. **Clapeyron's Equation.** This name is given to an important relation between the latent heat of steam or any other vapour, the change of volume which it undergoes in being vaporized, and the rate at which the saturation pressure varies with the temperature. To establish it we may revert to the ideally perfect steam engine of Art. 81, in which Carnot's cycle is followed with a liquid and vapour for working substance. We saw that this gave an indicator diagram (fig. 16) with two lines of uniform pressure (isothermals) connected by two adiabatic curves. The heat taken in was L per unit of working substance, and since the engine was reversible its efficiency was $(T_1 - T_2)/T_1$, from which it followed that the work done, or the area of the diagram, was $L(T_1 - T_2)/T_1$. This is in thermal units: to reduce it to units of work we multiply by J. Now suppose that the engine works between two temperatures which differ by only a very small amount. We may call the temperatures T and $T - \delta T$, δT being the small interval through which the engine works. The above expression for the work done becomes

$$\frac{JL\delta T}{T}.$$

The indicator diagram is now a long narrow strip (fig. 29). Its length ab is $V_s - V_w$, V_s being the volume of unit mass of the vapour and V_w the volume of unit mass of the liquid. Its height is δP, where δP is the difference between the pressure in ab and that in cd. In other words, since the vapour is saturated in cd as well as in ab, δP is the difference in the pressure of saturated vapour due to the difference in temperature δT. When

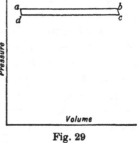

Fig. 29

δP is made very small, the area of the diagram becomes more and

---

* See Kearton on "The possibilities of mercury as a working substance for binary fluid turbines," *Proc. Inst. Mech. Eng.* 1928, p. 895; Emmet, *Mechanical Engineering*, May 1924, and *J. Franklin Inst.* Feb. 1925; also F. Bernhardt, *Phys. Zeit.* March 1925.

† See Dow on "Diphenyl-oxide Power Plants," *Mechanical Engineering*, Aug. 1926.

more nearly equal to the product of the length by the height, namely $\delta P (V_s - V_w)$. This is equal to the work done, whence

$$\delta P (V_s - V_w) = \frac{JL\delta T}{T}.$$

This equation is only approximate when the interval δP (or δT) is a small finite interval. In the limit, when the interval is made indefinitely small, it becomes exact, and may then be written

$$V_s - V_w = \frac{JL}{T} \left(\frac{dT}{dP}\right)_s,$$

where $\left(\frac{dT}{dP}\right)_s$ means the rate at which the temperature of saturated vapour changes relatively to the pressure: in other words it is the slope of the saturation curve of temperature and pressure. This is Clapeyron's Equation. It may be applied to find the volume of a saturated vapour, at any temperature, when the volume of the liquid, the latent heat, and the rate of change of temperature with pressure along the saturation curve are known. The values of pressure, volume and latent heat given in steam tables in relation to temperature must, if the tables are accurate, be such as will satisfy this equation. Take, for example, steam at 150° C., the tables at the end of the book would give dP/dT at 150° as 1·851. V_w is 0·01745 cub. ft. and the latent heat is 504·8 calories. Hence by Clapeyron's formula we should find, for the saturation volume in cub. ft.,

$$V_s = 0·0175 + \frac{1401 \times 504·8}{423·1 \times 1·851 \times 144} = 6·288;$$

the tabulated value is 6·287.

We might have obtained the Clapeyron Equation by considering that the entropy-temperature diagram corresponding to the indicator diagram of fig. 29 is a long narrow strip, whose length $\phi_s - \phi_w$ is L/T and height is δT. Its area is the thermal equivalent of the work done; hence $\delta P (V_s - V_w) = JL\delta T/T$, as before.

In the vaporization of a liquid V_s is greater than V_w, and heat is taken in; hence by Clapeyron's Equation $\left(\frac{dT}{dP}\right)_s$ is positive, which means that increasing the pressure raises the boiling point. When the change of volume $V_s - V_w$ is known for any substance, the equation may evidently be used to find the amount by which the boiling point is raised per unit increase of pressure.

99. **Further Applications of Clapeyron's Equation. The Triple Point.** The reasoning by which Clapeyron's equation was arrived

at was general: it applies to any reversible change in the state of aggregation of any substance, to the change from solid to liquid as well as to the change from liquid to vapour. The engine whose indicator diagram was sketched in fig. 29 may have anything for working substance, and the isothermal line in the first operation, during which heat is taken in, may be drawn to represent the change of volume corresponding to any change of phase. In the example already dealt with, the change was from liquid to vapour. But we might begin with a solid substance previously raised to the temperature at which it begins to melt (under a given pressure), and draw the line to represent the change of volume that occurs in melting, while the pressure remains constant. The substance does external work against that pressure if it expands in melting; or has work spent upon it if (like ice) it contracts in melting. The steps of the argument are not affected, and hence the equation may be written thus, with reference to any such transformation of state,

$$V'' - V' = \frac{J\lambda}{T} \frac{dT}{dP},$$

where V' is the volume of the substance (per unit of mass) in the first state, V'' its volume in the second state, λ is the heat absorbed in the transformation, and dT/dP is the rate at which the temperature of the transformation (say the melting point or boiling point) is altered by altering the pressure.

If a solid body expands in melting V'' is greater than V' and (since the latent heat λ is positive) it follows that dT/dP is positive: in other words the melting point is raised by applying pressure.

On the other hand if the body contracts in melting $V'' - V'$ is negative and dP/dT is negative: in other words the melting point is lowered by applying pressure. This is the case with ice. From the known amount by which ice contracts when it melts, James Thomson (in 1849) applied this method of reasoning to predict that the melting point of ice would be lowered by about 0·0074° C. for each atmosphere of pressure, and the result was afterwards verified experimentally by his brother, Lord Kelvin*.

* See Kelvin's *Mathematical and Physical Papers*, vol. I, pp. 156 and 165. The numerical result stated in the text is obtained as follows:—A pound of water changes its volume in freezing from 0·0160 to 0·0174 cub. ft. and gives out 80 calories. Hence

$$\frac{dT}{dP} = \frac{0 \cdot 0014 \times 273}{80 \times 1400} = 0 \cdot 00000341,$$

and if δP be one atmosphere or 2160 pounds per sq. ft., δT is $2160 \times 0 \cdot 00000341$ or 0·0074° C.

The lower of the two fixed points used in graduating a thermometer (Art. 15) is the temperature at which ice melts under a pressure of one atmosphere. If this pressure were removed—as it might be by putting the ice in a jar exhausted of air by means of an air-pump—the temperature of melting would be raised. The water-vapour given off at the melting point has a pressure of only 0·09 pound per square inch, and consequently if no air were present, and if the only pressure were that of its own vapour, ice would melt at approximately 0·0074° C., for the pressure would be reduced by nearly one atmosphere. The temperature at which ice melts under these conditions is called the *Triple Point*, because (in the absence of air) water-stuff can exist at that particular temperature in three states, as ice, as water, and as vapour, in contact with one another and in equilibrium.

At any pressure below that of the triple point a substance can exist only in the two states of solid and vapour, and when heat is applied it will vaporize without melting. Thus carbon dioxide, which has a triple point at a pressure of 75 pounds per square inch and a temperature of $-56\cdot5°$ C., and may be liquefied in bottles at high pressures, will freeze itself by partial vaporization when allowed to escape into the atmosphere, and the CO_2 ice or snow so formed will assume and keep a much lower temperature until all is vaporized. At atmospheric pressure the temperature of its vapour is only about $-79°$ C.: it will sink to that at least, and in freely moving air it will become still colder so that it may remain in equilibrium with the partial pressure of its own vapour.

100. **Entropy-Temperature Chart of the Properties of Steam.** Besides serving to illustrate the operations of ideal engines, a diagram in which the coordinates are the entropy and the temperature may be used as a general chart for exhibiting graphically the properties of steam or of any other fluid. The student will find it instructive to draw for himself a chart for steam, on section paper, to a scale large enough for reasonably accurate measurement.

The general character of such a chart will be apparent from fig. 30. It includes the boundary curves already described, which represent the relation of entropy to temperature in saturated steam and in water at the same temperature and pressure. Between these is the wet region, where the substance can exist in equilibrium only as a mixture of liquid and vapour. Beyond the steam boundary, to the right, is the region of superheated vapour.

Now let a system of *Lines of Constant Pressure* be drawn. Each of these shows the relation of ϕ to T while the substance changes its state in the manner of the imaginary experiment of Art. 56.

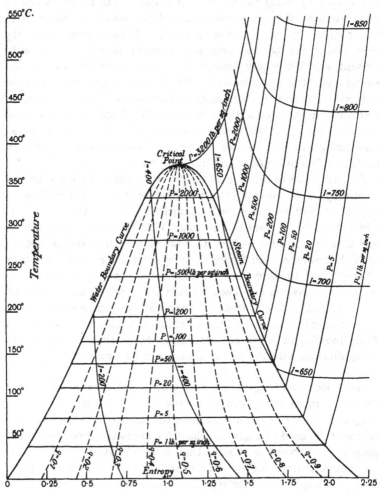

Fig. 30. Entropy-temperature Chart for Water and Steam.

Starting from the extreme left, a line of constant pressure for water is practically indistinguishable from the boundary curve; strictly it lies a little to the left of that curve, reaching it only when the temperature is such that steam begins to form. Then it crosses the wet region as a horizontal straight line, T being constant during the conversion of the substance from liquid into vapour. After reaching the steam boundary the line of constant pressure rises rapidly during the process of superheating. In the figure, a few lines of constant pressure are drawn, namely those for $P = 1, 5, 20,$ 50, 100, 200, 500, 1000 and 2000 pounds per square inch. When a sufficient number of such lines are drawn it is easy, by graphic interpolation, to mark on the chart the position of a point corresponding to any assigned condition of the substance as to pressure and temperature, and to trace, by measurement instead of by calculation, the changes which ensue during adiabatic expansion.

The convenience of the chart for such purposes is increased by including a system of *Lines of Constant Total Heat*. Examples of these lines are shown in the region of superheat for $I = 850, 800,$ 750, 700 and 650 calories. The line for 650 passes into the wet region and out again, undergoing a sharp change of direction. It is easy to map out the whole wet region by lines of constant I: one has only to calculate at any temperature, from the known values of I_w and I_s, what fraction of the substance must be vapour in order that the mixture may have the assigned total heat. As examples the lines for $I = 400$ and $I = 200$ have been drawn. Each of the lines of constant I represents what occurs in a throttling process. They tend, at the extreme right, to become nearly straight lines of constant temperature: this is because the vapour behaves more nearly like a perfect gas the more the pressure is reduced. In a perfect gas, as we saw in Art. 19, T is constant when the expansion is of such a nature as to keep I constant.

Another useful addition is a set of *Lines of Constant Dryness* in the wet region. These are drawn in the figure for $q = 0.1, 0.2, 0.3, 0.4, 0.5,$ 0.6, 0.7, 0.8 and 0.9. They divide each horizontal width between the two boundary curves into equal parts (see Art. 84). *Lines of Constant Volume* may also be drawn in the manner already described (Art. 95).

With the aid of such a chart one may find, for example, by drawing the appropriate adiabatic (vertical) line, that steam with a pressure of 200 pounds per square inch, superheated to 400° C., becomes saturated when its pressure falls, by adiabatic expansion. to 17 pounds. Continued into the wet region the adiabatic cuts the constant dryness line $q = 0.9$ at 52° C., showing that there is 10 per cent. of water present when the pressure has fallen to 2 pounds. The heat-drop may be inferred, but for its measurement a better form of chart is one which will be described in the next Article.

By drawing a vertical line to represent the adiabatic expansion of a mixture of steam and water, it is easy to trace the changes that occur in the proportion of water to steam. In the region of ordinary working pressures the line for $q = 0.5$ is nearly vertical. Hence if there is about 50 per cent. of water present at the beginning of adiabatic expansion, nearly the same percentage will be found as the expansion goes on. When the steam is much wetter than this to begin with, adiabatic expansion makes it drier.

The water and steam boundary curves are connected by a rounded top, the summit of which (at $T = 874°$) is the critical point. At that point there is no distinction between ϕ_w and ϕ_s or between I_w and I_s: the horizontal intercept between the water and steam boundaries, which corresponds to the taking in of latent heat, vanishes. In the figure a constant-pressure line of superheating is drawn for the critical pressure of 3200 pounds per sq. inch; it is tangent to the boundary curve at the critical point. At higher pressures the lines of constant pressure would pass, in the form of continuous curves, clear of the rounded top, from the region of water to that of superheated steam.

The water boundary curve is concave on the left for temperatures below 200° C., because the rise of entropy per degree, which is σ/T, where σ is the specific heat of water, becomes less as T increases, σ being nearly constant at low temperatures. But at higher temperatures the specific heat of water increases so fast as to make σ/T increase with rising temperature: the curve accordingly bends over to the right as it approaches the critical point.

In the next chapter we shall have occasion to refer to examples of entropy-temperature charts for other fluids. In one of these—carbonic acid—the region which is practically important, in connection with refrigerating processes, includes the rounded top whose summit is the critical point. In that diagram the lines of constant pressure in the liquid are clearly distinguishable from the boundary curve of the liquid state.

101. **Mollier's Chart of Entropy and Total Heat.** While the entropy-temperature diagram is invaluable as a means of exhibiting thermodynamic cycles and as a help towards understanding them, another diagram, introduced in 1904 by Dr R. Mollier*, is of greater service in the solution of practical problems. By taking for co-

* R. Mollier, "Neue Diagramme zur technischen Wärmelehre," *Z.V.D.I.* 1904, p. 271. Charts of this type have proved so useful that they are now issued with nearly all steam tables.

ordinates the entropy and the total heat, Mollier constructs a chart which has advantages that entitle it to the first place among devices for representing graphically the thermodynamic action of steam in steam-engines. Its application to refrigerating machines will be dealt with in the next chapter. As regards steam it furnishes the most convenient way to measure the heat-drop in adiabatic expansion, whatever be the initial state as to superheat, and consequently to find the greatest theoretical output that is attainable when the initial pressure and temperature, and the final pressure, are assigned. We have seen that this can be calculated without difficulty when complete tables are available; and also that it can be found by the aid of an entropy-temperature chart on which lines of constant total heat have been drawn. But the Mollier chart allows a graphic solution to be obtained with great directness and ease.

For practical purposes the Mollier $I\phi$ chart is drawn so as to show only the steam boundary curve and the region immediately above and below it, but it is instructive to consider the complete chart for water and steam, which is sketched in skeleton form and to a small scale in fig. 81. There abc is the water boundary curve and cde is the steam boundary curve, c being the critical point. The straight lines drawn between them, such as bd, are constant-pressure lines representing the process of vaporization. When these lines are continued across the boundary, as at dd', they become curved and represent the process of superheating under constant pressure. The slope of any line of constant pressure is a measure of the temperature, for at constant pressure $dI = dQ = Td\phi$, and consequently $dI/d\phi = T$. In the wet region the temperature along any line of constant pressure is constant, being the temperature of saturation for that pressure, and therefore any constant-pressure line in that region is straight*. It crosses the steam boundary without change of slope, but gradually bends upwards in the region of superheat as the temperature rises, for its slope continues to be a measure of the temperature.

Each constant-pressure line in the wet region may have its length between the two boundary curves divided into parts which express the dryness q at successive stages in the process of vaporization, just as in the $T\phi$ chart. Since the heat taken in up to any stage of that process is proportional to q (Art. 70), equal distances

---

* Since G is constant in an equilibrium process of evaporation or condensation, and $G = T\phi - I$, for any such process $I = T\phi -$ constant, hence the constant temperature or pressure lines on an $I\phi$ chart must be straight.

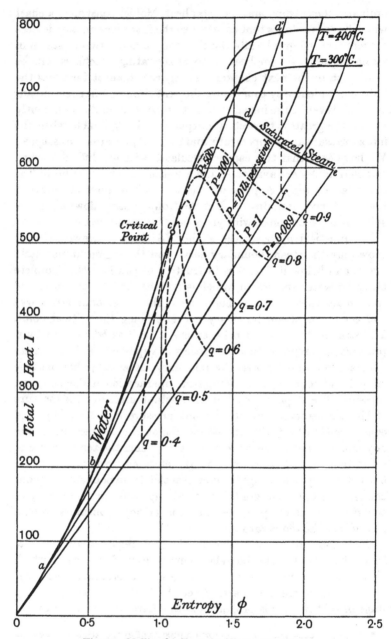

Fig. 31. Mollier $I\phi$ Chart for Water and Steam.

along the line, corresponding as they do to equal increments of total heat, correspond to equal changes of dryness. In this way lines of constant dryness are determined, some of which are shown in the sketch.

It is useful to have a system of lines of constant temperature drawn in the region of superheat: two such lines are shown in fig. 31. When they and the constant-pressure lines in that region have been drawn it is easy to mark the point which corresponds to any assigned condition of the steam as to temperature and pressure. Thus d' is the point corresponding to steam with a pressure of 100 pounds, superheated to 400° C. Then by drawing a vertical line such as $d'f$ through the point so found, we exhibit the process of adiabatic expansion. The length of that line, down to the final pressure, measures the adiabatic heat-drop, and therefore gives a very direct means of finding the greatest amount of work ideally obtainable from a pound of the working substance. In this example the final pressure, at f, is taken as 1 pound per sq. inch. The position of f among the lines of constant dryness shows how much of the steam is condensed by this adiabatic expansion. The advantage of a high vacuum, to which attention was drawn in Art. 94, will be obvious from the effect of the final pressure on the length of $d'f$.

A throttling process is represented by a horizontal straight line, since I is constant. Lines of constant temperature in the superheated region become nearly straight and horizontal at low pressures, for the behaviour of the vapour then approximates to that of a perfect gas.

A complete Rankine cycle is shown by the closed figure $abdd'fa$, where ab is the heating of the feed-water, bd the formation of steam in the boiler, dd' its subsequent superheating, $d'f$ its adiabatic expansion to the pressure of the condenser, and fa its condensation at that pressure. For the practical use of the diagram, however, there is no need to include the whole cycle. What is wanted is the region to the right, where the quality of the steam before and after expansion is exhibited, especially the region from $\phi = 1.3$ to 2.1 and from $I = 400$ to 850; and by restricting the chart to this region open scales may be used without making it unduly large.

Fig. 32 gives, in miniature form, a Mollier Chart for the useful region, showing a few lines of constant pressure, also lines of constant temperature in the region of superheat, and lines of

constant dryness in the wet region. In large published charts of
this type lines of constant volume are sometimes added. These
have been omitted in fig. 32, to avoid confusing the figure.

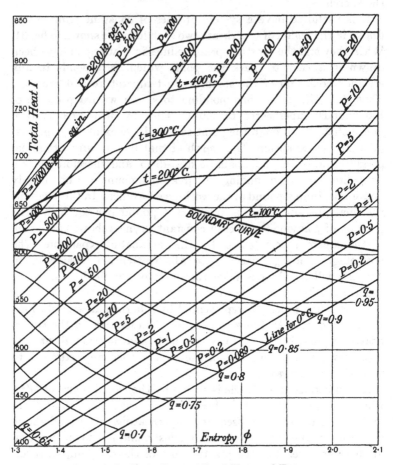

Fig. 32. Mollier's Chart of Total Heat and Entropy.

102. **Other Forms of Chart.** Besides the foregoing diagram
Mollier introduced another in which the coordinates are the
pressure and the total heat. This *PI* chart has not come into general
use in dealing with steam. It is, however, often employed by refrigerating engineers for fluids with a lower critical point which
serve as working substances in the vapour-compression process,
such as carbon dioxide, sulphur dioxide and methyl chloride.

An example of it as applied to carbon dioxide will be found in Chapter IV.

Many published charts for such substances employ log P instead of P as one of the coordinates, the other being I. This modification allows quantities to be measured from the chart with a more uniform degree of accuracy throughout the whole pressure range, but has the drawback of making the chart less suitable for exhibiting a thermodynamic process.

Other charts may be devised by selecting for the two coordinates other pairs of properties from the list given in Art. 75. In any such chart the characteristics of the fluid are exhibited by drawing systems of curves, each of which represents the relation that holds between the two properties chosen for coordinates when the state alters in such a manner that some third property is kept constant. By drawing several such systems of curves a comprehensive graphical substitute for numerical tables may be constructed. Of such devices it may be said that, so far as steam is concerned, the publication of full tables, which include the region of superheat, render graphic tabulation less necessary. It is now comparatively easy to find any required quantities directly from the tables, or by interpolation from them, with greater accuracy than is reached in measuring from a chart. But for certain purposes the graphic process is sufficiently exact and more convenient. Students should in any case make themselves familiar with the entropy-temperature chart, and also with the Mollier chart of entropy and total heat: the former because it will help them to understand cyclic processes; the latter as an instrument for dealing with practical problems in steam engineering and mechanical refrigeration.

102a. **Entropy-Temperature Diagrams of Various Fluids.** For most fluids the entropy-temperature diagram has a form which resembles its form for water; the vapour boundary curve slopes continuously to the right from the critical point showing a progressive increase of entropy as the temperature of saturation is reduced. When this is the case it is obvious that adiabatic expansion of the saturated vapour produces a wet mixture. But with some fluids the boundary curve takes such shapes as are illustrated in figs. 33 and 33a*. It overhangs on the vapour side, with the result that (within a certain range of temperature) adiabatic expansion of the saturated vapour will produce superheating instead of

---

* Ewing, *Phil. Mag.* vol. xxxix, June 1920, p. 633.

wetness, and conversely adiabatic compression will cause part of the vapour to condense instead of superheating it*.

103. **Effects of Throttling.** We have already seen (Art. 72) that when a throttling process is carried out under conditions that prevent heat from entering or leaving the substance the total heat I does not change. Lines of constant total heat on any of the diagrams accordingly show the changes in other quantities which are brought about by throttling. It is the process that

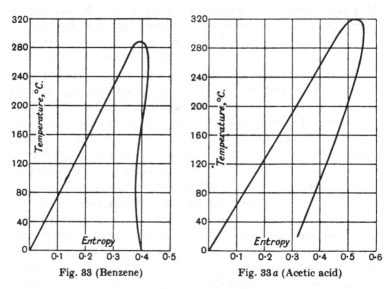

Fig. 33 (Benzene)   Fig. 33 a (Acetic acid)

occurs when a fluid passes through a "reducing valve" or other constricted orifice such as the porous plug of the Joule-Thomson experiment (Art. 19). It is *not* what occurs when a jet is formed, as in the nozzle of steam turbines. In that process, which will be dealt with later, the stream of vapour acquires kinetic energy that may be turned to useful account; whereas in throttling, any kinetic energy acquired in passing through the constriction is immediately dissipated by internal friction.

In a perfect gas throttling produces no change of temperature (Art. 19), but in steam and other vapours it produces a cooling

* Experiments by Cazin (*Ann. de Chim. et de Phys.* vol. xiv, 1868, p. 374) show that in benzene and some other vapours this does occur. At a particular temperature he observed a reversal of the effect, which agrees with the form of the benzene curve.

effect which is measured as the fall of temperature per unit fall of pressure under the condition that I is constant, or

$$\text{Cooling effect} = \left(\frac{dT}{dP}\right)_I.$$

Values of this quantity for steam under various conditions can be deduced from the steam tables. In steam that is highly super-heated, especially at low pressure, it is small, for the condition of the steam then approaches that of a perfect gas, but if the steam is saturated or only slightly superheated the cooling effect of throttling is much greater. The cooling effect can be measured experimentally: it plays an important part in determining the total heat and other properties of steam.

When a saturated vapour is throttled it becomes superheated as well as cooled. From the steam tables, the total heat of saturated steam at 180 pounds pressure is 665·2; if it is throttled down to a pressure of 20 pounds, the corresponding temperature of super-heated steam for the same total heat would be 156·3°. The satura-tion temperature for 20 pounds is 108·9°. The original temperature was 189·5°. Throttling has therefore cooled the steam by 33·2°; but at the same time it has caused it to become superheated to the extent of 47·4°. This apparent paradox is due to the fact that when the pressure is reduced by throttling the saturation temperature has fallen more than the actual temperature has fallen. Hence saturated steam is superheated by throttling, and steam that is initially superheated becomes more superheated. Similarly, a mixture of vapour and liquid is partially dried by throttling; it may be completely dried and even superheated if there is not much initial wetness and if there is a sufficient pressure-drop. This is illustrated in fig. 30 by the line of constant total heat for $I = 650$, which is drawn partly in the wet region and partly beyond it. It shows the effect of throttling on a wet mixture that contains about 4 per cent. of water at a pressure of 500 pounds; the steam becomes dry when the pressure is reduced to 39 pounds, corresponding to the temperature of 130° C. at which the line crosses the saturation curve, and with further throttling it is superheated.

The process of throttling is still more simply shown by horizontal lines ($I = $ constant) in the Mollier diagram (fig. 32). A horizontal line drawn through that point on the saturation curve which corresponds to the final pressure after throttling will show by its intercepts on lines of higher pressure what percentage of initial wetness the throttling process will remove. Similarly, it is easy to trace the extent to which liquid will evaporate in escaping through a throttle-valve from a region of high pressure to a region of lower pressure.

Drying by throttling has been used as a method of measuring wetness in a supply of steam, a sample being diverted from the steam-pipe into what is called a "throttling calorimeter." But the method cannot be relied on, for there is no security that the sample will be properly representative of the steam from which it is drawn.

104. The Heat-Account in a Real Process. The processes which have been considered in this chapter as going on in a steam-engine are ideal in the sense that they have been assumed to be *adiathermal*: that is to say, there is no transmission of heat to or from the working substance except what is originally taken from the source or finally rejected to the receiver; in all the intermediate operations the working substance has been enclosed in vessels that are assumed to transmit no heat. The assumed processes are also ideal in the sense that they are internally reversible. The process of throttling, which is a typically irreversible process, did not occur in the ideal engine cycles. In dealing with it also, however, we postulated adiathermal conditions; it was assumed in the argument of Art. 72 that no heat passed by conduction through the containing walls to or from the space outside.

Discarding these limitations we may now draw up, in general terms, a balance-sheet or heat-account for any real process, which will include thermal loss to the space outside and also irreversible actions within the engine or other apparatus.

Whether the apparatus considered be an engine cylinder, or the series of cylinders of a compound engine, or a turbine, or a throttling device, we may in all cases compare the state of the fluid at entry and at exit, as for example in the admission pipe of an engine and in the exhaust pipe. We imagine a steady flow of the working fluid through the apparatus. At entry let its pressure be P_1, its volume (per lb.) V_1, and its internal energy E_1. At exit let its pressure be P_2, its volume V_2 and its internal energy E_2. To make the comparison complete we may write K_1 for the kinetic energy (also per lb.) of the stream as it enters, and K_2 for its kinetic energy as it leaves. In passing through the apparatus the fluid will, in general, do external work, and also lose by conduction some heat to external space. Let W represent, in thermal units,

this output of work, and let Q_l represent the heat lost by conduction
to external space, both of these quantities (like the others) being
reckoned per pound of the fluid that passes through.

Each pound that enters the apparatus represents a supply of
energy equal to $K_1 + E_1 + AP_1V_1$, for E_1 is the internal energy it
carries, and P_1V_1 is the work done by the fluid behind in pushing
it in. But $E_1 + AP_1V_1$ is equal to I_1, the total heat per pound of
the fluid in its actual state at entry. Similarly, each pound that
leaves the apparatus represents a rejection of energy amounting to
$K_2 + E_2 + AP_2V_2$, for E_2 is the internal energy which the fluid
carries out, and P_2V_2 is the work spent upon it by the fluid behind
in pushing it out. $E_2 + AP_2V_2$ is equal to I_2, the total heat per
pound of the fluid in its actual state at exit. Hence, by the conservation of energy, for the apparatus as a whole,

$$K_1 + I_1 = K_2 + I_2 + W + Q_l.$$

The terms on the left of this equation represent the energy that
enters the apparatus; the terms on the right show how it is disposed
of in the issuing stream, in output of useful work, and in leakage
of heat.

The terms K_1 and K_2 are usually very small, except when the
apparatus is one for forming a steam jet, in which case K_2 is the
useful term: this will be considered in a later chapter. When the
change of kinetic energy in the stream is practically negligible,
as it is between the admission pipe and exhaust pipe of an engine,
we have
$$I_1 = I_2 + W + Q_l.$$

And when, in addition, the apparatus does not allow any appreciable amount of heat to escape to the outside ($Q_l = 0$), we have

$$I_1 - I_2 = W.$$

This means that when there is a steady flow of a working substance through any thermodynamic apparatus, the output of work
is measured by the *actual* Heat-Drop, *whether the internal action
is or is not reversible*, provided there is no loss of heat to the outside
by conduction through the walls.

The actual heat-drop must not be confused with the adiabatic
heat-drop, which is the difference between I_1 and that value
which the total heat would reach if there were adiabatic expansion
to the exit pressure P_2. The actual heat-drop $I_1 - I_2$ is identical
with the adiabatic heat-drop only when there is no loss of heat to

the outside and when, in addition, the internal action is wholly reversible.

Any irreversible feature in the internal action will increase I_2 above the value which would be reached by adiabatic expansion, and will consequently diminish the output of work.

In the extreme case of a throttling process there is no output of work, and therefore $I_2 = I_1$, provided there is no loss of heat to the outside. Any loss of heat to the outside in a throttling process will make I_2 correspondingly less, for we then have $I_2 = I_1 - Q_l$.

The losses of thermodynamic effect in a real engine, which make W less than the ideal output, namely the value corresponding to the adiabatic heat-drop, arise partly from loss of heat to the outside and partly from two kinds of irreversible internal action. One of these two kinds is mechanical; the other is thermal. In the mechanical kind, the action involves fluid friction within the working substance. It is of the same nature as that which occurs in throttling: there is irreversible passage of the working substance from one part of the engine to another where the pressure is lower, as for instance the passage of steam through somewhat constricted openings into the cylinder, or its passage, on release after incomplete expansion, into the exhaust pipe, with a sudden drop of pressure: or again, there is the same kind of irreversibility in a turbine in the frictional losses that attend the formation of steam jets or in the friction of the jets on the turbine blades. These are all instances of mechanical irreversibility. In the second kind of irreversible action there is exchange of heat between the working substance and the internal surface of the engine walls. The hot steam, on admission to a cylinder which has just been vacated by a less hot mixture of steam and water, finds the surfaces colder than itself. A part of it is accordingly condensed on them, which re-evaporates after the pressure has fallen through expansion. This alternate condensation and re-evaporation involves a considerable deposit and recovery of heat in a manner that is not reversible, for it takes place by contact between fluid and metal at different temperatures. The action may occur without loss of heat to the outside: it would occur, for instance, in an engine with a conducting cylinder covered externally with perfectly non-conducting material. Its effect, like that of throttling or fluid friction generally, is to reduce the output of work below the limit that is attainable only in a reversible process, and it does this by making the actual heat-drop $I_1 - I_2$ less than the adiabatic heat-drop.

The equation $W = I_1 - I_2$ takes account of both kinds of irreversibility—of the effect of thermal exchanges within the apparatus, as well as of any throttling or frictional effects in the action of the working substance. But it does not take account of heat lost to the outside, and for that the term Q_l has to be deducted, making

$$W = I_1 - I_2 - Q_l.$$

The full statement of the heat-account in a real process may be expressed as follows: When there is a steady flow of a working substance through any thermodynamic apparatus, the output of work is measured by the actual heat-drop from entrance to exit, less any heat that escapes by conduction to the outside, and less any gain of kinetic energy of the issuing stream over the entering stream; or, in symbols,

$$W = I_1 - I_2 - Q_l - (K_2 - K_1),$$

all these quantities being expressed in thermal units, and reckoned per unit quantity of the working substance.

This equation also applies to reversed heat-engines, or heat-pumps, which will be considered in the next chapter, but in them the quantity W is negative: work is expended on the machine instead of being produced by the machine. In such machines Q_l is also generally negative, for as a rule the apparatus is colder than its surroundings and the leakage of heat is inwards.

CHAPTER IV

THEORY OF REFRIGERATION

105. **The Refrigeration Process.** Refrigeration is the removal of heat from a body that is colder than its surroundings. In cold storage, for example, the contents of a chamber are kept at a temperature lower than that of the air outside, by extracting the heat which continuously leaks in through the imperfectly insulating walls. To maintain a refrigerating process requires expenditure of energy. It is generally done by means of a mechanically driven heat-pump, working on what is essentially a reversed heat-engine cycle. It may also be done by the direct use of high-temperature heat without intermediate conversion of that heat into work. We shall consider later the direct application of heat to effect refrigeration, but shall in the first instance treat of refrigerating machines driven by the expenditure of mechanical power.

Any process of refrigeration involves the use of a working substance which can be made to take in heat at a low temperature and discharge heat at a higher temperature. The heat is discharged by being given up to the air outside or to any water that is available to receive it. The process is a pumping-up of heat from the level of temperature of the cold body, at which it must be taken in, to the level at which it may be discharged. These levels should be as near together as is practicable, in order that no unnecessary work may be done: in other words the action of the working substance should be confined to the narrowest possible range of temperature. The temperature of discharge should be no higher than is necessary to get rid of the heat, and the lower limit should be no lower than will ensure transfer of heat into the refrigerating substance from the cold body whose heat is to be extracted.

Let T_1 be the temperature at which heat is discharged and T_2 the temperature at which it is taken in from the cold body. Consider a complete cycle in the action of the working substance. Let Q_1 be the quantity of heat which is discharged and Q_2 the quantity which is taken in from the cold body; and let W be the thermal equivalent of the work spent in driving the refrigerating machine. Then, by the conservation of energy,

$$Q_1 = Q_2 + W.$$

The useful refrigerating effect is measured by Q_2, and the "coefficient of performance," which is the ratio of that effect to the work spent in accomplishing it (Art. 4) is $\dfrac{Q_2}{W}$.

106. Reversible Refrigerating Machine. We have first to enquire what is the highest possible coefficient of performance when the limits of temperature T_1 and T_2 are assigned. We know by the principle of Carnot (Arts. 38, 39) that when heat passes down from T_1 to T_2 through a heat-engine, the ideally greatest efficiency in the conversion of heat into work is obtained when the engine is thermodynamically reversible. In that case

$$\frac{Q_1}{T_1} = \frac{Q_2}{T_2}.$$

The output of work W is $Q_1 - Q_2$. Hence the ideally greatest output of work is related to Q_2, the heat rejected at the lower limit of temperature, by the equation

$$W = \frac{Q_2 (T_1 - T_2)}{T_2}.$$

A corresponding proposition in the theory of refrigeration is that the ideally greatest coefficient of performance of a refrigerating machine, working to pump up heat from T_2 to T_1, is obtained when the machine is thermodynamically reversible. In that case the same relation holds, namely

$$\frac{Q_1}{T_1} = \frac{Q_2}{T_2},$$

and the amount of work W which is spent in driving the machine (and is equal to $Q_1 - Q_2$) is related to Q_2 by the equation

$$W = \frac{Q_2 (T_1 - T_2)}{T_2}.$$

In other words, the *greatest* amount of work that is theoretically obtainable in letting heat pass down through a given range of temperature is the *least* amount of work that will suffice to pump up the same quantity of heat through the same range.

To show that no refrigerating machine can be more efficient than one that is reversible, we shall use an argument like that of Art. 33. Let E, fig. 34, be a reversible refrigerating machine, reversed and therefore serving as a heat-engine. It takes a quantity of heat, say Q_1, from the hot body and delivers a quantity Q_2 to the cold body, converting the difference into work. Let all the work W which it develops be employed to drive a refrigerating machine R; and assume that there is no loss of power in the connecting mechanism. Accordingly the two machines, thus coupled, form a self-acting combination.

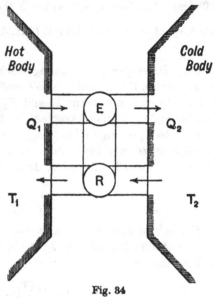

If it were conceivable that the machine R could have a greater coefficient of performance than the reversible machine E, that

Fig. 84

would mean that the ratio of Q_2 to W would be greater in R than in E. Hence (W being the same for both) R would take more heat from the cold body than E gives to it, and R would also give more heat to the hot body than E takes from it. The result would be a continuous transfer of heat from the cold body to the hot body by means of a purely self-acting agency. This would be contrary to the Second Law of Thermodynamics: we conclude therefore that no refrigerating machine can have a higher coefficient of performance than a reversible machine working between the same limits of temperature.

It follows that all reversible refrigerating machines, working between the same limits of temperature, have the same coefficient of performance. It also follows that the value of this coefficient is that which would be found in a reversed Carnot cycle, namely

$$\frac{Q_2}{W} = \frac{T_2}{T_1 - T_2}$$

This is the ideally highest coefficient—it measures the performance of what may be called a perfect refrigerating machine. The coefficient of performance in any real machine is necessarily less, for the cycle of a real machine falls short of reversibility.

107. **Conservation of Entropy in a Perfect Refrigerating Process.**
We saw in Art. 45 that a perfect, or reversible, heat-engine, such as Carnot's, may be regarded as a device which transfers entropy from a hot body to a cold body without altering the amount of the entropy so transferred, although the amount of heat which enters the engine is greater than the amount of heat which leaves the engine. The entropy taken from the hot body, namely Q_1/T_1, is equal to the entropy given to the cold body, namely Q_2/T_2; it may be said to pass through the engine without change, though the heat that passes through is reduced in the process by the amount which is converted into work, namely, by the amount $Q_1 - Q_2$.

Similarly a perfect, or reversible, refrigerating machine or heat-pump may be regarded as a device which transfers entropy from a cold body to a hot body without altering the amount of the entropy so transferred, although the amount of heat which enters the machine is less than the amount which leaves the machine. The action is in every particular a reversal of that of the perfect heat-engine. Entropy to the amount Q_2/T_2 is taken from the cold body, and entropy to the equal amount Q_1/T_1 is given to the warmer body to which heat is discharged. The amount of heat which is pumped up increases from Q_2 to Q_1 in the process, because an amount of work equivalent to $Q_1 - Q_2$ is expended in driving the machine and is converted into heat within the machine.

108. **Ideal Coefficients of Performance.** The following table shows the values of the coefficient of performance in a perfect or reversible refrigerating process, for various ranges of temperature.

Coefficients of Performance of a Perfect Refrigerating Machine.

Lower limit of temperature (Centigrade)	Upper limit of temperature (Centigrade)				
	10°	20°	30°	40°	50°
−20°	8·4	6·3	5·1	4·2	3·6
−15°	10·3	7·4	5·7	4·7	4·0
−10°	13·1	8·8	6·6	5·3	4·4
− 5°	17·9	10·7	7·7	6·0	4·9
0°	27·3	13·6	9·1	6·8	5·5
5°	55·6	18·5	11·1	7·9	6·2
10°	—	28·3	14·1	9·4	7·1

These are ideal figures, representing a theoretical limit which cannot be reached in practice. Though they relate to conditions of reversibility which are not fully attainable in a real machine, they illustrate clearly the practical importance of making the range of temperature as small as possible, by taking in the heat at a temperature no lower than can be helped and by discharging it after the least practicable rise.

The importance of a narrow range of temperature in refrigeration is further illustrated by fig. 35. It gives the entropy-temperature diagrams of three reversible refrigerating processes, in all of which the upper limit of temperature (T_1) is the same, and the same amount of work is spent. Each of the three supposed processes is

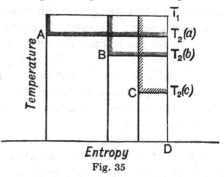

Fig. 35

ideally efficient: it is a reversed Carnot cycle, and its entropy-temperature diagram is a rectangle. The area of the rectangle represents the work spent, and the area under it, down to the absolute zero of temperature, represents the amount of heat that is taken from the cold body, and therefore measures the refrigerating effect. The three processes for which the diagram is sketched differ only in the temperature T_2 of the cold body from which heat is extracted. That temperature is relatively high in the first case (*a*), lower in case (*b*) and lower still in case (*c*). The refrigerating effect is measured by the area *AD* in the first case, by *BD* in the second, and by *CD* in the third. The result of lowering T_2 is very apparent, in reducing the amount of refrigeration that is ideally capable of being done by a given expenditure of work.

109. The Working Fluid in a Refrigerating Process. The working substance in a refrigerating cycle may be a gas which remains gaseous throughout, such as air, but generally it is a fluid which is alternately condensed and evaporated. During evaporation at a low pressure the fluid takes in heat from the cold body: it is then compressed and gives out heat in becoming condensed at a relatively high pressure. The selection of the fluid is governed by practical considerations. Water may be made to serve when the low temperature is not too low, but a serious drawback is the large volume and low pressure of the vapour. There are obvious advantages in using a fluid whose vapour-pressure is neither inconveniently small at the lower limit of temperature nor inconveniently large at the upper limit. The fluids most commonly used are ammonia and carbonic acid. Ammonia has a very convenient range of vapour-pressure throughout the range of temperature with which we are concerned in practical refrigeration. With carbonic acid the vapour-pressure is considerably higher, the critical point is reached at a temperature that may come within the range of operation, and the thermodynamic efficiency is somewhat less. Notwithstanding these objections carbonic acid is frequently preferred, especially on board ship, where it is more harmless should any of the fluid escape by leakage into the room containing the machine. For use on land, especially where the highest thermodynamic efficiency is aimed at, ammonia is usually chosen. Other fluids with lower vapour-pressures are extensively used for small-power plants, especially sulphur dioxide and methyl chloride (CH_3Cl).

110. **The Actual Cycle of a Vapour-Compression Refrigerating Machine.** If the reversed Carnot cycle were actually followed, the choice of working fluid would make no difference to the efficiency: the coefficient of performance for any fluid would have the value shown in Art. 106, namely $T_2/(T_1 - T_2)$. But a part of the reversed Carnot cycle is omitted in practice, with the result that the coefficient is reduced, and the extent of the reduction depends on the nature of the fluid; it is greater in carbonic acid than in ammonia.

To carry out a reversed Carnot cycle, with separate organs for the successive events which make up the cycle, would require:

(1) A compression cylinder in which the vapour is compressed from the pressure corresponding to T_2 to the pressure corresponding to T_1.

(2) A condenser in which it is condensed at T_1. A typical form of this organ would be a surface condenser in which the working fluid gives up its heat to circulating water.

(3) An expansion cylinder in which it expands from T_1 to T_2.

(4) An evaporator in which it takes up heat at T_2 from the cold body from which heat is to be extracted. This vessel is sometimes called the "refrigerator."

In nearly all refrigerating machines the expansion cylinder is omitted for reasons of practical convenience, and the fluid streams from (2) to (4) through a throttle-valve with an adjustable opening, called the "regulator" or "expansion-valve." In passing the expansion-valve the fluid does not change its total heat I: its pressure falls to that of the evaporator: its temperature falls to T_2 and part of it becomes evaporated before it begins to take in heat from the cold body.

The omission of an expansion cylinder, with the substitution for it of an expansion-valve, reduces the coefficient of performance for two reasons. The work which would be recovered in the expansion cylinder is lost; and also the refrigerating effect in the evaporator is reduced, for more of the liquid is vaporized in the act of streaming through the expansion-valve than would be vaporized in adiabatic expansion, consequently less is left to be evaporated by subsequently taking in heat from the cold body. The loss of efficiency from these two causes is not, however, very important under

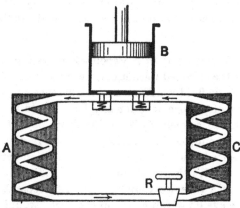

Fig. 86. Organs of a Vapour-Compression Refrigerating Machine.

ordinary conditions. To omit the expansion cylinder is a considerable simplification of the machine, all the more as the effective volume of such a cylinder would need adjustment relatively to that of the compression cylinder in order to secure the best effect under varying conditions as to the limits of temperature. Rather than introduce this complication it is worth while to make a slight sacrifice of thermodynamic efficiency.

In the usual type of vapour-compression refrigerating machine, accordingly, the expansion cylinder is omitted, and the organs are those shown diagrammatically in fig. 86. They are, (1) the compression cylinder B; (2) a condenser A such as a coil of pipe, cooled by circulating water, in which the working substance is condensed under a relatively high pressure and at the upper limit of temperature T_1; (8) an expansion-valve or regulator R through which it streams from A to C; (4) the evaporator C, in which it is vaporized at a low pressure by taking in heat from the cold body at the lower limit of temperature. The evaporator may for instance be a coil of pipe taking in heat from the surrounding atmosphere of a cold chamber; often it is a coil surrounded by cold circulating brine which serves as a vehicle for conveying heat to the working substance from a cold chamber or from cans for ice-making or other objects that are to be refrigerated.

The action of the compression cylinder is shown by the indicator diagram, fig. 37, for an ideal case which ignores effects of clearance and of throttling in the cylinder valves. During the forward stroke of the compressor the valve leading to A is shut and that leading from C is open. A volume V_1 of the working vapour is taken in from C at a uniform pressure corresponding to the lower limit T_2.

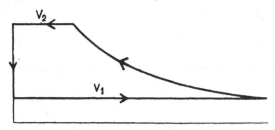

Fig. 37. Indicator Diagram of Compression Cylinder.

In some actual cases what is taken in is not dry-saturated vapour but a wet mixture, the wetness of which is regulated by adjusting the expansion-valve R. When this is done the subsequent compression may not produce much (if any) superheating. In modern practice, however, the tendency is to regulate the action so that what enters the cylinder is dry or very nearly dry saturated vapour. At the end of the forward stroke the valve leading from C closes and the piston is forced to move back compressing the vapour or wet mixture in the cylinder until its pressure becomes equal to that in A. This compression reduces the volume of the fluid in the cylinder to V_2. The valve leading to A then opens, and the back-stroke is completed under a uniform pressure while the working substance is discharged into A and condensed there. The admission and exhaust valves of the compressor are spring valves which open and close automatically and are so placed as to make the clearance exceedingly small. To complete the cycle, the same quantity of working substance is allowed to pass directly from A to C through the expansion-valve R. This step is not reversible (Art. 22).

The temperature T_1 at which condensation takes place, is in practice necessarily a good deal higher than that of the circulating water by which the condenser is kept cool, for a large amount of heat has to be discharged from the condensing vapour in a limited

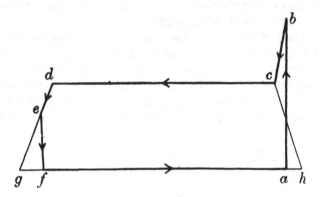

Fig. 88. The Vapour-Compression Cycle, using Ammonia.

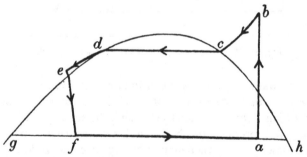

Fig. 89. The Vapour-Compression Cycle, using Carbonic Acid.

time. But it is important that the condensed liquid should be no warmer than is unavoidable before it passes the expansion-valve. Accordingly the condenser is arranged (sometimes by the addition of a separate vessel called a "cooler") so that the condensed liquid is brought as nearly as possible to the lowest temperature of the available water-supply before it passes the valve, though it may have been condensed at a considerably higher temperature. The advantage of this will be obvious when we consider, in the next article, the thermal effects of each step in the cycle.

111. Entropy-Temperature Diagram for the Vapour-Compression Cycle. The complete cycle is exhibited in the entropy-temperature diagram of fig. 88, which is drawn for ammonia as working substance, and fig. 89, which is drawn for carbonic acid. There dg and ch are portions of the boundary curves. The point a represents the condition of the mixture which is drawn into the compression cylinder, when compression is about to begin; its wetness is measured by the ratio ah/gh. The line ab represents adiabatic compression to the pressure of the condenser. The next process consists of cooling and condensation at this constant pressure: it is made up of three stages, bc, cd and de. In the first stage, bc, the superheated vapour is cooled to the temperature at which condensation begins; in the next stage, cd, the vapour is completely condensed; in the third stage, de, the condensed liquid is cooled to the lowest available temperature before it passes the expansion-valve. The lines bc, cd and de form parts of one line of constant pressure. In fig. 88 de is practically indistinguishable from the boundary line, but in fig. 89 the distinction is very apparent because we are there dealing with a liquid that is highly compressible in consequence of its nearness to the critical state. The line ef represents the process of passing through the expansion-valve, in which the pressure falls to that of the

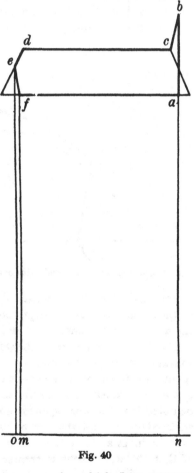

Fig. 40

evaporator. This is a throttling process, for which I is constant (Art. 72): ef is therefore a line of constant total heat; its direction changes in fig. 89 in crossing the boundary curve. By passing the

expansion-valve the working substance comes into the condition
shown by the point f. The proportion which is converted into
vapour by the mere act of passing the valve is shown by the ratio
gf/gh. Lastly we have the process of effective evaporation when
the substance is usefully extracting heat from the brine or other
cold body by evaporating in the refrigerator. This is represented by
the line fa, during which the dryness changes from gf/gh to ga/gh.

The refrigerating effect, that is to say, the amount of heat taken
in from the cold body, is represented by the area under the line fa,
measured down to a base-line corresponding to the absolute zero
of temperature, namely the area $mfan$ (fig. 40).

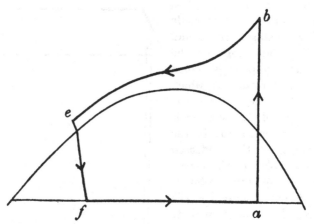

Fig. 41. Cycle for Carbonic Acid, with compression above the Critical Pressure.

The amount of heat rejected during cooling and condensation
of the vapour and subsequent cooling of the condensed liquid, is
the area under the lines bc, cd and de, namely the area $nbcdeo$.

The thermal equivalent of the work spent in carrying the working
substance through the complete cycle—which is simply the work
spent on it in the compressor—is the difference between those two
quantities, namely the area $nbcdeo$ *minus* the area $mfan$. It should
be noted that the work spent is not measured by the area $abcdefa$,
enclosed by the lines which represent the complete cycle, because the
cycle includes an irreversible step ef (see Art. 51). In consequence
of that the work spent is greater than the enclosed area by the
amount $oefm$.

As a further example we may take a compression process (fig. 41),
with carbonic acid for working substance, in which the temperature

of the cooling water is so high that the pressure during cooling is above the critical pressure. The line be is accordingly a continuous curve lying entirely outside of the boundary curve. The working substance passes from the state of a superheated vapour at b to the state at e without any stage corresponding to cd in fig. 39, in which it is a mixture of liquid and vapour. As before, the refrigerating effect is measured by the area under fa: the heat rejected to the cooling water is measured by the area under be: the difference between these two quantities measures the work spent, and is greater than the area of the closed figure $abefa$ by the area under the irreversible step ef.

112. **Refrigerating Effect and Work of Compression expressed in Terms of the Total Heat.** While it is instructive to state, as in the preceding Article, the refrigerating effect, the work of com-pression, and the heat rejected, in terms of areas on the entropy-temperature diagram, it is much more useful, for purposes of practical calculation, to express these as follows in terms of the total heat of the substance at the various stages of the operation.

The refrigerating effect, that is to say the amount of heat taken in from the cold body, is $I_a - I_f$, where I_a is the total heat at a and I_f is the total heat at f. This is because the (reversible) opera-tion fa is effected at constant pressure (Art. 68). For the same reason the amount of heat rejected to the condenser and cooler is $I_b - I_e$, where those quantities designate the total heat at b and at e respectively. Further, in the process ef of passing the expansion-valve there is no change of total heat, by the principle proved in Art. 72. Consequently, $I_f = I_e$. We may therefore state the amount of heat rejected as $I_b - I_f$.

Again, the work spent in the compressor is (in thermal units) $I_b - I_a$. It is the thermal equivalent of the area of the indicator diagram in fig. 37, namely $A \int_a^b V dP$, which is equal to $I_b - I_a$ by the general principle proved in Art. 80. We are dealing here with the increase of total heat in adiabatic compression instead of its decrease in adiabatic expansion.

That these results are in agreement with one another is seen by considering the heat-account of the cycle as a whole:

Work spent = Heat rejected − Heat taken in.

$$I_b - I_a = (I_b - I_f) - (I_a - I_f).$$

The *coefficient of performance*, which is the ratio of the heat taken in from the cold body to the work spent in the compressor, is

$$\frac{I_a - I_f}{I_b - I_a}.$$

It will be obvious that the numerical value of this coefficient would be reduced if we were to omit the cooling after condensation, which is represented by the line de. For in that case f would be shifted to the right, to a point on a line of constant total heat through d, and I_f would be increased. The refrigerating effect would be lessened; but the work spent in producing it would be the same as before, for the indicator diagram of the compression process, which is measured by $I_b - I_a$, is not affected. The values of I_a and I_b depend only on the state of the substance at a and at b respectively, and are the same as before.

113. Charts of Total Heat and Entropy for Substances used in the Vapour-Compression Process. The above results will show that calculations of performance are easy when we have tables or charts showing the properties of the working fluid under conditions of superheat as well as saturation. Such data are now available for a number of substances used in practice For ammonia, carbon dioxide and sulphur dioxide Mollier charts of entropy and total heat will be found in a Report of the Refrigeration Research Committee of the Institution of Mechanical Engineers*.

* *Min. Proc. Inst. Mech. Eng.* 1914. The charts given there were drawn by C. F. Jenkin. (See also Jenkin and Pye on carbon dioxide, *Phil. Trans. R.S.* vol. ccxv, A, 1915 and *Proc. R.S.* A, 1921, p. 352.) Charts for ethyl chloride and methyl chloride by Jenkin and Shorthose were published by the Food Investigation Board of the Department of Scientific and Industrial Research in 1921.

Numerical tables of the properties of saturated and superheated ammonia, carbon dioxide, sulphur dioxide, methyl chloride, and dichlordifluormethane or "peon" (CCl_2F_2) will be found in the *Refrigerating Data Book* of the American Society of Refrigerating Engineers, 1932. The figures given there for ammonia are chiefly based on researches by the Bureau of Standards, Washington, along with earlier work by Goodenough and Mosher, and those for carbon dioxide on the work of Plank and Kuprianoff, *Zeit. f. d. gesammte Kälte-Industrie*, 1929, whose tables are remarkably complete, extending as they do below the Triple-point to a temperature of $-100°$ C. The properties of CO_2 in that region have become practically important because it is now largely used as a portable refrigerant in the form of solid blocks, generally made by compressing CO_2 "snow." A block of solid CO_2 has the great advantage that it keeps perfectly dry as it passes into the gaseous state (Art. 99). While evaporating in the open it continues to give off gas at a temperature of $-80°$ C. or under, until it disappears.

For further reference to places where $I\phi$ and $T\phi$ charts of various refrigerants may be found, see Keesom, *Actes du Cinquième Congrès International du Froid* (Rome, 1928), vol. II, p. 146.

In drawing these charts a geometrical device is resorted to for the purpose of making the diagrams at once open and compact, with the effect that the quantities may be measured with sufficient accuracy on a chart of reasonable size. This device, which Mollier originally adopted in drawing his $I\phi$ chart for carbonic acid, is to use oblique coordinates, as illustrated in fig. 42. The lines of constant I are horizontal: the lines of constant ϕ instead of being perpendicular to them are inclined at a small angle. The result is

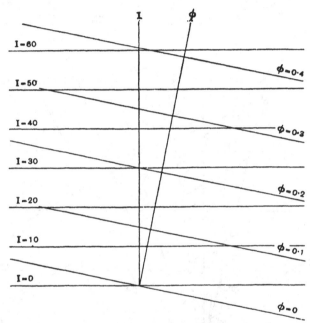

Fig. 42. Use of oblique coordinates in the $I\phi$ chart.

that when the chart is drawn the curves on it are sheared over, as compared with the form they would take on a chart with rectangular axes. Any figure which when drawn with rectangular coordinates is relatively long in one diagonal direction may with advantage be opened out by the use of oblique coordinates. This is true of $I\phi$ charts; as applied to them, the device gives a better separation of lines that run more or less diagonally across the sheet, like the lines in fig. 31 (Art. 101). There is consequently a great gain in clearness and in the power of accurately measuring those changes of I that take place in refrigerating processes. The inclination selected for the oblique axis will depend on the degree of opening out that is convenient in any particular chart. In the case of fig. 42 it is 5 along the slope to 1 vertically, and hence a

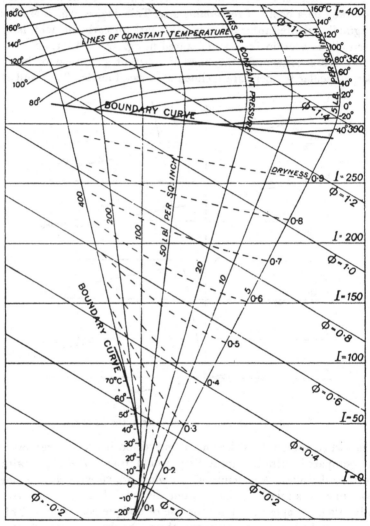

Fig. 43. $I\phi$ chart for Ammonia.

measurement of I if made along a line of constant ϕ would have to be interpreted on a scale five times as coarse as the normal scale for I.

An $I\phi$ chart for ammonia drawn with oblique coordinates is shown (on a small scale*) in fig. 43. In this case the amount of shearing is moderate, for the slope of the lines of constant entropy is only two to one. The diagram, for the useful region, consists of a fan-like group of lines of constant pressure extending as straight lines through the region of wetness from the liquid boundary to the vapour boundary or saturation curve, and then as curves into the region of superheat. Lines of constant temperature are also drawn in the region of superheat, and lines of constant dryness (shown as broken lines in the chart) are drawn by dividing the straight portion of each line of constant pressure into a number of equal parts. This chart should be compared with that shown for water and steam in fig. 31 (Art. 101) in which, however, there was no shearing, for rectangular coordinates were employed. Allowing for that difference the remarks made in Art. 101 apply here. The slope of any constant-pressure line, when properly interpreted with reference to the coordinates used in the drawing, measures the temperature, for $T = dI/d\phi$. Hence there is no abrupt change of direction between the straight part of any such line and its curved continuations into the liquid region at one end and into the region of superheat at the other. This of course applies to any substance. The $I\phi$ chart for sulphur dioxide is generally similar to the chart for ammonia.

The $I\phi$ chart for carbon dioxide is shown on a small scale in fig. 44. It shows the region round about the critical point. That point lies where a line of constant pressure would be tangent to the boundary curve. Constant-pressure lines are drawn for pressures that are higher than the critical pressure as well as for the wet region. The principle already stated applies to these lines, that the slope at any point (due regard being had to shearing) measures the temperature. In passing up along any line of constant pressure above the critical pressure, the slope, which measures the temperature, increases continuously. The straight portions of the constant-pressure lines, within the boundary curve, are divided by broken lines which are lines of constant dryness. Lines of constant temperature are also drawn in the region outside the boundary curve. In the region within the boundary, where the state is that of a mixture of saturated vapour and liquid, these lines would of course be straight, and would coincide with lines of constant

---

* For similar charts drawn in fuller detail and on a scale large enough for use in solving problems, reference should be made to the publications cited above.

pressure. To avoid confusion the straight portions of the constant-temperature lines are omitted in the figure.

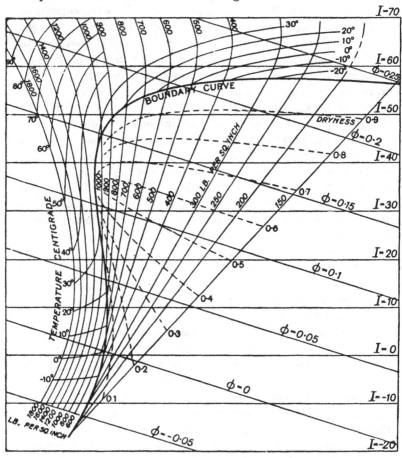

Fig. 44. $I\phi$ chart for Carbon Dioxide.

114. Applications of the $I\phi$ Chart in studying the Vapour-Compression Process. We are now in a position to represent the vapour-compression refrigerating process by diagrams which exhibit the changes of total heat in relation to entropy. With the help of $I\phi$ charts numerical values of the total heat are readily found by measurement at each stage in the assumed cycle.

To trace a refrigerating cycle on the appropriate chart, begin as before at a point a (fig. 45) which represents the state of the substance when it is about to enter the compressor. This point is on

the constant-pressure line corresponding to the process of evaporation in the cold body or evaporator (fig. 36), and its distance from
the two boundary curves corresponds to the proportion of vapour
to liquid in the mixture. If the compression is to be completely
"dry" it will begin at a_1, but for greater generality we take it as
beginning with a slightly wet mixture at a. The straight line ab,

drawn parallel to the lines of
constant entropy on the chart,
is the process of adiabatic compression. The position of b is
determined by the intersection
of this line with a line of constant pressure corresponding to
the known upper limit of pressure at which condensation
is to occur. The temperature
reached in the process of compression is seen by the position
of b among the lines of constant
temperature. In general there
will be some superheating. But
if the mixture is so wet to begin
with that the adiabatic line
through a does not cross the
boundary curve during compression before the upper limit
of pressure is reached there
is none, and in that case the
process is spoken of as "wet"
compression. This would be the
case if compression had begun
at a_c instead of a. By beginning

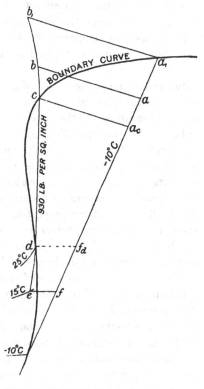

Fig. 45. Refrigeration cycle traced on
the $I\phi$ chart for Carbon Dioxide.

at a it carries the substance into the region of superheat before
compression is completed at b. Next we have the constant-pressure
process of cooling and condensation and further cooling, represented in its three stages by the lines bc, cd and de, the position of
e being fixed by the temperature to which the liquid is known to
be cooled before it reaches the expansion-valve. Then a horizontal
straight line through e (a line of constant total heat) represents the
process of passing through the expansion-valve, and determines a
point f, on the evaporation line, which exhibits the condition in

which the substance enters the evaporator. The process of evaporation fa, which is the effective refrigerating process, completes the cycle. The values of I_a, I_b, I_e and I_f (which is the same as I_e) are read directly by measurement from the chart. As has been already pointed out, the work spent in compressing the substance is $I_b - I_a$, and the refrigerating effect is $I_a - I_f$.

We may illustrate the use of the chart by some examples. Take first a case in which the working substance is carbonic acid, with $-10°$ C. as the temperature of evaporation, $25°$ C. as the temperature of condensation, and $15°$ C. as the temperature to which the substance is cooled before passing the expansion-valve. The diagram for the performance of an ideal machine under these conditions is sketched in fig. 45, assuming various degrees of dryness at the beginning of the compression. If the substance is then entirely dry the operation starts at a_1, namely, the end of the evaporation line for $-10°$ C., and compression brings it to b_1 which is on a line of constant pressure equal to the pressure of saturated vapour at $25°$ C., or about 930 pounds per sq. inch. But the vapour is considerably superheated at b_1, its temperature there (as the lines of constant temperature show) being $58°$ C. The work spent in compression, which is most accurately found by reading off the length of the line $a_1 b_1$ on a scale which makes that length a direct measure of the change of I, is 8·7.

We next trace the process of condensing and cooling, under the constant pressure of the condenser. From b_1 to c the gas is losing its superheat; from c to d it is being condensed; and from d to e it is being cooled as a liquid. The point e is found by the intersection of the line of constant pressure under which the process is carried out with the line of constant temperature for $15°$ C. Next draw ef parallel to the lines of constant total heat to meet the evaporation line for $-10°$ C. The refrigerating effect $I_{a_1} - I_f$ is 47·9. The coefficient of performance is therefore 5·5. This cycle corresponds to completely dry compression.

Suppose on the other hand that the compression is just wet enough to avoid any superheating. In that case it must commence at a_c in order that the adiabatic line representing the compression may pass through c on the boundary curve. Then the work done in compression is smaller than before, for $a_c c$ is smaller than $a_1 b_1$. The refrigerating effect is also smaller, for fa_c is smaller than fa_1. The coefficient of performance is almost unchanged.

Between these two there is a certain degree of dryness which (under the assumed ideal condition of adiabatic compression) would give a slightly higher coefficient of performance than either. This may be shown by taking a succession of points for various states of dryness between a_0 and a_1 as the starting point of the cycle. But we may reach the same conclusion more directly as follows, by a general method which is applicable to any $I\phi$ chart:

The refrigerating effect for any state of initial dryness, a, is proportional (on some scale) to the length fa. The work done is proportional (on another scale) to the length ab. Hence the position of b which will give the highest coefficient of performance is that which gives the smallest ratio of ab to fa. This is found by drawing a tangent from f to the line of constant pressure on which b lies. By applying this method the point b has been determined in the figure, and hence the point a is found at which compression should begin if the coefficient of performance is to have its maximum value. In the example that value is fully 5·7, and is obtained when the initial dryness is about 0·87.

As another example, still with carbonic acid, take the same conditions as before, except that the condensed liquid, instead of being cooled to 15° C. before expansion, reaches the valve at the temperature of condensation, namely 25° C. In that case the process of expansion corresponds to the line df_d in fig. 45. For maximum coefficient of performance, under these conditions, compression should no longer start from a but from a point so chosen that the adiabatic line through it reaches the constant-pressure curve b_1c at the point where the tangent from f_d meets that curve. This corresponds to a coefficient of only 4·4, as compared with the maximum of 5·7 found above, which shows how much loss results from omitting to cool the condensed liquid before it reaches the expansion-valve.

A further example will serve to illustrate the application of the $I\phi$ chart to carbonic acid working under tropical conditions, so that the higher limit of pressure is above the critical pressure of the substance. Still taking −10° C. as the temperature of evaporation, we shall suppose the pressure in the condenser to be 1200 pounds per sq. inch, and the temperature to which the liquid is cooled before expansion to be 30° C. With these data the diagram takes the form shown in fig. 46, where a_1b_1 represents a process of completely dry compression, and ab a process of compression in which the position of a has been so chosen as to give the maximum co-efficient of performance. The line ab consequently cuts the curve of constant pressure for 1200 pounds per sq. inch at the place where a tangent from f would meet that curve. The point e is determined by following the curve of constant pressure till it cuts the line of temperature for 30° C. The maximum coefficient of performance is obtained when the dryness before compression is about 0·95. Its value is 3·1, and under these conditions the vapour is superheated to 70° C. at the end of compression. The coefficient calculated for completely dry compression, when the compression line is $a_1 b_1$, has almost the same value.

In all these examples it is interesting, and practically important, to notice how little the coefficient of performance in the theoretical cycle is affected even by considerable changes in the dryness before compression. This is true not only of carbonic acid but of any working substance.

In practice dry compression is preferred. By reducing exchanges of heat between the working substance and the metal of the cylinder it gives a closer

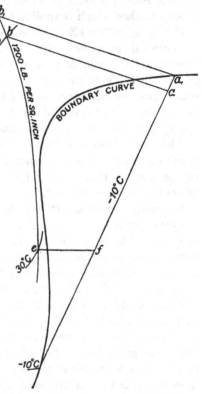

Fig. 46. Refrigeration cycle with Carbon Dioxide when the upper limit of pressure exceeds the critical pressure.

approximation to the adiabatic ideal than wet compression would give.

114a. Multiple-Effect Compression and Precooling of the Liquid by Partial Throttling. Attention has been drawn to the importance of cooling the condensed liquid before it passes the expansion-valve, a point which is seen in fig. 45 when the throttling lines ef and df_d are compared. In many CO_2 machines a practice is adopted, especially when the cooling water is comparatively warm, of precooling the liquid by means of an auxiliary process which, though it complicates the action, may have a thermodynamic advantage, as was first pointed out by Linde. In any such process the condensed refrigerant is cooled before it passes the expansion-valve to a temperature below that of the circulating water. One way of doing this is to employ an independent auxiliary circuit to refrigerate the liquid in the main circuit after it has been condensed and before it is throttled by passing the valve. Another way is to divide the throttling process itself into two steps by employing an intermediate receiver in which the pressure is lower than that of the condenser but higher than that of the evaporator. In the first step all the condensed fluid passes through a throttle-valve into that vessel, and is thereby chilled. Part of it becomes gaseous as a result of this throttling: the gas so formed is removed, recompressed, and recondensed. The remainder, which is now a precooled liquid, passes on to the second stage of the throttling process and so serves as the effective refrigerant in the main operation.

The compressing mechanism has therefore to deal with gas at two pressures, which involves some form of what is called "multiple effect compression." In an arrangement introduced by Windhausen the compressing process is made compound and the gas from the intermediate receiver is taken in to be compressed in the second stage. In another device, by Voorhees, a single cylinder is made to serve, but part of its charge enters after the first part has been inhaled. In the practical application of such arrangements for precooling by preliminary throttling, the regulation of the whole process is made automatic by providing a float in the intermediate receiver, which controls the comparative amounts of throttling in the first and second steps*.

115. Entropy-Temperature Chart for Carbon Dioxide in the Gaseous, Liquid and Solid States. The extended use of carbon dioxide not only as a working substance in the compression process but as a portable refrigerant or conveyor of cold in the form of blocks of "dry ice," gives special interest to a comprehensive $T\phi$ chart published by Plank and Kuprianoff†, which includes the region below the triple-point and therefore shows what happens

---

* For a discussion of the thermodynamic effects of multiple-effect compression, see a paper by Professor E. H. Lamb in *Ice and Cold Storage* for February and March, 1927.

† See footnote on p. 144.

when the substance passes direct from the solid to the gaseous phase by what is called sublimation.

A miniature version of the chart, showing a few representative lines, is given in fig. 47. There the entropy and the total heat are taken as zero for the liquid at 0° C. and consequently have negative values at lower temperatures. Lines of constant pressure are drawn, ranging from 120 At. downwards, the word "atmosphere" being used here in a technical sense to imply a pressure of 1 kg. per sq. cm. Lines of constant total heat are drawn, ranging from $I = 90$ down to $I = -80$. Both of these are here shown as broken lines in the region below the boundary curve. In the original chart there are also lines of constant volume, but these are omitted in this miniature version. Here c is the triple-point ($T = -56 \cdot 6°$ C., $P = 5 \cdot 28$ At.) from which the liquid boundary curve cd starts; d is the critical point; and de is the boundary curve for saturated vapour produced from the liquid. The continuation ef is for saturated vapour formed by sublimation from the solid state. A point to be noted is the change of direction in the boundary curve at e. The curve ab shows the relation of entropy to temperature in the solid phase when sublimation is about to begin. The horizontal distance from ab to ef at any level of temperature measures the change of entropy that occurs during sublimation. Thus at the triple-point the change of entropy in sublimation from solid to vapour, namely the length be, is 0·600, and since the absolute temperature is 216·5° K. this corresponds to $0 \cdot 6 \times 216 \cdot 5$ or 129·9 thermal units as the latent heat of sublimation. For lower pressures the latent heat of sublimation is somewhat greater, becoming 136·9 thermal units at one atmosphere and 139·8 when the vapour-pressure is so low as to bring the temperature down to $-100°$ C. It is this high latent heat (in combination, of course, with its low temperature) which gives to solid CO_2 its great practical value as a portable refrigerant.

The lines of constant pressure which lie to the left of the boundary curve cd relate to the liquid under conditions of pressure higher than those of saturation.

115a. **Mollier's *PI* Diagram for Carbon Dioxide.** Mention was made in Art. 102 of a type of diagram, due to Mollier, in which the coordinates are the pressure and the total heat. A chart of this type for CO_2 is given by Plank and Kuprianoff, part of which is copied in skeleton form in fig. 48, namely the part above the triple-point. Lines of constant volume are drawn, as well as lines of constant temperature and constant entropy. The volumes are expressed in cub. decimetres per kg.; to reduce to cub. ft. per lb. multiply by 0·016018. The pressures are in "atmospheres," that is kg. per sq. cm.

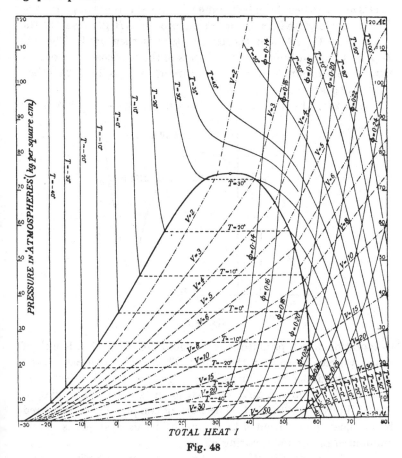

TOTAL HEAT I

Fig. 48

For exhibiting a cyclic process of refrigeration this diagram has the advantage that three out of the four operations are represented by straight lines parallel to the axes and the lengths of two of these lines, namely the two at constant pressure, directly measure the quantities of heat that are taken from the cold body and rejected to the hot body respectively. A third straight line represents the throttling process by passage through the regulating valve at constant I, and the diagram is completed by a curve of constant ϕ which, in the ideal case, represents the process of compression. Its projection on the base of I measures the work spent in compressing the substance.

The use of the PI diagram is illustrated in fig. 48a where two cycles are sketched, both beginning with dry saturated vapour at A. In the cycle $ABEF$, a compression process AB goes on till the pressure is 65$\frac{1}{2}$ At. and the temperature is then 55° C. The line $BCDE$ is the constant-pressure process in the condenser and cooler, namely first cooling to the temperature of condensation (25°) then condensing at that temperature, and then cooling the condensed liquid to 20°. The line EF is the throttling process, and FA is the

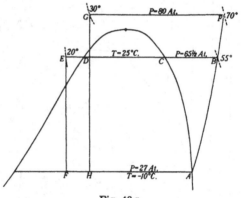

Fig. 48a

evaporation which produces the refrigerating effect. The length of FA is a direct measure of that effect. The length BC measures the heat rejected before condensation begins, CD that rejected in condensation, and DE that rejected in the subsequent cooling of the liquid from 25° to 20° C. The work of compression is measured by the change of I from A to B, which in this example is barely one-fourth of the refrigerating effect.

The other cycle $AF'GH$ illustrates what may occur with warmer cooling water, when heat is rejected at a pressure above the critical pressure and there is consequently no change from vapour to liquid in the "condenser." Here the compression AF' is carried to 80 At. and approximately 70° C. $F'G$ is the process of cooling the compressed gas at that pressure to 30° which is taken here as the lowest temperature the cooling water will admit of. The throttling process is GH. The effective refrigeration HA is substantially less than in the first example and the work of compression is greater. In conditions such as these, some thermodynamic advantage would result from dividing the throttling process into two stages and applying "multiple-effect compression" as described in Art. 114*a*.

116. Vapour-Compression by means of a Jet. Water-Vapour Machine. Whatever be the working substance, a vapour-compression refrigerating machine acts by pumping the vapour from the low-pressure region in which it has been evaporated to the high-pressure region in which it is to be condensed. So long as the vapour-pressure is moderately high the use of a compressing piston for this purpose is satisfactory. But when the pressure is very low, as it would be if water were used for the working substance, the volume to be swept through by a compressing piston would be inconveniently large, and the waste through friction would be excessive. At 0° C., for example, the density of water-vapour is so small that about 365 cubic feet of it are required to absorb as much latent heat as one cubic foot of ammonia vapour. Hence to use water-vapour as a refrigerating agent some appliance must be resorted to which will avoid the bulk and frictional waste of an ordinary compression pump. One such appliance is a centrifugal pump or reversed turbine: another is an ejector or jet pump, in which an auxiliary stream of vapour, supplied at a comparatively high pressure, forms a motive jet which drags with it the vapour to be "aspirated," namely the vapour which has been formed by evaporation at low pressure, so that both pass on together to be condensed. The vapour of the motive jet necessarily mixes with the vapour to be aspirated and both are condensed together: there are thus two circuits which coalesce in the condenser. Part of the condensed liquid returns through the expansion-valve to the cold evaporator, and acts as the effective working substance in producing refrigeration: the other part is forced by a feed-pump into a boiler where it is vaporized at a relatively high pressure, so that it may act as the motive jet: the two then meet again in the ejector on their way to the condenser. The thermodynamic efficiency of this device is not high, but it has the merit of simplicity. It is specially applicable where only a moderate degree of refrigeration is desired, as in "air-conditioning," when a fresh-air supply has to be cooled before its admission to rooms.

117. The Step-down in Temperature. Use of an Expansion Cylinder in Machines using Air. So long as the working substance in a refrigerating machine is a vapour which becomes liquefied during the operation, it is practicable, as we have seen, to dispense with an expansion cylinder and still have a large amount of refrigerating effect. The step-down in temperature, which is necessary in any refrigeration cycle, occurs as a consequence of the process of throttling, while the substance passes the expansion-valve. This is true also of a gas near its critical point, and hence a machine using carbonic acid under tropical conditions can be effective without an expansion cylinder although the substance may not undergo liquefaction. A gas near its critical point is very far from perfect and does not even approximately conform to Joule's Law. A gas which conforms to that law would suffer no step-down of temperature in passing an expansion-valve (Art. 19). With a gas such as air, which is nearly perfect at the temperatures and pressures that occur in ordinary refrigeration, the step-down would be too small to serve the desired purpose. Hence with air for working substance an expansion cylinder becomes an essential element of the machine. Refrigerating machines using air, and cooling it by means of expansion in a cylinder in which it does work against a piston, are amongst the oldest effective means of producing cold by mechanical agency. It was through their development that the cold-storage industry was created and the business was established of conveying refrigerated cargoes overseas. The type has become obsolete because machines using a condensable vapour are not only more compact but give a better thermodynamic return for the work spent, but it has enough scientific as well as historical interest to deserve description.

An air-machine operates by taking in a portion of air from the chamber that is to be kept cold, compressing it more or less adiabatically with the result that its temperature rises considerably above that of the available water-supply, then extracting heat from it in the compressed state by means of circulating water, then expanding it in a cylinder in which it does work, with the result that its initial pressure is restored and its temperature falls greatly below the initial temperature. It is then returned into the atmosphere of the cold chamber, with which it mixes; the object being either to lower the temperature in the chamber or to keep it from rising through leakage of heat from outside. This type was known as the Bell-Coleman air-machine.

The cycle is a reversal of that of Joule's Air-Engine, described in Art. 54. As applied in refrigeration the apparatus took the form shown diagrammatically in fig. 49. In the phase of action shown there, the pistons are moving towards the left. Air from the cold chamber C is being drawn into the compression cylinder M. In the return stroke it will be compressed from one atmosphere to about four, with the result that its temperature may be raised to 130° C. or higher. It is delivered under this pressure to the cooler A where it gives up heat to the circulating water and comes down to near atmospheric temperature. It then passes, still at high pressure, to the expansion cylinder N, where it does work in expanding to the initial pressure of one atmosphere and thereby

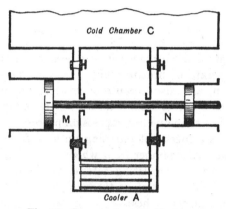

Fig. 49. Organs of an Air-Machine.

becomes very cold, reaching a temperature of perhaps −60° C. or −70° C., in which condition it is returned to the cold chamber. An ideal indicator diagram for the whole cycle is given in fig. 49 a, where $fcbe$ shows the action of the compression cylinder and $eadf$ shows that of the expansion cylinder. The area $abcd$ measures the net amount of work that is expended. In the diagram the compression and expansion are both treated as adiabatic and the volume of A as well

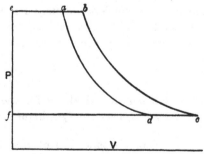

Fig. 49 a. Indicator Diagram of Air-Machine.

as that of C is assumed to be so large that during delivery of the air its pressure does not sensibly change. Writing T_a, T_b, T_c and

T_d for the temperature of the working air, at the points a, b, c and d of the diagram, we have $Q_A = C_p (T_b - T_a)$ for the heat rejected to the cooling water, and $Q_C = C_p (T_c - T_d)$ for the heat usefully extracted from the cold chamber. The net amount of work expended is equal to $Q_A - Q_C$. The coefficient of performance is

$$\frac{Q_C}{Q_A - Q_C}.$$

For the reason explained in Art. 54

$$\frac{T_b}{T_c} = \frac{T_a}{T_d}; \text{ from which } \frac{T_c - T_d}{T_b - T_a} = \frac{T_c}{T_b}.$$

Hence $\qquad \dfrac{Q_C}{Q_A} = \dfrac{T_c}{T_b}$, and $\dfrac{Q_C}{Q_A - Q_C} = \dfrac{T_c}{T_b - T_c}.$

This coefficient of performance is low because of the very large range of temperature through which the working air is carried. For this reason, and also because of greater frictional losses, an actual air-machine gives results that compare unfavourably with those obtained in the vapour-compression process.

Considered as a means of pumping up heat from T_c, the temperature of the cold chamber from which heat is taken in, to T_a, the temperature of the circulating water to which heat is discharged, the air-machine has two serious thermodynamic defects. There is an irreversible transfer of heat when the working air, after being heated by compression to T_b, comes into thermal contact with the circulating water at T_a; and there is another irreversible transfer when the working air, chilled by expansion to T_d, mixes with the less cold atmosphere of the chamber at T_c. An ideally efficient refrigerating machine, namely a reversed Carnot engine, working between T_a and T_c as upper and lower limits would have (Art. 105) a coefficient of performance equal to

$$\frac{T_c}{T_a - T_c}.$$

The coefficient found above for the reversed Joule cycle is substantially less, because T_b is higher than T_a.

118. **Direct Application of Heat to produce Cold. Absorption Machines.** In another class of refrigerating appliances there is no application of mechanical power: the agent is heat, which is supplied from a high-temperature source, and is employed in such a way as to cause another quantity of heat to pass from a cold body

and to be discharged at a temperature intermediate between that of the cold body and the hot source. In such machines the efficiency of the action from the thermodynamic point of view is measured by the heat ratio $\dfrac{Q_2}{Q}$, where Q_2 is the heat extracted from the cold body, and Q is the high-temperature heat which is supplied to carry out the operation.

A typical example is the ammonia-absorption refrigerating machine. Essentially this is a device in which the vapour of ammonia is alternately dissolved by cold water under a relatively low pressure, and distilled from solution in water under a relatively high pressure by the action of heat. The ammonia vapour, driven off by applying heat to the solution, is condensed in a vessel that is kept cool by means of circulating water. This gives anhydrous liquid ammonia at high pressure which (just as in a compression machine) is allowed to pass through an expansion-valve, into a coil or vessel forming the evaporator. A low pressure is maintained in the evaporator by causing the evaporated vapour to pass into another vessel, called the absorber, where it comes into contact with cold water in which it becomes dissolved. When the water in the absorber has taken up a sufficient proportion of ammonia it in turn is heated to give off the vapour again under high pressure. In the simplest form of the apparatus the same vessel serves alternately as absorber and as generator or distiller. For continuous working there are separate vessels, and the rich solution is transferred from the absorber to the generator by a small pump, while the water from which ammonia has been expelled flows back to the absorber to dissolve more ammonia. The scheme of such an apparatus is shown in fig. 50. Heat is applied to the solution in the generator by means of a steam-coil. The gas passes off at top to the condenser, then through the expansion-valve to the evaporator, and then on to the absorber, where it meets a current of water or very weak solution that has come over from the bottom of the generator. Between the generator and absorber is the interchanger, a device for economizing heat by taking it from the water that is returning to the absorber, and giving it to the rich solution that is being pumped into the generator. This rich solution is delivered at the top of the column in the generator; as the liquid parts with the ammonia it becomes denser and falls to the bottom, where it escapes to the absorber through an adjustable valve. When water absorbs ammonia a large amount of heat is given out. Hence the absorber as well as the condenser has to be kept cool by means of circulating water or otherwise. Under the most favourable conditions the quantity of heat which such a machine takes in from the cold body is considerably less than the quantity of high-temperature heat that has to be supplied, for it needs more thermal units to separate ammonia gas from solution in water than simply to evaporate the same amount of liquid ammonia.

An ingenious type of ammonia absorption machine has lately come into extensive use for domestic and other small-scale refrigeration under the name of *Electrolux*. In this machine

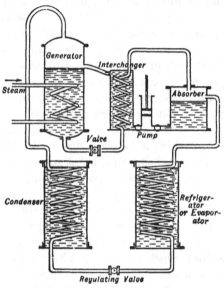

Fig. 50. Organs of an Ammonia Absorption Machine.

a continuous action is secured without any pump, and without duplicating the vessels. This is done by causing the ammonia as it evaporates after condensation to mix with a neutral and insoluble gas, namely hydrogen, so that although the whole circulating system is kept at a fairly high pressure the *partial* pressure of the ammonia (Art. 62), which determines its evaporation, is sufficiently small to allow the ammonia to assume the low temperature that is desired. The mixture of hydrogen and ammonia gas passes on to an absorber where the ammonia is taken up by water and the hydrogen is left. The hydrogen then passes back to the evaporator to resume its action as a diluent of the evaporating ammonia. The water in which the ammonia has been dissolved passes on by gravity to a "boiler" where heat is applied to it so that the dissolved ammonia is driven off to a condenser where it liquefies, the condenser being kept cool by circulating water. There is no pump and no expansion-valve. Except for small differences due to gravity the whole circulating system is at a uniform pressure of about 180 lb. per sq. inch: once charged it is kept sealed up. The hydrogen circulates continuously between the evaporator, where it mixes with the ammonia gas, and the absorber where the ammonia gas separates from it by being absorbed. Between the two there is a heat-exchanger which cools the hydrogen on its way back to the evaporator. The water which acts as absorbent circulates continuously from the absorber to the boiler and back, and between the two there is a heat-exchanger which cools the water on its way from the boiler to the absorber. Both circulations are maintained by gravity, the flow of the rich solution to the boiler (which it enters at the top) being assisted by a vertical stand-pipe in which bubbles are formed by a heating-coil at the bottom. The gaseous circulation also depends on gravity: the absorber is at a lower level than the evaporator so that the hydrogen rises and the heavier mixture of hydrogen with ammonia gas comes down. The evaporator is placed in the chamber which is to be kept cold. The apparatus is automatic in its action, easily regulated, and compact*.

In another type of absorption machine water-vapour is the substance which is absorbed: it is taken up by sulphuric acid, from which it may again be separated by the agency of heat. Such a machine has been used for ice-making, the evaporation of part of the water serving to freeze the rest. In this case also the heat ratio, namely the ratio of heat usefully extracted to heat supplied, is less than unity, for it takes more heat to separate the vapour of water from a sulphuric-acid solution than from pure water. It is a familiar fact that when water is mixed with sulphuric acid much heat is given out.

In other refrigerating appliances, essentially of the same class, the working vapour is "adsorbed" into the pores of a solid body and no solvent liquid is required. An interesting device of this kind, which has been applied to the cooling of railway waggons in the United States, uses Silica Gel as the adsorbing substance. Silica gel (SiO_2) is a hard solid with very minute pores that will

---

* For particulars of the Electrolux device and a discussion of its action, see a paper by H. Mawson, *Proc. Inst. Mech. Eng.* May 1931, p. 680.

take up large quantities of water-vapour or other vapour to which it is chemically inert, and give it off again when heat is applied. When this property is utilized for refrigeration the adsorbed vapour is generally sulphur dioxide (SO_2). Silica gel is obtained as a gelatinous mass which after drying becomes a sort of sponge that may be broken into granules. In practice the granules will take up a quarter or even a third of their weight of SO_2. During this adsorption the gas gives out latent heat which has to be carried away by causing air or water to circulate over the vessel in which the granules are contained. For the refrigeration of railway waggons the granules are carried in pipes which are stacked vertically in a compartment at one end of the car; these pipes can be exposed either to heat from a burner or to circulating air. When the stack of pipes is heated the adsorbed fluid is given off under pressure and passes as a gas into air-cooled tubes on the roof of the car where it is condensed. The condensed refrigerant, which may be stored for a time, afterwards passes through an expansion-valve to a set of evaporating pipes within the car, where it exerts its cooling effect by evaporating under low pressure, and then returns to the silica-gel granules to be again adsorbed, the pipes in which they are placed being kept cool by air circulation at this stage of the process. By having two stacks of such pipes, alternately acting as adsorbents and desorbents, the action may be made practically continuous, but a single set controlled automatically by a thermostat in the car is usually sufficient. It will be obvious that the process as a whole is precisely like that of an absorption machine except that adsorption by a porous solid takes the place of solution in water.

Silica gel is also employed to extract moisture from air as a means of air-conditioning. To dry the gel by driving off the water adsorbed in it takes about $1\frac{1}{2}$ times the normal latent heat of water-vapour at the same pressure.

The adsorption of vapour by a solid with minute pores occurs, apart from any chemical action, as an effect of surface tension in modifying the relation of vapour pressure to temperature on a curved liquid surface (see Art. 222).

119. Limit of Efficiency in the Use of High-temperature Heat to Produce Cold. Any appliance, such as an absorption machine, for the direct production of cold by the agency of heat, requires a supply of heat at a temperature higher than that of the surroundings.

There are necessarily three temperatures to be considered: (1) the low temperature T_2 of the cold body from which heat is being extracted; (2) the intermediate temperature T_1 of the available condensing water or other "sink" into which heat can be rejected; and (3) the high temperature T of the source from which heat is supplied to perform the operation. Any such appliance may be regarded as equivalent to the combination of a motor or heat-engine driving a refrigerator or heat-pump (fig. 51). A quantity Q of high-temperature heat goes in at one place, and thereby causes a quantity Q_2 of low-temperature heat to go in at another place. Heat is rejected at the intermediate temperature T_1, and the heat so rejected is equal to the sum of Q and Q_2, for no work is done

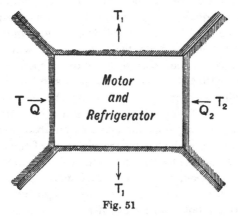

Fig. 51

by the appliance or spent upon it, as a whole. This description applies whether the appliance is actually a mechanical combination of a heat-engine with a heat-pump, or is an absorption machine with no conversion of heat into work and work into heat. In either case we have to consider what is the ideally greatest ratio of the low-temperature heat Q_2, which is extracted from the cold body, to the high-temperature or driving heat Q, when the three temperatures T_2, T_1 and T are assigned.

Suppose, first, that the machine consists of a perfect (reversible) heat-engine driving a perfect (reversible) heat-pump. Then it is easy to calculate the ratio of the heat extracted Q_2 to the heat supplied Q. Writing W for the heat-equivalent of the work developed in the heat-engine and employed to drive the heat-pump, we have (by Art. 38)

$$W = \frac{Q(T - T_1)}{T},$$

since the heat-engine is reversible. Again, since the heat-pump is also reversible,

$$W = \frac{Q_2 (T_1 - T_2)}{T_2}$$ (Art. 106).

Hence

$$\frac{Q_2}{Q} = \frac{T_2 (T - T_1)}{T (T_1 - T_2)},$$

which gives the required ratio of heats.

The importance of this result lies in the fact that no other method of applying heat to produce cold can give a higher ratio of Q_2 to Q, the three temperatures T, T_1 and T_2 being assigned. To prove this, imagine the combination of reversible heat-engine and reversible heat-pump to be reversed: it will then give out an amount of heat equal to Q to the hot body and an amount equal to Q_2 to the cold body, and it will take in an amount equal to $Q + Q_2$ from the intermediate body at T_1. It will still develop no work as a whole, nor require work to be spent in driving it. Imagine further that between the hot body and the cold one there are two appliances working—both using the same intermediate temperature—one of which is this reversed combination, and the other is a refrigerating machine (such as an absorption machine) whose efficiency we wish to compare with that of the combination. Then if it were possible for that machine to have a higher efficiency than the combination, it would extract more heat than Q_2 from the cold body for the same expenditure of high-temperature heat Q. Hence, when both work together, namely the combination working reversed and the other machine working direct, the cold body would lose heat while on the whole the hot body would lose none. In other words we should then have an impossible result, namely a simple transfer of heat, by a purely self-acting agency, from the cold body at T_2 to a warmer body at T_1, the intermediate temperature. The agency would be self-acting in the sense of being actuated by no form of energy, mechanical or thermal. Such a result would be a violation of the Second Law (Art. 81). The conclusion is that no means of employing heat to produce cold, whether directly, as in an absorption machine, or indirectly, as in a compression machine driven by an engine, can be more efficient (for the same three temperatures) than the combination of a reversible heat-engine driving a reversible heat-pump. Hence the expression

$$\frac{Q_2}{Q} = \frac{T_2 (T - T_1)}{T (T_1 - T_2)}$$

measures the ideally greatest ratio of heat extracted to heat supplied. Any real appliance will show a smaller heat ratio in consequence of irreversible features in its action. The action of an ammonia-and-water absorption machine, for example, is very far from being reversible: the heat ratio in it is much less than unity. But, as the above expression shows, when T_2 is not much lower than T_1 and T is much higher, Q_2 may be much greater than Q in the ideal use of heat to produce cold.

120. **Expression in Terms of the Entropy.** The above expression for the ideal performance under reversible conditions may be written

$$\frac{Q}{T}(T - T_1) = \frac{Q_2}{T_2}(T_1 - T_2),$$

from which

$$\frac{Q + Q_2}{T_1} = \frac{Q}{T} + \frac{Q_2}{T_2}.$$

This expresses the conservation of entropy for the complete reversible operation. The entropy of the system as a whole does not change. For the term on the left is the gain of entropy by the body at T_1 to which heat is rejected: the two terms on the right are the losses of entropy by the hot body and cold body respectively. The whole action may be regarded as a transfer of entropy from two sources at T and T_2, to an intermediate sink at T_1. So long as the action is reversible this transfer occurs without affecting the aggregate entropy, but if it is not completely reversible the aggregate entropy will increase; in that case the term on the left becomes greater than the sum of the terms on the right.

Again, the equation shows that, under reversible conditions, the product of the entropy lost by the hot source (through the removal of the heat Q) and the drop in temperature which that heat undergoes, namely from T to T_1, is equal to the product of the entropy lost by the cold body and the rise of temperature of the abstracted heat Q_2. Each of these products is in fact a measure of W, the work which the heat-engine produces and the heat-pump consumes, in the ideal combination of reversible engine with reversible pump.

A mechanical analogue is illustrated in fig. 52. Here a quantity of water M, supplied at a high level H, descends to a lower level H_1 and serves to raise another quantity M_2 from a still lower level H_2 up to H_1. Both quantities are discharged at the level H_1. The operation is reversible, and the energy equation may be written

$$M(H - H_1) = M_2(H_1 - H_2).$$

On comparing this with the equation given above, for a corresponding reversible thermal operation, it will be noticed that the analogue of weight (of water) is not heat but entropy, namely the quantity of heat divided by the temperature of supply.

The reversible thermal operation may be represented on the entropy-temperature diagram as in fig. 53. There the area *abon*

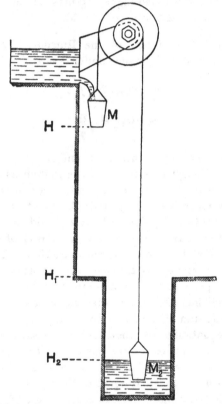

Fig. 52. Mechanical analogue of the use of heat to produce cold.

represents the high-temperature heat which is supplied at temperature T, and the area *abcd* represents the work which would be done in a perfect heat-engine by letting down that quantity of heat from T to the lower level T_1. Between the given levels of temperature T_1 and T_2 draw a rectangle *defg* whose area is equal to the area *abcd*, and produce *ef* to meet the base-line for zero temperature in *m*. Then the area *fgnm* represents the refrigerating effect, namely the heat extracted from the cold body at T_2. The amount of heat discharged at the intermediate level T_1 is equal to the area *ecom*, which is equal to the sum of the areas *abon* and *fgnm*.

121. The Refrigerating Machine as a means of Warming. In any such appliance, whether reversible or not, the quantity of heat delivered at the intermediate temperature T_1 is greater than the quantity supplied at T by the amount of the heat raised from T_2, and may, as we have seen, be much greater. This fact is the basis of an interesting suggestion made by Kelvin in 1852, that in the warming of rooms it would be thermally more economical to apply the heat got from burning coal in this indirect way than to discharge it into the room to be warmed. The thermodynamic value of high-temperature heat is wasted if we allow it directly to enter a comparatively cold substance. That value might be better utilized by employing

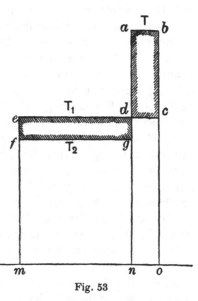

Fig. 53

the heat to pump up more heat, taken in from say the outside atmosphere, to the level to which the room is to be warmed. By using, for example, an efficient steam-engine to drive an efficient heat-pump, a small quantity of heat supplied at a high temperature will suffice to raise a much greater quantity of heat through the small range that is required, and consequently to produce a much greater warming effect. Similarly, if a supply of power from any source is available as a means of warming to a moderate temperature, it will be turned to better account if we set it to drive a heat-pump than if we simply convert it into heat. The suggestion that some of the coal which is used for heating rooms might be saved by applying heat in this indirect manner has at present no more than a theoretical interest*.

* See a paper by T. G. Haldane, *J. Inst. Elect. Eng.* 1930, for a revival of the suggestion.

121 a. The "Cold Multiplier." It is relevant in this connection to mention, for its thermodynamic interest, a refrigerating device which has the object of making the most effective use of a very cold body when what is desired is the removal of heat from a chamber at a moderately low temperature. Suppose for example that we have a supply of solid CO_2 to be used as a cooling agent, which will act as a sink of heat at a very low temperature T_2; also that the temperature of the available water-supply is T; and that our object is to extract heat from a chamber at some intermediate temperature T_1. If we simply bring the solid CO_2 into thermal contact with the contents of the chamber there is thermodynamic waste; heat flows irreversibly from T_1 to T_2; hence we are not utilizing the very cold sink's full value. The problem is like that dealt with in Kelvin's suggestion for warming rooms, and it may be solved by an *inversion* of his process. Take the reversible system of fig. 51 and imagine the action shown there to be reversed in all respects. Heat Q_2 will then flow into the cold sink at T_2 and heat Q will flow into the comparatively warm water-supply at T. Thus a quantity of heat equal to $Q + Q_2$ is extracted from the chamber at T_1. In other words the refrigerating effect of the combined process is greater than that which would be given by direct contact with a very cold sink, in the ratio of $Q + Q_2$ to Q_2.

In this scheme the temperature T_1 of the chamber may be regarded as the upper limit of a heat-engine whose sink is at T_2, and also as the lower limit of a heat-pump whose upper limit is the temperature T of the available water-supply. In the ideal reversible operation the power developed by the heat-engine serves to drive the heat-pump.

The ratio of $Q + Q_2$ to Q_2, which measures the theoretical limit of advantage, may be worked out for particular temperatures by the formulas of Art. 119. These apply only when the processes are reversible; but even under practical conditions a considerable increase of effective refrigeration may be realized.

A device for carrying out this idea in the utilization of solid CO_2 was shown at The Refrigeration Exhibition in London in 1934 under the name of the "Cold Multiplier." It works by absorption of ammonia in water and resembles the Electrolux contrivance in dispensing with a pump by causing the ammonia to evaporate under a low partial pressure of its own vapour, in a mixed atmosphere containing an inert gas. The inert gas is nitrogen, and the gases circulate continuously by gravity under a constant total pressure, for the mixture of ammonia and nitrogen is lighter than the nitrogen which has undergone rectification.

122. The Attainment of Very Low Temperature. Cascade Method. Another part of the science of refrigeration deals with methods of producing cold so extreme as to liquefy air and other so-called permanent gases. This is now the basis of an important industry, which employs the liquefaction of air as a step towards the separation of its constituents, in order to obtain supplies of oxygen, nitrogen, argon and neon. To liquefy any gas the temperature must be reduced below the critical point (Art. 77), and for nitrogen this means a cooling below $-146°$ C. Temperatures much lower than this have been reached by the methods which will now be described. Hydrogen, whose critical temperature is $-240°$ C., has not only been liquefied but solidified: its melting point under atmospheric pressure is about $-259°$ C. or $14°$ K. Even helium, the most refractory of gases, with its critical point at $5·2°$ K., has been liquefied and solidified under conditions that brought it within two or three degrees of the absolute zero*.

In the scientific study of low temperatures and their physical effects reference should be made to the important work carried on at the University of Leiden under the direction of the late Professor Kamerlingh Onnes and his successors, Drs W. H. Keesom and W. J. de Haas. Among the many publications of the Leiden Laboratory are $T\phi$ charts (and in some cases $I\phi$ charts) of methane, ethylene, nitrogen, hydrogen and helium, exhibiting the properties both above and below the critical point†.

One way of reaching a very low temperature, called the "cascade" method, is to have a series of compression refrigerating machines so connected that the working substance in one, when cooled by its own evaporation, acts as the circulating fluid to cool the condenser of the next machine of the series, and so on. Different working fluids are selected for the successive machines, so that each in turn reaches a lower temperature than its predecessor. The general idea of the method is illustrated in fig. 54. In that diagram the first working substance is carbon dioxide, which is represented in the sketch as supplied from a reservoir on the left, into which it has been compressed. It expands through a throttle-

---

* See Keesom (*Nature*, vol. CXVIII, p. 81, 1926). Helium was solidified at $4·2°$ K. at a pressure of 140 atmospheres, the temperature of solidification diminishing with the pressure.

† See Leiden *Communications*, Supplement No. 65, 1928.

valve into the vessel A, from which it escapes at atmospheric
pressure (this part of the apparatus might be completed by a
compression pump restoring the substance to the reservoir). The
effect is that the vessel A is kept at a temperature of about $-80°$ C.
Within it is another vessel which serves as the condenser of a
machine using ethylene as working substance. Ethylene has a
critical temperature about $10°$ C., and needs only moderate pressure
to liquefy it at $-80°$ C. It is pumped into the inner part of the
condenser A, is there liquefied, and passes on through an expansion-
valve to the outer part of the vessel B in which it evaporates,

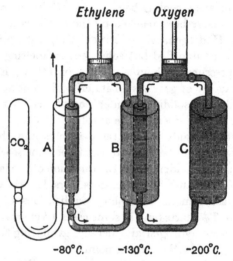

Fig. 54. Cascade Method of reaching very Low Temperatures.

producing a temperature of say $-180°$ C. at the low pressure which
is maintained by the pump. This cools the vessel B below the
critical point of oxygen (namely $-119°$ C.); accordingly oxygen
may be used as the working substance of the next machine. It is
condensed in the inner part of the condenser B, and after passing
through an expansion-valve it may produce a temperature of
$-200°$ C. or less in the vessel C by evaporating there under a low
pressure. Each machine of the series is a vapour-compression
machine, working on the principle already described, and made up
of an evaporator, a compressing pump, a condenser, and an expansion-valve. The essential feature in the combination is that the
working substance in any one machine must be evaporated at a

temperature that is lower than the critical point of the working substance of the next machine in the series.

123. Regenerative Method. But it is in a different way that low temperatures are now attained for the commercial liquefaction of air. The usual process is a regenerative one, first successfully developed by Linde, in which the Joule-Thomson effect of irreversible expansion in passing a constricted orifice serves as the step-down in temperature, and a cumulative cooling is produced by causing the gas which has suffered this step-down to take up heat in a thermal interchanger from another portion of gas that is on its way to the orifice.

Consider first what would happen if there were no such thermal interchange. Imagine a gas such as air to have been compressed to a high pressure P_A, and to have had the heat developed by compression removed by circulating water or otherwise, so that its temperature is that of the surroundings. Call this initial temperature T_1. Let the compressed gas at that temperature enter an apparatus in which it expands irreversibly (through an expansion-valve or plug or constricted orifice of any kind) to a much lower pressure P_B, at which pressure it leaves the apparatus. If the gas were an ideal perfect gas this irreversible expansion would cause no fall in temperature. In a real gas there is in general a fall, from T_1 to some lower temperature T'. The fall $T_1 - T'$ measures the Joule-Thomson cooling effect of the given drop in pressure. In Joule and Thomson's experiments on air it was about a quarter of a degree for each atmosphere of drop in pressure*.

The cooling effect of the drop in pressure may be measured by the quantity of heat which would have to be supplied to the gas, per lb., after expansion, to restore it to the temperature at which it entered the apparatus. Call that quantity Q, then

$$Q = C_p (T_1 - T'),$$

where C_p is the mean specific heat of the gas between these temperatures, at the lower pressure P_B.

We may define Q as the quantity of heat which each lb. of the gas would have to take up within the apparatus if its temperature

---

* According to their results for air, the fall of temperature expressed in degrees centigrade is

$$0 \cdot 275 (P_A - P_B) \left(\frac{273}{T_1}\right)^2,$$

where P_A and P_B are the pressures in atmospheres. For more recent results, see Hoxton, *Phys. Rev.* vol. xiii, p. 488, 1919, or Roebuck, *Proc. Amer. Acad.* vol. lx, p. 537.

on leaving the apparatus were made equal to its temperature on entry. It measures the available cooling effect due to each lb. of gas that passes through the apparatus.

So long as there is no communication of heat to the gas, by thermal interchange or otherwise, while it is passing through the apparatus, the gas simply passes off at a lower temperature T'. The gas that passes off has the same total heat I as the gas that enters (Art. 72), though its temperature has dropped. If we were to restore it to the original temperature T_1 before letting it pass off, it would take away more total heat than it brings in, the difference being equal to Q. Its total heat I at exit would then be greater than its total heat on admission by the quantity Q, though its temperature would be the same. The existence of a Joule-Thomson cooling effect in any real gas depends on the fact that the total heat I is a function of the pressure: for a given temperature the total heat is greater when the pressure is low.

Suppose now that there is a counter-current interchanger by means of which the stream of gas which has passed the orifice takes up heat from the stream that is on its way to the orifice, with the result that the outgoing stream, before it escapes, has its temperature restored to T_1 or very near it. This is easily accomplished by having, within the apparatus, a long approach pipe or worm through which the compressed gas passes before it reaches the orifice, and round the outside of which the expanded gas passes away, so that there is intimate thermal connection between the two streams. For simplicity we may assume the interchanger to act so perfectly that when the outgoing gas reaches the exit it has acquired the same temperature T_1 as the entering gas. Each lb. of it will therefore have taken up a quantity of heat equal to Q as defined above.

124. Regenerative Method. First Stage. Under these conditions the apparatus will steadily lose heat at the rate of Q units for every lb. of gas that passes through. If we suppose the apparatus as a whole to be thermally insulated against leakage of heat into it from outside, there will consequently be a continuous reduction of the stock of heat that is held by the pipes and the gas in them. The result is a progressive cooling which constitutes the first stage of the action.

It may help to make the action clear if we draw up an account of the energy received and discharged by the apparatus. Gas enters at A (fig. 55) under the pressure P_A and at the temperature T_1. Gas leaves the apparatus at B under the pressure P_B and at the same temperature T_1, having taken up, through the action of the interchanger, a quantity of heat equal to Q. The pipes and expansion orifice are not shown in the sketch: they are within the enclosing case, which is assumed to be a perfect non-conductor of heat. During the first stage of the action the stop-cock C is closed, and all the gas that has gone in at A goes out at B; it is only by the entry of gas at A and by its escape at B that energy enters or leaves the apparatus.

Each lb. of entering gas contains a quantity of internal energy E_A, and the work that is done upon it as it goes in is $P_A V_A$. Each lb. of outgoing gas contains a quantity of internal energy E_B, and does work, against external pressure, equal to $P_B V_B$. Hence, for

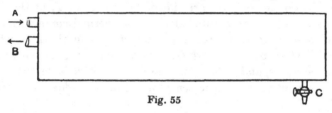

Fig. 55

each lb. that flows through, the net amount of heat which the apparatus loses is

$$E_B + P_B V_B - (E_A + P_A V_A), \text{ or } I_B - I_A.$$

But the amount so lost is Q, namely the heat that is required to restore the gas to the temperature at which it makes its exit. Hence

$$Q = I_B - I_A.$$

The contents of the apparatus become colder and colder in consequence of this continued abstraction of heat. But it is important to notice that their fall in temperature does not affect the value of Q. We assume that the action of the thermal interchanger continues to be perfect; in that case the exit temperature will still be equal to the initial temperature T_1 however cold the interior becomes in the neighbourhood of the expansion-valve. There will be no change in the value of either I_B or I_A, and consequently no change in Q. The value of Q, as the above expression shows, depends entirely on the conditions at A and at B; with perfect interchange this means that it depends only on P_A, P_B and T_1. It is independent of any temperature conditions within the apparatus.

It is therefore not affected by the progressive cooling, and retains the same value as the action proceeds*.

This stage of progressive cooling continues until the temperature of the gas at the place where it is coldest, namely on the low-pressure side of the expansion-valve, falls not only below the critical point, but to a value T_2 which is low enough to let the gas begin to liquefy under the pressure P_B. In other words T_2 is the boiling point corresponding to P_B. This is the lowest temperature that is reached.

A continuous gradient of temperature has now become established along the flow-pipe within the apparatus from the point of entrance, where it is T_1, to the high-pressure side of the expansion-valve, where it exceeds T_2 by the amount of the Joule-Thomson drop. There is also a continuous gradient along the return pipe from T_2, on the low-pressure side of the valve, to T_1 at the exit. The flow and return streams are in close thermal contact, and at each point there is an excess of temperature in the flow which allows heat to pass by conduction into the return, except at the entrance where, under the ideal condition which we have postulated of perfect interchange, the temperature of both flow and return is T_1.

This state of things is diagrammatically represented in fig. 56. There the flow and return are represented as taking place in straight pipes, one inside the other to provide for interchange of heat. Entering along the inner pipe A the compressed gas expands through a constricted orifice E (equivalent to an expansion-valve) into a vessel from which it returns by the outer pipe B. The vessel is provided with a stop-cock C by which that part of the fluid which

---

* It will be shown in Chapter VIII that the quantity Q, which measures the available cooling effect within the apparatus when the pressures P_A and P_B and admission temperature T_1 are assigned, can be calculated if we know the co-efficient of expansion of the gas under constant pressure for various pressures, and also the volume (per lb.) for various pressures, at the temperature T_1.

Writing V for the volume at any pressure, and $\left(\dfrac{dV}{dT}\right)_P$ for the coefficient of expansion, namely the rate of change of volume per unit of change of temperature when the pressure is constant, we shall see there (Art. 189) that

$$Q = \int_{P_B}^{P_A} \left[T_1 \left(\frac{dV}{dT}\right)_P - V \right] dP,$$

the temperatures being taken as T_1 throughout.

Q may also be found experimentally, by observing the drop of temperature $T_1 - T'$ which takes place when the gas expands from P_A to P_B through a Joule-Thomson orifice without any interchange of heat.

is liquefied can be drawn off when the second stage of the operation
has been reached. In the temperature diagram (fig. 56 a) MN represents the length of the interchanger, DM is the initial (and final)
temperature T_1, GN is T_2 and FG is the Joule-Thomson drop.
DF is the gradient for the flow-pipe, and GD for the return.

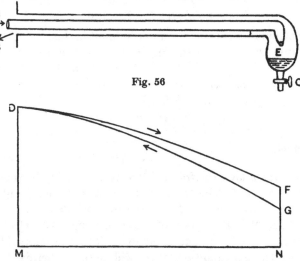

Fig. 56

Fig. 56 a.   Ideal process of Regenerative Cooling.

125.   Regenerative Method. Second Stage. When this gradient
has become established the gas begins to liquefy, the apparatus
does not become any colder, and the action enters on the second
stage, which is one of thermal equilibrium. A certain small fraction
of the gas is continuously liquefied and may be drained off as a
liquid through the stop-cock C. The larger fraction, which is not
liquefied, continues to escape through the interchanger and to
leave the apparatus at the same temperature as before, namely
the temperature T_1 equal to that of the entering gas. Call this
unliquefied fraction q; then $1-q$ represents the fraction that is
drawn off as a liquid at the temperature T_2. Since the apparatus
is now neither gaining nor losing heat on the whole, its heat-account
must balance; from which

$$I_A = qI_B + (1-q)\,I_C,$$

where I_A is the total heat per lb. of the gas entering at A, I_B is the
total heat per lb. of the gas leaving at B, and I_C is the total heat
per lb. of the liquid leaving at C. In this steady working the

aggregate total heat of the fluid passing out is equal to that of the fluid passing in. The fluid, as a whole, takes up no heat in passing through the apparatus.

Suppose now that the liquid leaving at C were evaporated at its boiling point T_2, and then heated at the same pressure from T_2 to T_1. The heat required to perform that operation would be

$$(1-q)\,[L+C_p\,(T_1-T_2)].$$

But that hypothetical operation would result in this, that the whole of the fluid then leaving the apparatus would be restored to the temperature of entry, namely T_1, since the part which escapes at B is already at that temperature. Hence the heat required for it is equal to the quantity Q as defined in Art. 123. We therefore have

$$(1-q)\,[L+C_p\,(T_1-T_2)]=Q,$$

from which
$$1-q=\frac{Q}{L+C_p\,(T_1-T_2)}.$$

This equation allows the fraction that is liquefied to be calculated when Q is known*. The fraction so found is the ideal output of liquid, for we have assumed that there is no leakage of heat from without, and that the action of the interchanger is perfect in the sense that the outgoing gas is raised by it to the temperature of entry. Under real conditions there will be some thermal leakage, and the gas will escape at a temperature somewhat lower than T_1: the effect is to diminish the fraction actually liquefied.

The fraction $1-q$ is increased by using a larger pressure-drop. It is also increased by reducing the initial temperature T_1; thus the output of a given apparatus can be raised by using a separate re-frigerating device to pre-cool the gas. Pre-cooling is indispensable if the method is to be applied to a gas in which, like hydrogen, the Joule-Thomson effect is a *heating* effect at ordinary temperatures, but becomes a cooling effect when the initial temperature is suffi-ciently low. This reversal occurs in hydrogen at about 190° K. With helium the Joule-Thomson effect becomes a cooling only when the temperature is below 40° K.

126. **Linde's Apparatus.** The principle of regenerative cooling described in the preceding article was first successfully applied by Linde in 1895 for the production of extremely low temperatures, and for the liquefaction of air, by means of an apparatus shown

---

* The specific heat of the vapour is here treated as constant from T_2 to T_1, which is very nearly true at low pressures.

diagrammatically in fig. 57. It consists of an interchanger *CDE* formed of two spiral coils of pipes, one inside the other, enclosed in a thermally insulating case. A compressing pump *P* delivers air under high pressure through the valve *H* into a cooler *J* where the heat developed by compression is removed by water circulating in the ordinary way from an inlet at *K* to an outlet at *L*. The highly compressed air then passes on through the pipe *BC* to the inner worm and after traversing the worm it expands through the throttle-valve *R* into the vessel *T*, thereby suffering a drop in temperature. Then it returns through the outer worm *F* and, being in close contact with the inner worm, gives up its cold to the gas that is still on its way to expand. Finally it reaches the compression cylinder *P* through the suction-valve *G*, and is compressed to

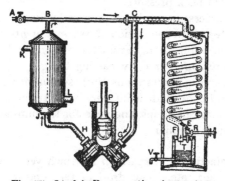

Fig. 57. Linde's Regenerative Apparatus.

go again through the cycle. During the first stage it simply goes round and round in this way; but when the second stage is reached and condensation begins, the part that is liquefied is drawn off at *V* and the loss is made good by pumping in more air through the stop-valve at *A* by means of an auxiliary low-pressure pump, not shown in the sketch, which delivers air from the atmosphere to the low-pressure side of the circulating system.

Linde showed that by keeping this lower pressure fairly high, it is practicable to reduce the amount of work that has to be spent in liquefying a given quantity of air. He pointed out that while the cooling effect of expansion depends upon the difference of pressures P_A and P_B on the two sides of the expansion-valve, the work done in compressing the air in the circulating system depends on the ratio of P_A to P_B. It is roughly proportional to the logarithm of that ratio, for it approximates to the work spent in the isothermal compression of a perfect gas, which (by Art. 28) is $RT \log_e r$, where r is the ratio of the volumes or of the pressures. If, for example, P_A is 200 atmospheres and P_B is one atmosphere, the cooling effect is proportional to 199 and the work of the compressing pump is roughly proportional to $\log 200$. If on the other hand the back pressure P_B is 50 atmospheres, the cooling effect is proportional to 150 and the work of the main compressing pump to $\log 4$. The cooling effect is reduced by only about one-fourth, while the work is reduced by nearly three-fourths. After allowing for the extra amount of work that has, in the second case, to be spent on the auxiliary pump in supplying air at 50 atmospheres to replace the fraction which is liquefied, there is still a marked advantage, in point of thermodynamic efficiency, in using a closed cycle with a moderately high back pressure.

The Linde process is employed on a commercial scale to liquefy air as a first step in the separation of its constituents, each of which finds industrial application. From the nitrogen a fertilizer called cyanamide is manufactured by passing the gas over hot calcium carbide. The oxygen serves many purposes, in medicine as well as for welding and other industrial purposes. Argon is used for the filling of incandescent lamps, and neon for the luminous shop-signs where red glow is familiar.

127. Liquefaction by Expansion in which Work is done. Apparatus of Claude, etc. The drop in temperature which a gas undergoes in passing from a region of high pressure to a region of low pressure would be greater if the process were conducted reversibly, as by expansion in a cylinder in which the gas does mechanical work. We should still have the small Joule-Thomson cooling effect, but in addition there would be the (generally much larger) cooling effect that is due to the energy which the gas loses in doing work. Early attempts made by Siemens, Solvay and others to reach very low temperatures by applying a thermal interchanger to an expansion cylinder, failed mainly because the cylinder soon reached a temperature at which the lubricant froze. This difficulty was successfully overcome in 1902 by Claude, who found that the difficulties attendant on expansion in a working cylinder down to a temperature below the critical point of air could be overcome by using certain hydrocarbons as lubricants. A hydrocarbon such as petroleum-ether does not solidify but remains viscous at a temperature as low as $-160°$ C. Using a lubricant of this kind Claude succeeded, as an experimental *tour de force*, in liquefying air in an expansion cylinder furnished with a regenerative counter-current thermal interchanger: the expansion cylinder simply taking the place of the expansion-valve in an apparatus such as that of Art. 126. He also found that the liquid, once it begins to form, serves itself as a lubricant, and no other need then be supplied. Under these conditions, however, there is little if any advantage in using an expansion cylinder, for the volume of the fluid at the lowest extreme of temperature is so small as to make the work of expansion insignificant. There is not much additional cooling: at the same time it is far less practicable to secure thermal insulation with an expansion cylinder than with a Joule-Thomson orifice. Claude

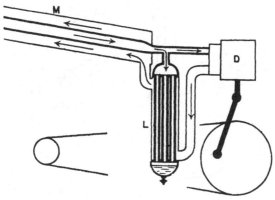

Fig. 58. Claude's method.

subsequently obtained a more economical result by giving the apparatus the modified form shown in fig. 58. In that arrangement part of the compressed air expands in a working cylinder to a temperature which may be just below the critical temperature, and the air which is cooled (but not liquefied) by that expansion is used as a cooling agent on the remainder of the air, with the result that some of the latter is liquefied under the higher pressure at which it is supplied. The supply comes in, at a pressure of 40 atmospheres or so, through the central pipe of the counter-current interchanger *M*. Part of it passes into the expansion cylinder *D* where it expands doing work, and is then discharged through the condensing vessel *L*, where it serves as the cooling agent to maintain a temperature somewhat lower than $-140°$ C., the critical temperature of air. The remainder of the compressed air enters the tubes of *L* and is condensed there, under pressure, dropping as a liquid into the chamber below, from which it can be drawn off.

In a further development of this invention Claude made the expansion compound, and caused the expanded gas to act as a cooling agent after each stage, becoming itself warmed up in the process. The expanded gas is thereby prepared to suffer further expansion without an excessive fall of temperature. During its expansion the gas in the cylinder is not so near the liquid state as to make expansion in a working cylinder of little use. The arrangement with compound expansion is illustrated in fig. 59. Air under pressure

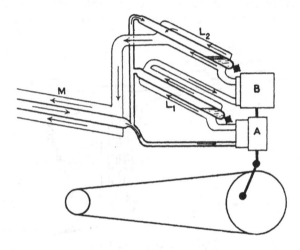

Fig. 59. Claude's later method with compound expansion.

enters, as before, through the central pipe of M. Part of it goes to the first expansion cylinder A, does work there, and proceeds at reduced pressure, and at a temperature below the critical point, through the outer vessel of the condenser L_1, in the inner tube of which some of the compressed air is being condensed. This warms up the expanded air to some extent, and it then passes on to complete its expansion in B, which again brings its temperature down sufficiently to allow it to act as condensing agent for the remaining portion of the air under pressure, in the second condenser L_2. This division of the expansion into two (or it may be more than two) stages is equivalent to making the process as a whole more nearly isothermal, so that the air need not at any stage deviate very widely from a temperature which is just sufficiently below the critical point to allow liquefaction to go on under the pressure at which the air is supplied*.

In a modification of Claude's method, introduced and practically developed on a large scale by C. W. P. Heylandt, the air, compressed to 200 atmospheres or so and cooled only by circulating water, is divided into two streams. About half of it is expanded in a working cylinder which lowers the temperature by 150°, and the portion so chilled is then brought into thermal contact, through an interchanger, with the other half, which is thereby much pre-cooled. The pre-cooled portion then passes an expansion-valve where a large part of it is liquefied by the Joule-Thomson effect.

Professor Kapitza has succeeded in liquefying helium mainly by adiabatic expansion in a cylinder where the need for lubricant is avoided by fitting the piston loosely with a small clearance round it through which some of the gas may leak†. The piston moves rapidly and the amount of leakage is too small to have much effect on the efficiency of the process. The helium is compressed to 25 or 30 atmospheres and is pre-cooled by liquid air. It is then cooled by the use of the expansion cylinder and a regenerative spiral to about 8° K., after which the final liquefaction takes place through the Joule-Thomson effect in an expansion-valve. With helium there is great advantage in applying the method of adiabatic expansion because the Joule-Thomson effect is negligible or even negative except at very low temperatures.

Professor Simon (*Z. f. Phys.* p. 816, 1933 and *Proc. Roy. Inst.* p. 28, 1934) employs a simple method for liquefying small quantities of helium which depends on the heat which is taken from it in expansion against the atmosphere. Two small metal chambers are separated by a partition, the one contains helium at 100 atmospheres, the other liquid hydrogen. The vessels are cooled in liquid hydrogen, the space between the vessels and the liquid hydrogen container being at first filled with helium. Then the hydrogen in the upper vessel is pumped off, so as to lower the temperature. The helium in the lower smaller chamber is allowed to expand through the fine connecting tube which ends in a valve communicating with another tube and balloon in which the evaporated helium can be stored. The helium then cools itself until it liquefies. By further evaporation at reduced pressure temperatures as low

---

* G. Claude, *Comptes Rendus*, June 11 and Oct. 22, 1906. See also his book on *Air liquide*, 2nd edition, 1926.

† *Nature*, May 12, 1934, p. 708.

as 1° K. are reached. The method by which the lowest temperatures noted in Art. 16 have been obtained depends on the decrease of the entropy of certain paramagnetic substances by magnetization*, and the temperatures are measured by taking advantage of the fact that the magnetic susceptibility of these substances varies inversely as the temperature to a first order of approximation. Vapour pressures at temperatures in the neighbourhood of absolute zero become so remarkably small, that the thermal insulation is very good and these low temperatures can be maintained for a long time. The specific heats are also very small and large quantities of metal are cooled by the abstraction of very small quantities of heat.

128. Separation of the Chief Constituents of Air. The liquefaction of air enables the constituent gases to be separated because in re-evaporation they have different boiling points. The boiling point of nitrogen, under atmospheric pressure, is about −196° C. or 13° lower than that of oxygen, which is −183° C. When a quantity of liquefied air evaporates freely both gases pass off, but not in the original proportion in which they are mixed in the liquid. The nitrogen evaporates more readily, and the liquid that is left becomes richer in oxygen as the evaporation proceeds. This difference in volatility between oxygen and nitrogen makes it possible to carry out a process of *rectification* analogous to the process which is used by distillers for extracting spirit from the "wash" or fermented wort, which is a weak mixture of alcohol and water, by means of a device known as the Coffey Still.

In the still patented by Aeneas Coffey in 1830 there is a rectifying column consisting of a tall chamber containing many zig-zag shelves or baffle plates. The wash enters at the top of the column and trickles slowly down, meeting a current of steam which is admitted at the bottom and rises up through the shelves. The down-coming wash and the up-going steam are thereby brought into close contact and an exchange of fluid takes place. At each stage some of the alcohol is evaporated from the wash and some of the steam is condensed, the heat supplied by the condensation of the steam serving to evaporate the alcohol. The condensed steam becomes part of the down-coming stream of liquid: the evaporated alcohol becomes part of the up-going stream of vapour. Finally at the top a vapour

---

* First suggested by Debye (*Ann. d. Phys.* vol. LXXXI, p. 1154, 1926) and by Giauque (*J.A.C.S.* vol. XLIX, p. 1864, 1927).

comparatively rich in alcohol passes off: at the bottom a liquid accumulates which is water with little or no alcohol in it. A temperature gradient is established in the column: at the bottom the
temperature is that of steam, and at the top there is a lower temperature approximating to the boiling point of alcohol. The wash
enters at this comparatively low temperature, and takes up heat
from the steam as it trickles down.

Linde applied the same general idea in a device for separating
the less volatile oxygen from the more volatile nitrogen of liquid
air. In this device, the primary purpose of which was to obtain
oxygen, there is a rectifying column down which liquid air trickles,
starting at the top at a temperature a little under 79° absolute,
which may be taken as the boiling point of liquid air under atmospheric pressure. As the liquid trickles down it meets an up-
going stream of gas which consists (at the bottom) of nearly pure
oxygen, initially at a temperature of about 90° absolute, that being
the boiling point of oxygen under atmospheric pressure. As the gas
rises and comes into close contact with the down-coming liquid,
there is a give and take of substance: at each stage some of the
rising oxygen is condensed and some of the nitrogen in the down-
coming liquid is evaporated; the liquid also becomes rather warmer.
By the time it reaches the bottom it consists of nearly pure oxygen:
the nitrogen has almost completely passed off as gas, and the gas
which passes off at the top consists very largely of nitrogen. More
precisely it consists of nitrogen mixed with about 7 per cent. of
oxygen: in other words, out of the whole original oxygen content
of the air (say 21 per cent.) two-thirds are brought down as liquid
oxygen to the bottom of the column, while one-third passes off
unseparated along with all the nitrogen. The oxygen that gathers
at the bottom is withdrawn for use, and is evaporated in serving to
liquefy fresh compressed air, which is pumped into the apparatus
to undergo the process of separation. The cold gases that are leaving
the apparatus, namely the oxygen which is the useful product, and
the nitrogen which passes off as waste gas at the top of the column,
are made to traverse counter-current interchangers on their way
out, so as to give up their cold to the incoming compressed air that
is on its way to be liquefied.

In the diagram, fig. 60, these counter-current interchangers are
omitted for the sake of clearness, but the essential features of the
condensing and rectifying apparatus are shown. The figure is based
on one given in Linde's patent of June 1902, which describes the

invention by which a process of rectification has been successfully applied in the extraction of oxygen from air.

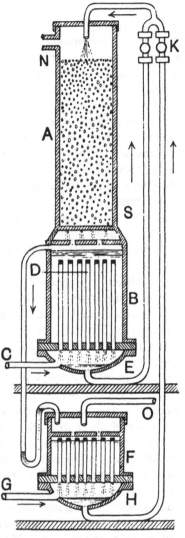

There *A* is the rectifying column, consisting in this instance of a vertical chamber stacked with glass balls, through the interstices of which the liquid trickles down. The lower part *B* contains an accumulation of fluid which, when the apparatus has been at work long enough to establish a uniform *régime*, consists of nearly pure liquid oxygen. Compressed air, which has been cooled by passing through a counter-current interchanger, enters at *C*, becomes liquefied in the vertical condenser pipes *D*, which are closed at the top, and drops down into the vessel *E*. It gives up its latent heat to the oxygen in *B*, thereby evaporating a part of that, and so supplying a stream of gaseous oxygen which begins to pass up the rectifying column. On its way up, this stream of gas effects an exchange of material with the liquid air which is trickling down: gaseous oxygen is condensed and returns with the stream to the vessel *B*, while nitrogen is evaporated and passes off at the top of the column, at *N*, mixed with some oxygen. The escaping gas goes through an interchanger, taking up heat from the incoming compressed air.

Fig. 60. Linde's apparatus of 1902 for extracting oxygen by rectification.

The accumulation of nearly pure liquid oxygen in *B* overflows into the lower vessel *F*, where a supplementary supply of compressed air entering at *G* is employed to evaporate it by means of a similar arrangement of condenser tubes open at the bottom and closed at the top, this air becoming itself condensed in the process, and falling as a liquid into the vessel H. The liquefied air from E and from H is still under pressure: it passes up through expansion-valves K to the top of the rectifying column, where it is discharged over the glass balls at a pressure not materially above that of the atmosphere. This secures the necessary difference in temperature between the bottom and top of the column. The compressed air plays the part of heater and evaporator of the liquid oxygen at the bottom, at the comparatively high temperature of about 91° absolute, before it undergoes rectification. In other words, it not only corresponds to the "wash" of the Coffey still, but it also serves as the equivalent of the heater by which the liquid at the bottom of the still gives off an upward current of steam. Gaseous oxygen, the desired product in this case, passes off at O, and like the waste gas, consisting mainly of nitrogen, which escapes at N, goes through a counter-current interchanger, taking up heat from the compressed air which enters partly at C and partly at G. It is the waste gas in this process that forms the analogue of the rectified spirit which is the useful product of the Coffey still.

At first, when the machine begins working, the air is highly compressed, but after the operation has gone on for some time, and a steady state is approached, a much lower pressure is sufficient. It must be high enough to make the air liquefy at the temperature of the liquid oxygen bath, say 91° absolute, and in practice it is kept higher than this to ensure that the drop in temperature at the expansion-valve may be sufficient to make good any losses due to leakage of heat from outside, and to imperfect interchange in the counter-current apparatus. For some time after the apparatus is first started the rectifying action is imperfect, but as the process goes on the liquid contents of the vessel B become richer and richer in oxygen, the rectification becomes more complete, and the pressure may be reduced. Under practical conditions it is easy to secure that the gaseous product shall be pure to the extent of containing 98 per cent. of oxygen.

129. Baly's Curves. The action of the rectifying column will be made more intelligible if we refer to the results of experiments published in 1900 by Baly*, which deal with the nature of the evaporation in mixtures of liquid oxygen and nitrogen. Given a

* E. C. C. Baly, *Phil. Mag.* vol. XLIX, p. 517, 1900.

mixture of these liquids in any assigned proportion, equilibrium between liquid and vapour is possible only when the vapour contains a definite proportion of the two constituents, but this proportion is not the same as that in the liquid mixture. Say for example that the liquid mixture is half oxygen and half nitrogen, then according to Baly's experiments the vapour proceeding from such a mixture will consist of about 22 per cent. of oxygen and

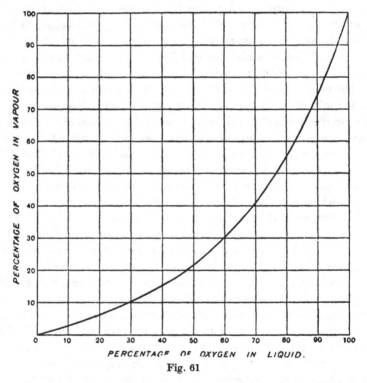

Fig. 61

78 per cent. of nitrogen. With these proportions there will be equilibrium. If however a vapour richer than this in oxygen be brought into contact with the half-and-half liquid, part of the gaseous oxygen will condense and part of the liquid nitrogen will be evaporated, until the proportion giving equilibrium is reached. The curve, fig. 61, shows, for each proportion in the mixed liquid, what is the corresponding proportion in the vapour necessary for equilibrium: in other words what is the proportion which the constituents have in the vapour, when that is being formed by evaporation of the mixed liquid, in the first stages of such an evaporation,

before the proportion in the liquid changes. In this curve the base-line specifies the proportion of oxygen in the liquid mixture, from 0 to 100 per cent., and the ordinates give the proportion of oxygen in the corresponding vapour, when the vapour is formed under a pressure equal to that of the atmosphere. Much the same general relation will hold at other pressures. It will be seen from the curve that when the evaporating liquid mixture is liquid air (oxygen 21 per cent., nitrogen 79 per cent.), the proportion of oxygen present in the vapour that is coming off is about 7 per cent. or a little less.

This is what occurs at the top of the rectifying column in the apparatus of fig. 60. The liquid that is evaporating there is freshly formed liquid air, and hence the waste gases carry off about 7 per cent. of oxygen. Coming down the column of liquid finds itself in contact with gas containing more oxygen than corresponds to equilibrium. Accordingly oxygen is condensed and nitrogen is evaporated at each stage in the descent, in the effort at each level to reach a condition of equilibrium between the liquid and the vapour with which it is there in contact.

Fig. 62 is another form of Baly's curve, the form, namely, in which the results of the experiments were originally shown. There the ordinates represent the absolute temperature (in centigrade degrees) at which, under atmospheric pressure, the mixed liquid boils, and two curves are drawn which show by means of the scale on the base-line the percentage of oxygen in (1) the liquid, (2) the vapour, when the condition of equilibrium between liquid and vapour is attained. A horizontal line drawn across the curves at any assigned level of temperature shows the composition of vapour and liquid respectively for that temperature, when the two are in equilibrium. Taking an intermediate point between the top and bottom of the rectifying column, and drawing the line for the corresponding temperature, we should find the respective compositions of liquid and vapour there to approximate to the values found from the two curves, this approximation being closer the more slowly the liquid trickles down, and the more intimate the contact between liquid and gas.

If a similar condition of equilibrium holds at each stage in the process of liquefying a mixture of the gases, these curves may also be taken as showing what is the proportion of the constituents in the mixed liquid at each stage while condensation of the mixed gas proceeds. Thus when air containing 21 per cent. of oxygen begins to liquefy, the liquid initially formed should, under equilibrium conditions, be much richer in oxygen: the proportion of oxygen in it, according to the curve, is 48 per cent.

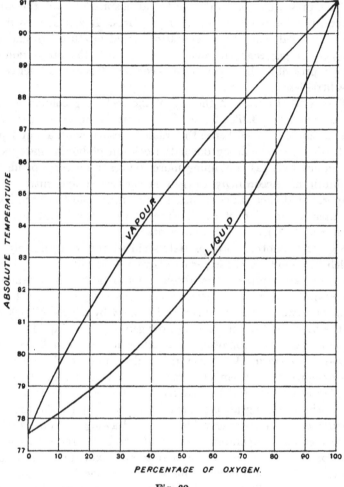

Fig. 62

These conditions are approximately realized when the process known as "scrubbing" is resorted to in the liquefaction of air. By this process, which will be presently described as carried out by Claude, a partial separation between the two constituents is effected during the act of liquefaction.

130.   **More Complete Rectification.**   In Linde's invention of 1902
the rectification is imperfect, for although it yields oxygen that is
free from nitrogen, the nitrogen which passes off is by no means
free from oxygen. A modified process makes the separation much
more complete.

The modification consists in extending the rectifying column
upwards and in supplying it at the top with a liquid rich in nitrogen.
A fractional method of liquefaction is adopted, which separates the
condensed material at once into two liquids, one containing much
oxygen and the other little except nitrogen. The latter is sent to the
top of the rectifying column, while the former enters the column
at a lower point, appropriate to the proportion it contains of the
two constituents. Nearly pure nitrogen passes off as gas at the top,
and oxygen free of nitrogen from the bottom.

Fig. 63 is a diagram showing this modified process in a form given
to it by Claude. The counter-current interchangers which are of
course part of the actual apparatus are omitted from the diagram.

Compressed air, cooled by the interchanger on its way, enters the
condenser at A. The condenser consists of two sets of vertical pipes,
communicating at the top, where they all open into the vessel B,
but separated at the bottom. The central pipes, which open from
the vessel A, are one set: the other set form a ring round them and
drain into the vessel C. Both sets are immersed in a bath, S, of
liquid which, when the machine is in full operation, consists almost
entirely of oxygen. The condensation of the compressed air causes
this oxygen to be evaporated. Part of it streams up the rectifying
column D, to be condensed there in carrying out the work of rectification and consequently to return to the vessel below. The rest of
the evaporated oxygen, forming one of the useful products, goes
off by the pipe E at the side. In these features the apparatus is
substantially the same as Linde's, but there is a difference in the
mode of condensation of the compressed air. Entering at A it first
passes up the central group of condenser pipes, and the liquid which
is formed in them contains a relatively large proportion of oxygen.
This liquid drains back into the vessel A, where it collects, and the
gas which has survived condensation in these pipes goes on through
B to the outer set of pipes, is condensed in them, and drains into
the other collecting vessel C. It consists almost wholly of nitrogen.
Then the liquid contents of C are taken (through an expansion-
valve) to the top of the rectifying column, while those of A enter
the column lower down, at a level L, chosen to correspond with the

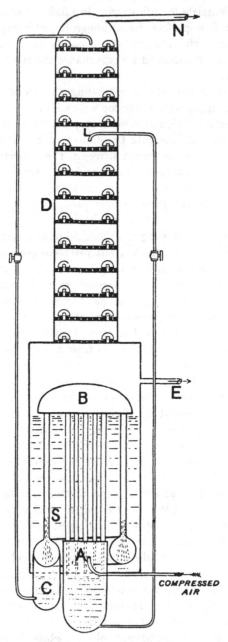

Fig. 68. Claude's apparatus for the complete separation of oxygen and nitrogen.

proportion of the constituents. The result is to secure practically complete rectification, and the second product of the machine—commercially pure nitrogen—passes off at the top through the pipe N and may be collected for use.

130 a. **Separation of Other Constituents of Air.** When the liquefaction of air was first employed as a means of separating its oxygen from its nitrogen no attention was paid to the inert constituents that are present in small quantities, except that argon was looked on as an embarrassing impurity, difficult to separate from oxygen because it has nearly the same boiling point. Argon is now extracted as a valuable product. It occurs to the extent of nearly 1 per cent. by volume; the proportions of the other inert gases are very much less. The boiling points of the several constituents of air at atmospheric pressure are approximately:

Xenon	164° K.	Nitrogen	77° K.
Krypton	121° K.	Neon	27° K.
Oxygen	90° K.	Helium	4° K.
Argon	87° K.		

The boiling point of argon comes between that of oxygen and that of nitrogen. Accordingly in the rectifying column of fig. 63 there is an intermediate level, not far from the bottom (the oxygen end), where argon tends to accumulate. In other words the mixture there contains a larger proportion of argon than is found either lower or higher in the column. If the argon there tries to rise, as a gas, it becomes liquefied by the colder nitrogen above: when it descends in the liquid it finds the oxygen below sufficiently warm to vaporize it. The result is that at the favourable level the liquid may contain 5 or even 10 per cent. of argon, along with very little nitrogen but much oxygen*.

This fact is made the basis of a separating process. From that level some of the liquid relatively rich in argon passes into an auxiliary rectifying column, where oxygen gathers at the bottom and is returned to the main column, while a gaseous mixture rises, becoming richer in argon. Finally the residue of oxygen is removed by chemical action.

Neon, of which there are only one or two parts in 100,000 of air, being much more volatile than nitrogen, tends to accumulate above the liquid nitrogen in the vessel C of fig. 63. This is also true of

---

* Claude, *loc. cit.* p. 414.

helium, which is present to a still smaller extent. Hence the vapour space there contains a mixture of these gases along with much nitrogen. Most of the nitrogen may be removed by applying a degree of cold which liquefies it but leaves the neon and helium in the gaseous state. In this way Claude succeeded in obtaining a gas of which some 50 per cent. consisted of mixed neon and helium*. The rest, chiefly nitrogen with a little oxygen, is readily absorbed by charcoal cooled in liquid air. The neon and helium of the purified mixture may then be separated by cooling it to about 20° K. with boiling hydrogen, which solidifies out the neon, leaving the helium gaseous. It may be added that the object of the process is to produce neon. The atmosphere is not a serious source of helium, which is obtained in much greater quantity from natural gas in Canada and the United States.

* Claude, *loc. cit.* p. 407.

CHAPTER V

JETS AND TURBINES

131. Theory of Jets. We have now to consider the manner in which a jet is formed in the discharge, through an orifice, of steam or any other gas under pressure. To simplify matters it will be assumed that the fluid takes in no heat and gives out no heat to other bodies during the operation; in other words that the jet is formed under adiathermal conditions. Suppose a gas to be flowing through a nozzle or channel of any form, from a region where the pressure is relatively high to one where it is lower. Each element of the stream expands, and the work which it does in expanding gives energy of motion to the element in front of it. The whole stream therefore acquires velocity in the process and also increases in volume. Let A and B (fig. 64) be imaginary partitions, across which it flows, taken at right angles to the direction of the stream lines, A being in the region of higher pressure. Let P_a be the pressure

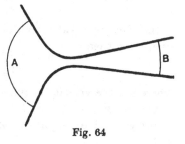

Fig. 64

at A, v_a the velocity there, and V_a the volume which unit quantity of the gas has as it passes the imaginary partition at A. Similarly let P_b, v_b and V_b be the pressure, velocity and volume of unit quantity at B. Let E_a and E_b be the internal energy of the gas at A and B respectively. In flowing from A to B the velocity changes from v_a to v_b and there is consequently a gain of kinetic energy amounting, per unit of mass, to $\dfrac{v_b{}^2 - v_a{}^2}{2g}$.

Each unit quantity of gas that enters the space between A and B has work done upon it by the gas behind, amounting to $P_a V_a$. In passing out of this space at B it does work on the gas in front amounting to $P_b V_b$. In flowing from A to B it loses internal energy amounting to $E_a - E_b$. Hence by the principle of the conservation of energy, since by assumption no heat is taken in or given out,

$$\frac{v_b{}^2 - v_a{}^2}{2g} = E_a - E_b + P_a V_a - P_b V_b \quad \ldots\ldots\ldots\ldots(1).$$

But $E_a + P_a V_a$ is I_a, the total heat at A, and $E_b + P_b V_b$ is I_b, the total heat at B, and the equation may consequently be written

$$\frac{v_b^2 - v_a^2}{2g} = I_a - I_b \quad \dots\dots\dots\dots\dots(2).$$

The gain in kinetic energy is therefore equal to the loss of total heat, or what is commonly called the "heat-drop." We are treating E and I as if they were expressed in work units: when expressed in heat units they have to be multiplied by the mechanical equivalent J.

The equation applies as between any two places in the flow, and taking the process as a whole, from the initial condition in which the velocity is v_1 and total heat I_1 to the final condition in which the velocity is v_2 and total heat I_2, we have

$$\frac{v_2^2 - v_1^2}{2g} = I_1 - I_2 \quad \dots\dots\dots\dots\dots(3).$$

In many practical cases the initial velocity is zero or negligibly small, and then

$$\frac{v^2}{2g} = I_1 - I_2 \quad \dots\dots\dots\dots\dots(4),$$

where v is the velocity acquired in consequence of the heat-drop.

This is the fundamental equation from which to calculate the velocity which an expanding fluid acquires in a jet, starting from rest.

So far there has been no assumption as to absence of losses through friction or eddy currents. If we assume, as an ideal case, that in the formation of the jet the fluid is expanding under such conditions that there is no conduction of heat to or from or within the fluid, and also no dissipation of energy through friction or eddies, the heat-drop in the equation

$$\frac{v_2^2 - v_1^2}{2g} = I_1 - I_2$$

is that which occurs in expansion with constant entropy. We have already seen (Art. 80) that this heat-drop is equal to the area $ABCD$ of the ideal indicator diagram (fig. 65) for adiabatic expansion from the initial to the final state, or $\int_{P_2}^{P_1} V dP$.

Hence $\dfrac{v_2^2 - v_1^2}{2g} = \text{area } ABCD = \int_{P_2}^{P_1} V dP \dots\dots\dots\dots(5).$

This result might also be inferred from the fact that, under the assumed conditions, the gas is doing all the work of which it is ideally capable, as it expands from the first to the second state, in giving kinetic energy to its own stream. The gain of kinetic energy is, therefore, equal to the area of the ideal indicator diagram.

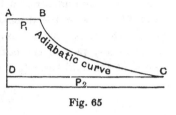

Fig. 65

Assume that we may, with sufficient accuracy, express the expansion in the ideal indicator diagram by a formula of the type $PV^m =$ constant. Then the area of the diagram, namely

$$\int_{P_2}^{P_1} V dP = \frac{m}{m-1} (P_1 V_1 - P_2 V_2)$$

$$= \frac{m}{m-1} \left[1 - \left(\frac{P_2}{P_1}\right)^{\frac{m-1}{m}} \right] P_1 V_1.$$

Hence when the expanding fluid starts from rest, at pressure P_1, to form a jet, we have

$$\frac{v^2}{2g} = \frac{m}{m-1} \left[1 - \left(\frac{P}{P_1}\right)^{\frac{m-1}{m}} \right] P_1 V_1 \dots\dots\dots\dots\dots(6),$$

or

$$v = \sqrt{\frac{2gm}{m-1} \left[1 - \left(\frac{P}{P_1}\right)^{\frac{m-1}{m}} \right] P_1 V_1} \dots\dots\dots(6a),$$

as an equation from which to find the velocity v when the pressure has fallen to any lower pressure P, under the assumed conditions of flow without friction or eddies and with no conduction of heat. Equation (6) is a particular case of equation (4), namely the case where the expansion is isentropic and where the relation of pressure to volume in isentropic expansion admits of being expressed by the formula $PV^m =$ constant.

132. **Form of the Jet (De Laval's Nozzle).** As expansion of the fluid in a jet proceeds, the volume and velocity both increase. It is easy in frictionless adiabatic flow to calculate both, and in that way to determine the proper form to give to the nozzle or channel to make provision for the increased volume, having regard to the increased velocity. At any stage the area of cross-section of the channel required for each lb. of fluid discharged is equal to the volume per lb. divided by the velocity. It is convenient to reckon

the area of section per unit of mass in the discharge, and afterwards multiply by the number of lbs. or kilogrammes.

.Let M represent the discharge, namely the mass which passes through the nozzle per second, X the area of cross-section of the stream at any part of the nozzle, v the velocity there, and V the volume of the fluid there (per unit of mass); then

$$M = \frac{vX}{V} \text{ and } \frac{X}{M} = \frac{V}{v}.$$

On making the calculation for a gaseous fluid starting from rest and discharged into a region of much lower pressure, it will be found that in the earliest stages the gain of velocity is relatively great, but as expansion proceeds the increase of volume outstrips the increase of velocity. The result is that the ratio of volume to velocity at first diminishes, passes a minimum value, and then increases; and hence the channel to be provided for the discharge, after passing a minimum of cross-section, expands in the later stages. The proper form for the nozzle, to allow the heat-drop corresponding to a large drop in pressure to be utilized as fully as possible in giving kinetic energy to the stream, is therefore one in which the area of section at first contracts to a narrow neck or "throat" and afterwards becomes enlarged to an extent that is determined by the available fall of pressure.

It is on this principle that De Laval's "convergent-divergent" nozzle (fig. 66) is designed. The throat, or smallest section, is approached through a more or less rounded entrance which allows the stream lines to converge, and from the throat outwards to the discharge end the nozzle expands in any

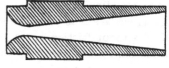

Fig. 66

gradual manner, generally in fact as a simple cone, until an area of section is reached which will correspond to the proper area of discharge for the final volume and velocity, the values of which depend upon the final pressure.

The divergent taper from the throat onwards is made sufficiently gradual to preserve stream-line motion as completely as is practicable, and so avoid the formation of eddies which would dissipate the kinetic energy of the stream. A very short rounded entrance to the throat is sufficient to guard against eddies in the convergent portion of the stream, but in the divergent portion a much more gradual change of section is required. The nozzle shown in the figure was designed for an initial pressure of 250 pounds per sq. inch and a back pressure of about $1\frac{1}{2}$ pounds. By the back pressure is meant the pressure in the space into which the fluid is discharged.

In the design of such a nozzle the purpose is (1) to make the discharge have a given value, and (2) to give the stream as high a final velocity as possible by utilizing completely the energy of the fluid in expanding down to the back pressure. The data for the design are the initial pressure, the back pressure, and the intended amount of the discharge. It will be shown as we proceed that the area of section at the throat depends only on the initial pressure and the intended discharge*; and that the enlargement from the throat to the final section depends further on the back pressure against which the stream is to escape.

At any place in the nozzle the discharge per unit area of cross-section is

$$\frac{M}{X} = \frac{v}{V}.$$

At the throat, where the cross-section is least, this is a maximum.

Consider now the ideal case of isentropic expansion in a nozzle when the fluid is one for which PV^m is constant during such expansion. Equation (6) is then applicable. The velocity at any point, the pressure there having fallen to P, is

$$v = \sqrt{\frac{2gm}{m-1}\left[1 - \left(\frac{P}{P_1}\right)^{\frac{m-1}{m}}\right]P_1 V_1},$$

and the volume is

$$V = V_1\left(\frac{P_1}{P}\right)^{\frac{1}{m}}.$$

Hence for the discharge per unit area of section at the place where the pressure is P, we have

$$\frac{M}{X} = \frac{1}{V_1}\left(\frac{P}{P_1}\right)^{\frac{1}{m}}\sqrt{\frac{2gm}{m-1}\left[1 - \left(\frac{P}{P_1}\right)^{\frac{m-1}{m}}\right]P_1 V_1} \quad \ldots\ldots(7).$$

This may be applied to calculate the proper section X for a given discharge M when the pressure has fallen from the initial pressure P_1 to any assigned lower pressure P. For the purpose of designing a nozzle there are only two places where this calculation has to be made, namely at the throat, and at the end where the fluid escapes against the assigned back pressure. When the throat-section X_t

---

* It should also be noted that at the entry to the nozzle the area varies directly as $\sqrt{T}$.

and the final section X_f have been calculated, a suitable form for the nozzle is readily drawn; any smooth curve will serve for the convergent entrance, and any conical taper may be selected for the divergent extension from the throat to the end, provided it is neither so abrupt as to interfere with stream-line flow, nor so gradual as to make the nozzle unduly long and thereby introduce unnecessary friction. The appropriate divergence is of the order of, but not greater than, 10 per cent.

To calculate the final section X_f which will allow the energy of the fluid to be fully utilized by expansion down to the assigned back pressure, that pressure is to be taken for the value of P in Eq. (7). To calculate the section at the throat the pressure there has first to be found. The pressure at the throat is determined by the consideration that the discharge per unit of section (M/X) is there a maximum. If the expression for M/X in Eq. (7) is differentiated with respect to P/P_1 and the differential written equal to zero, the resulting value of P/P_1 will be that for which M/X is a maximum; in other words, it will be the value of P_t/P_1, where P_t is the pressure at the throat.

Eq. (7) may be written

$$\frac{M}{X} = \sqrt{\frac{2gm}{m-1} \cdot \frac{P_1}{V_1}} \sqrt{\left(\frac{P}{P_1}\right)^{\frac{2}{m}} - \left(\frac{P}{P_1}\right)^{\frac{m+1}{m}}} \quad \dots\dots\dots(7a).$$

The condition for a maximum is found by differentiating the quantity under the second root:

$$\frac{2}{m}\left(\frac{P_t}{P_1}\right)^{\frac{2-m}{m}} - \left(\frac{m+1}{m}\right)\left(\frac{P_t}{P_1}\right)^{\frac{1}{m}} = 0,$$

$$2\left(\frac{P_t}{P_1}\right)^{\frac{1-m}{m}} - (m+1) = 0,$$

from which
$$\frac{P_t}{P_1} = \left(\frac{m+1}{2}\right)^{\frac{m}{1-m}} = \left(\frac{2}{m+1}\right)^{\frac{m}{m-1}} \quad \dots\dots\dots\dots(8).$$

Further, by substituting this in Eq. (6a), we have for the velocity at the throat

$$v_t = \sqrt{\frac{2gmP_1V_1}{m+1}} \quad \dots\dots\dots\dots\dots\dots(9).$$

The volume (per lb.) of the fluid at the throat is

$$V_t = V_1\left(\frac{P_1}{P_t}\right)^{\frac{1}{m}} = V_1\left(\frac{m+1}{2}\right)^{\frac{1}{m-1}} \quad \dots\dots\dots\dots(10).$$

By combining these an equation is obtained for the discharge per unit of cross-section at the throat,

$$\frac{M}{X_t} = \frac{v_t}{V_t} = \left(\frac{2}{m+1}\right)^{\frac{1}{m-1}} \sqrt{\frac{2gmP_1}{(m+1)\,V_1}} \quad \ldots\ldots\ldots\ldots(11).$$

From this equation the cross-section at the throat is found which will give an assigned discharge when the initial pressure is known. The ratio of the cross-section at any place, where the pressure is P, to the cross-section at the throat, is readily found from Eq. (7a):

$$\frac{X}{X_t} = \frac{\sqrt{\left(\frac{P_t}{P_1}\right)^{\frac{2}{m}} - \left(\frac{P_t}{P_1}\right)^{\frac{m+1}{m}}}}{\sqrt{\left(\frac{P}{P_1}\right)^{\frac{2}{m}} - \left(\frac{P}{P_1}\right)^{\frac{m+1}{m}}}} \quad \ldots\ldots\ldots\ldots(12).$$

This expression is convenient in determining the proper amount of enlargement of the nozzle from the throat to the end when the back pressure is assigned.

133. **Limitation of the Discharge through an Orifice of Given Size.** It follows from these equations that the discharge through a given orifice under a given initial pressure P_1 depends only on the cross-section at the narrowest part of the orifice, and is independent of the back pressure, provided the back pressure is not greater than P_t as calculated by Eq. (8). By continuing the expansion in a divergent nozzle after the throat is passed, the amount of the discharge is not increased, but the fluid acquires a greater velocity before it leaves the nozzle, because the range of pressure which is effective for producing velocity is increased. To put it in another way, we may say that the heat-drop down to the pressure at the throat determines the amount of the discharge, and the remainder of the heat-drop, which would be wasted if there were no divergent extension of the nozzle, is utilized in the divergent portion to give additional velocity to the escaping stream. This velocity is given in a definite and useful direction, whereas if there were no divergent extension of the nozzle the fluid, after leaving the nozzle, would expand laterally, and its parts would acquire velocity in directions such that no use could be made of the kinetic energy so acquired.

Consider what happens with a nozzle such as that of fig. 67, which has no divergent extension. Fluid is expanding from a chamber where the pressure is P_1 into a space where the pressure is P_2. Assume the back pressure P_2 to be less than P_t as calculated by Eq. (8). In that case the pressure in the jet, where it leaves the nozzle, will be P_t, and the further drop of pressure to P_2 will occur through scattering of the stream. The discharge is determined by Eq. (11). *It is not increased by any lowering of the back pressure P_2,* because any lowering of P_2 does not affect the final pressure in the nozzle, which remains equal to P_t. Osborne Reynolds* explained the apparent anomaly by pointing out that the stream is then leaving the nozzle with a velocity equal to that with which sound (or any wave of extension and compression) is propagated in the fluid, and consequently any reduction of the pressure P_2 cannot be communicated back against the stream: its effects are not felt at any point within the nozzle. The pressure in the stream at the orifice therefore cannot become less, however low the back pressure P_2 may be. But if P_2 is increased so as to exceed P_t, the lateral scattering close to the orifice ceases, the velocity is reduced, the pressure at the orifice then becomes equal to P_2, the discharge is reduced, and its amount is to be calculated by writing P_2 for P in Eq. (7) or (7.a).

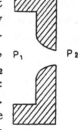

Fig. 67

In applying these results to a nozzle of any form, the least section is to be regarded as the throat: if there is a divergent extension beyond the least section the amount of the discharge is not affected, though the final velocity of the stream is increased. Taking a nozzle of any form, and a constant initial pressure P_1, if we reduce the back pressure P_2 from a value which, to begin with, is just less than P_1, the discharge increases until P_2 reaches the value $P_1 \left(\dfrac{2}{m+1} \right)^{\frac{m}{m-1}}$. After that, any further reduction of P_2 does not increase the discharge. But the velocity which the fluid acquires before it leaves the nozzle may then be augmented by lowering P_2 and adding to the divergent portion of the nozzle. The nozzle will be rightly designed when it provides for just enough expansion to make the final pressure equal to the back pressure; the jet then escapes as a smooth stream, and the energy of expansion is utilized to the full. If the nozzle does not carry expansion far enough; if in other words the final pressure exceeds the back pressure, energy will be wasted by scattering. If on the other hand the back pressure is too high for the nozzle, so that the nozzle

---

* *Phil. Mag.* March 1886; *Collected Papers*, vol. II, p. 311.

provides for more expansion than can properly take place, vibrations are set up which cause some waste*. We shall now consider the application of these general results to air and to steam.

134. Application to Air. In applying the above formulas to any permanent gas, such as air, the index m is γ, the ratio of the two specific heats (Art. 25). Its value for air may be taken as 1·40. Substituting this number in Eq. (8) we have, for a jet of air expanding under isentropic conditions,

$$\frac{P_t}{P_1} = \left(\frac{2}{2 \cdot 4}\right)^{3 \cdot 5} = 0 \cdot 528.$$

Hence if the jet is being delivered against a back pressure less than $0 \cdot 528 P_1$ a divergent extension of the nozzle is required to give the greatest possible velocity to the issuing stream, though the quantity delivered will be the same as that which would be delivered against a back pressure of $0 \cdot 528 P_1$. If the back pressure be increased it must exceed $0 \cdot 528 P_1$ before there is any diminution in the discharge.

135. Application to Steam. In applying the general equations to the expansion of steam jets we have to recognize an important distinction from the expansion which was treated of in Art. 78. Here, just as in that case, we assume adiabatic conditions: ϕ is constant. But in a steam jet, as Callendar† pointed out, the expansion occurs too fast to allow a state of equilibrium to be attained as regards condensation. The expanding jet is never an equilibrium mixture of saturated vapour and water: it is more or less supersaturated when the equilibrium condition would be one of wetness. At any stage of expansion, the steam, instead of being in the stable state corresponding to its pressure, is in what is called a *metastable* state, a state that cannot be permanent in any vapour. In the metastable state the steam is supersaturated; it may be completely dry, or it may have some water mixed with it, but necessarily less than there would be in a stable mixture at the same pressure. In other words a metastable state exists only before the proper fraction of the vapour has become liquid. In passing from the metastable state to the stable or equilibrium state, at the same pressure, part of the vapour is condensed; heat is accordingly given out, the

* For experiments on the effects of nozzles which carry expansion too far, or not far enough, see Stodola's book on the Steam Turbine.

† "On the steady flow of steam through a nozzle or throttle," *Proc. Inst. Mech. Eng.* Feb. 1915.

temperature rises, and the entropy of the fluid as a whole is increased.

If the steam is superheated to begin with, it behaves like a gas in the initial stages of the expansion, and its equilibrium at each stage is stable until it crosses the boundary or saturation line, that is to say, until its temperature falls to a value which corresponds to saturation at the pressure then reached. It is only in further expansion, beyond that stage, that a metastable condition can be produced. If the steam is initially saturated a metastable condition is produced as soon as expansion begins.

According to Callendar, the adiabatic expansion of superheated steam can be represented by the equation $P(V-b)^{1.3} = $ const., where b is a term so small that it may be omitted without material error*, making the formula $PV^{1.3} = $ const.

The same formula continues to apply in expansion beyond the saturation line provided no water condenses out, that is to say, provided the metastable condition of supersaturation is so complete that the steam remains quite dry. It also applies, under the same proviso, in the expansion of initially saturated steam.

The experiments of C. T. R. Wilson (already referred to in Art. 79) have shown that in the absence of foreign nuclei, such as dust particles, and of nuclei due to ionization†, water-vapour does not condense when it is suddenly expanded until its pressure is largely reduced, and then a cloud of small water-particles is observed. Even then, however, the conditions are not those of equilibrium, for when the expansion is continued a much denser cloud, composed of many more particles, appears at a later stage. Wilson's experiments were made by expanding air (or other gas) saturated with water-vapour, but the general conclusion would no doubt apply if water-vapour were expanded alone.

Given plenty of time, a condition of equilibrium would be reached by condensation of part of the vapour on the walls of the containing vessel, but in the very rapid expansion which occurs during the passage of steam through a nozzle, condensation on the inner surface of the nozzle can do little towards bringing it about. The effects of surface condensation are insignificant. Hence in the earlier stages of the expansion, as far as the throat and for some way beyond it, steam behaves like the vapour in Wilson's experiments before the cloud of water-particles appeared; it is supersaturated and practic-

---

* For Callendar's revised value of b, see *Proc. Inst. Mech. Eng.* 1929, p. 831.

† Wilson, *Phil. Trans. R.S.* A, vol. cxcii, p. 403, and vol. cxciii, p. 289.

ally dry. This is true of steam that is initially saturated, and *a fortiori* of steam that is initially superheated.

It follows that in calculating the discharge of steam through a nozzle with a given size of throat, or the size of throat required for an assigned amount of discharge, the proper formula to use, in the ideal case of isentropic expansion, is that which refers to supersaturated, as well as superheated, steam, namely $PV^{1\cdot 3}=$const. The equations already given will apply with the value $1\cdot 3$ for the index m.

Thus from equation (8) we have $P_t = 0\cdot 545P_1$. This applies whether the steam is saturated or superheated to begin with; in either case the steam is dry when it passes the throat, and it will be supersaturated there unless there has been much initial superheat.

Before attention was drawn to the true character of the action in a jet the index $1\cdot 135$, which is appropriate for the equilibrium type of adiabatic expansion, was usually taken as applicable, with the result that the calculated pressure at the throat was too high and the calculated discharge too small. The revised theory gives a calculated discharge slightly greater than the actual discharge, but with no more difference than can properly be ascribed to friction*.

Using the index $1\cdot 3$ in equation (11) we have, for the discharge per unit area of section at the throat,

$$\frac{M}{X_t} = \left(\frac{2}{2\cdot 3}\right)^{\frac{10}{3}} \sqrt{\frac{2\times 32\cdot 2\times 1\cdot 3}{2\cdot 8}} \sqrt{\frac{P_1}{V_1}} = 3\cdot 786 \sqrt{\frac{P_1}{V_1}},$$

with pounds and feet as units throughout. With the units more commonly employed this gives

$$\frac{M \text{ in lb. per sec.}}{X_t \text{ in sq. inches}} = 0\cdot 3155 \sqrt{\frac{P_1 \text{ in pounds per sq. inch}}{V_1 \text{ in cub. ft. per lb.}}},$$

as a formula for calculating the size of throat in a nozzle that is supplied with either saturated or superheated steam.

After passing the throat some condensation no doubt occurs in the form of a cloud of small water-particles, as in Wilson's experiments. But the process takes time, and the whole time occupied by the steam in passing through the nozzle is so short that it may

---

* Callendar, *Proc. Inst. Mech. Eng.* 1915, p. 53. For experiments on the discharge of steam through nozzles, see six Reports of a Committee of the Institution of Mechanical Engineers, *Proc.* Jan. and March 1923, May and Oct. 1924, May 1925, Jan. 1928, and Feb. 1930. See also a discussion on "Steam Nozzle Efficience," *Proc. Inst. Mech. Eng.* May 1931.

be doubted whether the condensation that occurs within the nozzle does much to restore equilibrium. It is probable that the steam is still to a large extent supersaturated when it escapes*. As regards calculation of the final area of cross-section no great error will be introduced if we consider the formula $PV^{1.3}$ to be applicable throughout, and this formula will also give a good approximation in estimating the final velocity of the jet.

136. **Comparison of Metastable Expansion with Equilibrium Expansion.** It may help to make this matter intelligible if we compare more fully the adiabatic expansion of steam under such conditions that it is a wet mixture in a state of equilibrium throughout, with its adiabatic expansion in a metastable state, in which it remains completely dry. Let steam expand from an initial state

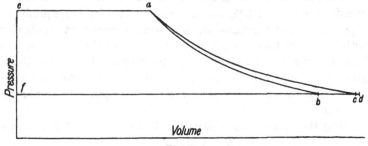

Fig. 68

represented by a (fig. 68), in which we will assume it to be dry and saturated. The curve ac represents adiabatic expansion of the type treated in Chapters II and III. At each stage of that process the fluid is a mixture, in stable equilibrium, of saturated vapour and water. Its volume at any pressure is determined by the method explained in Art. 78. The curve ab represents metastable adiabatic expansion during which the steam remains quite dry. Its form is determined by the equation $PV^{1.3} = \text{const.}$ In both cases the expansion is isentropic and therefore

$$\phi_a = \phi_b = \phi_c.$$

* Observations of the appearance of escaping jets support this conclusion. They show that when steam initially dry (but not necessarily superheated) escapes from a divergent nozzle in which it has expanded through a considerable ratio, no particles of water become visible until the steam has travelled some distance from the orifice. See Stodola, *Zeit. des Ver. deutsch. Ing.* 1913. Cf. Yellott, *Trans. Amer. Soc. Mech. Eng.* 1934, vol. LVI, p. 411, who finds that in a convergent-divergent nozzle condensation does not occur until the steam reaches a condition represented by a 3 per cent. moisture line on the Mollier diagram.

But though the entropy and the pressure are the same at b as at c, the fluid is in two very different states. At b it is a homogeneous gas; at c it is a wet mixture. At c its temperature is the temperature of saturation corresponding to its pressure there; at b its temperature is much lower, being determined by the equation $PV = RT$, which makes $T/P^{\frac{3}{13}} = \text{const.}$ The volume is of course less at b.

The heat-drop from a to c is the thermal equivalent of the work represented by the area $eacf$, and the heat-drop from a to b is the thermal equivalent of the work represented by $eabf$. Hence the heat-drop is less in the metastable expansion, by an amount that is equivalent to the area abc, and the total heat at b is therefore greater than the total heat at c by that amount.

Suppose now that after sudden expansion to b, along the curve ab, the metastable fluid at b is allowed to become stable by partially condensing under constant pressure, without any gain or loss of heat. Its temperature will rise to the saturation value for that pressure; it will, therefore, come to have the same temperature as the mixture at c, but it will be somewhat drier, because its total heat remains equal to I_b, which, as we have seen, is greater than the total heat I_c of the mixture at c. Its volume will, therefore, increase to a point d, which is beyond c.

If we write q_c for the dryness at c of steam that has expanded in a stable state, or state of equilibrium as a whole, from a to c, and q_d for the dryness at d of steam that has expanded in a metastable state to b and has subsequently attained equilibrium, by water separating out at constant pressure, without loss or gain of heat, the difference of total heats is

$$I_d - I_c = L\ (q_d - q_c).$$

But $I_d = I_b$ and $I_b = I_c + A$ (Area abc).

Hence $L\ (q_d - q_c) = A$ (Area abc).

In attaining equilibrium the fluid as a whole has gained entropy, for ϕ_d is greater than ϕ_b, or ϕ_a, or ϕ_c in the equilibrium state, by the amount that would convert the equilibrium mixture at c into the equilibrium mixture at d. Thus

$$\phi_d - \phi_b = \frac{L\ (q_d - q_c)}{T_d} = \frac{A\ (\text{Area } abc)}{T_d}.$$

This increase of entropy is not due to any gain of heat, for no heat has been communicated to the fluid; it is due to the fact that there has been an irreversible internal change in passing from the metastable to the stable state.

We may think of the substance as undergoing a cycle of changes. Starting from a let it expand suddenly and adiabatically to b; then let it change from b to d at constant pressure without taking in or giving out heat. Then let it be partially condensed, under constant pressure, from d to c; during this stage a quantity of heat must be given out equal to $L(q_d - q_c)$. Then let it be slowly compressed along the equilibrium adiabatic curve from c to a. This completes the cycle. Work has been expended, equal to the area abc, and a corresponding quantity of heat has been removed.

During its transition (along bd) from the metastable to the stable state, the fluid passes through a state in which its pressure and volume are the same as those of the equilibrium mixture at c. But its state in other respects is by no means the same; it is then a mixture of supersaturated vapour with some liquid, not in equilibrium; its temperature is lower and its total heat is greater.

It is scarcely necessary to add that the remarks which were made in Art. 75 about the specification of the state of a fluid assumed that the fluid as a whole was in equilibrium. They do not apply to metastable or transition states. The loss of available energy which occurs in a nozzle as a result of supersaturation, is distinct from and additional to any loss that may occur through friction.

137. Measure of Supersaturation. Supersaturation involves *supercooling*; that is to say, the vapour is cooled below the saturation point corresponding to its pressure*. Supersaturation also involves an excess of pressure, and a corresponding excess of density, when comparison is made between the pressure or the density of a supersaturated vapour and that of a saturated vapour at the same temperature. The ratio of densities is nearly the same as the ratio of pressures. Supersaturated vapour behaves like a gas with PV nearly constant at constant temperature. Either ratio serves as a convenient means of specifying the degree of supersaturation.

138. Retarded Condensation. Wilson found that when dust-free air, saturated with water-vapour at 20° C., was adiabatically expanded by suddenly enlarging the vessel in which it was contained, a cloud of fog particles did not form until the volume of the

* This is called "undercooling" by some writers, but the word "supercooling" is more appropriate as a description of cooling which is in excess of the normal amount.

vessel was increased in the ratio of 1·375 to 1. This corresponds to a nearly eight-fold supersaturation of the vapour; that is to say the ratio of the vapour densities, or of the actual vapour-pressure after expansion to the pressure of saturated vapour at the temperature reached by the expansion, was then nearly 8. He found that the time-rate of the expansion might be varied considerably without affecting this result; and also that when the expansion was carried further a much denser fog cloud was formed, containing many more particles. It follows from these results that the growth of those fog particles which were first formed did not go on fast enough to restore and maintain equilibrium in the expanding fluid.

Condensation of expanding steam, by the formation of water particles suspended in the vapour, is accordingly retarded in two different ways. There is what may be called a static retardation which does not depend on the time-rate of expansion, for the fog does not begin to form until the volume has increased by a definite and considerable amount. In addition, there is a time-lag which prevents equilibrium from being reached while the expansion continues. One reason for this is that the drops, once they have formed, must have time to cool in order that they may continue to act as centres for condensation. Hence the more rapid the expansion the less near will be the approach to equilibrium at any stage after condensation has begun.

It may be questioned whether, even in such slow expansion as occurs in steam-engines of the cylinder and piston type, equilibrium of the working fluid is approximately attained, notwithstanding the assistance which is given by condensation on the metal surfaces. It is quite possible that exhaust steam discharged to the condenser may consist in part of supercooled vapour though it also contains water*. Supersaturation in it would be readily detected if we could observe the temperature and compare that with the pressure; but attempts to measure the temperature of supersaturated steam directly, by means of a thermometer, give fallacious results on account of condensation of water on the bulb, or on the pocket in which it is enclosed, or on the wire if it is an exposed thermometer of the resistance type. The theory of ideal steam-engines using adiabatic expansion, which was discussed in Chapter III, and from which efficiencies of the "Rankine Cycle" were calculated, assumed

---

* Compare Callendar and Nicholson, "On the Law of Condensation of Steam," *Min. Proc. Inst. Civ. Eng.* vol. XXXI, pp. 171–4, where experimental evidence is mentioned of supersaturation during expansion and exhaust.

a condition of equilibrium on the part of the working fluid through-out the whole operation. So far as there is any departure from that condition in a real engine it makes for reduced efficiency: in this as well as in other respects the real performance of an engine falls short of the standard set by the Rankine Cycle.

In the more rapid expansion which steam undergoes while it passes through a turbine of any type, it appears that the state is far from being one of equilibrium even in the later stages. This view is supported by an examination of the results of trials of the performance of turbines, working under various conditions as to exhaust pressure and initial superheat[*].

The reason why drops of liquid do not form freely enough to prevent an expanding vapour from becoming supersaturated is because the vapour-pressure of a liquid is greater above a curved sur-face than a plane surface, and before very small droplets can form, the pressure must be higher than the normal saturation pressure.

139. Action of Steam in a Nozzle, continued. Returning now to the action of steam in a nozzle, we may note in passing how metastable expansion may be represented on the entropy-tempera-ture diagram, and on the Mollier diagram of entropy and total heat.

Taking first the entropy-temperature diagram (fig. 69), adiabatic expansion from an initial state a where the pressure is P_1, to any lower pressure P_2, under equi-librium conditions, is repre-sented by the isentropic ac, where c is on the equilibrium line of constant pressure for a wet mixture at P_2. But if the expansion is so sudden as to occur without condensation, it is represented by the isentropic ab, where b is a point on the constant-pressure curve bh for

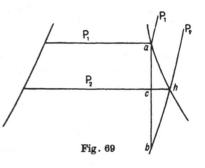

Fig. 69

supersaturated vapour. That curve is a continuation, below the boundary line ah, of the constant-pressure curve for superheated steam at P_2. The ultimate state d which would be reached if equilibrium were attained at constant I and constant P (as in fig. 68, Art. 136) may be calculated, but it would serve no useful

[*] See H. M. Martin, "A New Theory of the Steam Turbine," *Engineering*, vol. CVI, 1918.

purpose to attempt to represent on this diagram the irreversible transition from state b to state d. The diagram shows clearly the amount of supercooling cb.

In the $I\phi$ diagram (fig. 70) adiabatic expansion under equilibrium conditions is again represented by the isentropic line ac, the straight line ch being the equilibrium line of constant pressure for a wet mixture at pressure P_2. Adiabatic expansion under dry supersaturated conditions is represented by ab; b is again a point on the constant-pressure curve bh for supersaturated vapour, which is a continuation below the boundary curve ah of the constant-pressure curve for superheated steam at P_2. Here we may determine d graphically by drawing a horizontal straight line through b to meet the equilibrium constant-pressure line in d; the assumption being, as before, that the metastable vapour, after expansion, ultimately comes to a stable state in d without change of pressure and without gain or loss of heat. The horizontal straight line bd is a line of constant total heat.

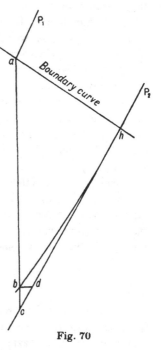

Fig. 70

In these diagrams, as well as in fig. 68, we have assumed that the steam is saturated to begin with. But the construction can obviously be modified to apply to steam with initial superheat; the point a may be anywhere in the constant-pressure line for P_1.

140. Effects of Friction. The losses that occur in jets or turbines through friction and through supersaturation cannot be separated in practice; but for the purpose of considering the thermodynamic effect of friction it will be convenient to treat that separately by imagining a case in which there is no supersaturation. Such a case may be realized by using steam that is highly superheated before expansion.

Let BC be the equilibrium adiabatic curve on the pressure-volume diagram drawn, say, for 1 lb. of steam; then the area $ABCD$ (fig. 71) represents the amount of work available for setting the steam in motion as a jet, or for getting mechanical effect out of it in any manner. The area $ABCD$ is equivalent to the whole heat-drop in adiabatic expansion under equilibrium conditions, and measures the utmost work obtainable in any method of utilizing the energy of the steam. It is on this basis that the work of the Rankine Cycle (Art. 87) is calculated, which forms an ideal standard with which the

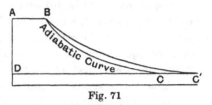

Fig. 71

actual output of any steam-engine or steam turbine may be compared. The actual output per lb. of steam is necessarily less in all cases than the area $ABCD$, and the ratio of the actual output to that area is called the "efficiency ratio" of the engine or turbine (Art. 94). In a steam jet the output is the kinetic energy of the jet itself.

Consider now the effect of friction in a nozzle. Assume the conditions to be adiathermal. If there were no friction (as well as no supersaturation) the whole work represented by $ABCD$ would be utilized in giving velocity to the jet and would appear in it as kinetic energy: in that case we should have

$$\frac{v^2}{2g} = \text{Area } ABCD = I_1 - I_2,$$

where v is the velocity produced in the jet (starting from rest) and $I_1 - I_2$ is (in units of work) the heat-drop in adiabatic expansion.

But in any real nozzle there is some friction between the fluid and the sides of the channel, and the flow is to some extent turbulent, which means that eddies are formed in which there is dissipation of the energy of flow through internal friction. There is also the considerable loss in traversing the bends of nozzles and blades. We shall apply the word friction broadly to all such losses. Their effect is as follows. On discharge, or at any stage during the expansion, the jet has less kinetic energy, and therefore less velocity, than it would have at the same stage if there were no friction. But its volume, after expansion to any given pressure, is greater than it would be if there were no friction, because the energy that has been dissipated through friction has taken the form of heat. Thus up to any stage in the expansion there has been a loss of kinetic energy, but there has also been a gain of heat. Consequently the fluid has a greater volume than it would have in the absence of friction. Moreover it has a greater stock of heat still available for conversion into work in the later stages of expansion, though that advantage cannot in any event compensate completely for the loss of energy to which the increased stock of heat is due. The heat that is restored at any stage as a result of friction has lost availability for conversion into a mechanical form, for the working substance then has a lower temperature than it had in the earlier stages when the mechanical energy was generated out of which that heat has been produced. Thus the net result is to reduce the kinetic energy of the jet below the standard for no friction, although part of the energy that has been lost through friction up to any stage is recovered in subsequent stages. The matter may be put in another way by saying that, in consequence of friction, the fluid, after expansion to any pressure, has suffered less drop of total heat than it would have suffered had there been no friction. There is less mechanical effect; but there is more heat left in it and its volume is greater, at each stage. A progressive increase of entropy occurs during expansion, as a result of the irreversible processes that are going on within the fluid, whereas with no friction the entropy would be constant.

Taking the pressure-volume diagram, fig. 71, the effect of friction is to give the actual expansion curve a form such as BC', in which the fluid has a greater volume at each stage than it would have in the adiabatic process represented by BC. But though this apparently implies a gain of work there is really a loss. The area $ABC'D$ does not measure an actual output of work, but an artificial quantity which we may call the "gross apparent work." Of this gross apparent work, a part is reconverted into heat, as the expansion proceeds, namely a quantity sufficient to supply enough heat at each stage to bring the expansion curve out from BC to BC'. At the end of the operation the net amount of work that is obtained, far from being greater than the adiabatic area $ABCD$, is less than that area by the equivalent of $I_2' - I_2$, where I_2' is the total heat at C' and I_2 is the total heat at C. In other words it is less by the quantity of heat which would be required to change the condition of the expanded fluid at constant pressure from C to C'.

To prove this we may think of what happens when the substance is carried through an imaginary cycle. Starting from state B let it expand, with friction, to C'. Let W' be the net amount of work actually done by it in this expansion. Then let it be changed from state C' to state C by removal of heat under constant pressure. The quantity of heat so removed is $I_2' - I_2$. Then let it be compressed adiabatically from C to B. The work W done upon it during the compression (which is reversible) is the same as the work that would be done by it in adiabatic expansion. The cycle is now complete, and by the conservation of energy we have

$$W' + (I_2' - I_2) - W = 0,$$

or

$$W' = W - (I_2' - I_2).$$

Hence also $\quad W' = I_1 - I_2 - (I_2' - I_2) = I_1 - I_2',$

or the net amount of work done is equivalent to the *actual* heat-drop, in agreement with Art. 104.

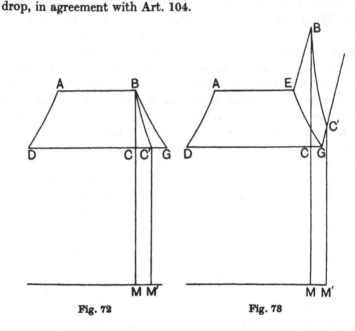

Fig. 72   Fig. 73

Turning to the entropy-temperature diagram (fig. 72), the ideal case, without friction, is represented by $ABCD$. Friction gives the expansion curve some such form as BC', in which the entropy increases progressively as the temperature falls. The area $MBC'M'$ represents the heat produced by friction; it is the heat required to give the expansion curve its actual form, and since no heat comes from outside sources it is supplied at the expense of the kinetic energy of the jet, by a conversion which is going on from the beginning to the end of the expansion. The gross apparent work is represented by the area $DABC'$, but from this we have to deduct the area $MBC'M'$ to find the net amount of work which finally appears as kinetic energy in the jet. Thus allowing for friction the net amount of work W' is

$$\text{Area } DABC' - \text{Area } MBC'M',$$

or $\qquad\qquad \text{Area } DABC - \text{Area } MCC'M'.$

Hence the net loss, as compared with the work W that would be got in adiabatic expansion (with no friction), is the area $MCC'M'$, which is $I_2' - I_2$, as above.

In fig. 72 the steam is initially saturated. If it be superheated, let B be the initial condition (fig. 73), AEB being the constant-pressure line for P_1 and DGC' the constant-pressure line for P_2. Frictionless expansion (in equilibrium) would be represented by BC. The actual expansion is along some such line as BC'. The gross apparent work is represented by the area $DAEBC'G$, and the net amount is found by deducting from that the area $MBC'M'$, which represents the heat developed through friction. The net effect of friction is to deduct an amount of work equal to the area $MCGC'M'$ from the ideal performance $DAEBC$. This deduction is equivalent to $I_2' - I_2$ as before.

For practical purposes it is more useful to represent the effects of friction on the Mollier diagram of entropy and total heat (fig. 74). Let B represent the initial state (in this example there is some superheat; the broken line is the boundary curve). BC represents an ideal adiabatic process of expansion and BC' the actual process. I_1 is the initial total heat, I_2 the total heat that would be left in the steam after adiabatic expansion to P_2, and I_2' is the total heat

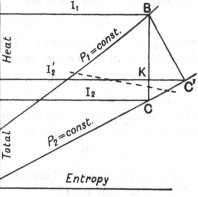

Fig. 74

actually left in the steam after expansion to that pressure. The actual heat-drop, to which the net amount of work done is equivalent, is BK or $I_1 - I_2'$, and the net loss resulting from friction is KC or $I_2' - I_2$.

When the proportion is known beforehand of the frictional loss KC to the total theoretically available heat-drop BC we can mark the point K on the adiabatic line through B and draw a horizontal line through K to find C'. When there are experimental data for estimating the frictional losses in expansion down to various intermediate pressures we can apply this construction to trace the actual expansion curve BC' in a series of steps. The method is applied to compound steam turbines, as a means of determining the state of the steam after each of a series of stages in the passage of steam through the turbine (Art. 145).

The student may find it useful to express the effect of friction thus. When there is no friction, and the expansion is adiabatic (Art. 80),
$$dI = VdP,$$
where dI represents (in units of work) the drop of total heat which takes place while the pressure drops by dP. When there is friction
$$dI' = V'dP - dQ,$$
where dI' represents the drop of total heat as affected by friction, and V' the volume as affected by friction, dQ being the quantity of heat generated by friction at the expense of the gross apparent work and restored to the fluid as heat. Hence
$$dI - dI' = dQ - (V' - V)\, dP.$$
Integrating between the limits (1) and (2)
$$I_1 - I_2 - (I_1' - I_2') = Q - \int (V' - V)\, dP,$$
where Q is the whole quantity of heat generated by friction. Since I_1 and I_1' are the same, this gives
$$I_2' - I_2 = Q - \text{Area } BCC' \text{ of fig. 71,}$$
which expresses the fact that in consequence of friction the net loss of mechanical effect is equal to the heat generated, less the work that is recovered through the augmentation of volume which friction brings about.

141. Application to Turbines. The above discussion of the effects of friction relates, in the first place, to jets, but it may readily be extended to turbines. When a jet is considered alone the "efficiency-ratio" (Art. 94) of the process, regarded as a conversion of energy, is measured by comparing the actual kinetic energy of the issuing stream with the energy obtainable under ideal conditions, that is to say with the adiabatic heat-drop. But in a turbine, of any type, the process of conversion goes further: the object is to do mechanical work on the rotor or revolving part of the machine, and the efficiency of the process as a whole is measured by comparing the work done by the fluid on the rotor with the adiabatic heat-drop.

142. Simple Turbines. In one type of turbine (De Laval's) the two operations are entirely distinct. The steam enters the turbine through a fixed nozzle of the convergent-divergent type (fig. 66) in which its pressure drops in a single step through the whole available range from the initial pressure P_1 to the back pressure P_2. Subject to frictional losses all the energy takes a kinetic form in the jet. The jet then impinges on blades that project from the circumference of a very rapidly revolving wheel, acting on them just as a jet of water acts on the blades of a Pelton water-wheel, with an impulse which is measured by the loss of momentum of the stream. The jet thereby converts its kinetic energy into work done on the wheel. This second conversion is a purely hydraulic process, a question of dynamics but not of thermodynamics. It involves frictional losses (distinct from the earlier frictional losses in the nozzle) as well as losses arising from the fact that the stream is not wholly deprived of momentum by its impact on and passage over the revolving blades.

The whole available heat-drop from such pressures as are usual in steam-supply down to the vacuum of a good condenser may easily be as much as 300 calories. If this were all utilized to give velocity to the steam in a single operation, the jet would leave the nozzle with a velocity of 5180 feet per second, since

$$v^2 = 2gJ\,(I_1 - I_2).$$

Making allowance for frictional losses we should still have to deal with jet velocities of 4500 feet per second or more. If the energy of the jet is to be extracted by making it impinge on the blades or buckets of a simple turbine wheel, the condition for efficiency is that the velocity of the blades should be not far short of one-half the velocity of the jet. Hence to get a good return from this heat-drop by means of a simple turbine we should require a speed of something like 2000 feet per second at the periphery of the wheel which carries the blades. This is an impracticable speed: there are no materials of construction fitted to withstand the forces it would involve. De Laval's turbine does indeed run at very high speeds, but they are far short of this and involve the sacrifice of a considerable part of the available energy of the jet. The remedy, as the late Sir Charles Parsons realized, is to divide the whole heat-drop into many stages, and so give the stream only a moderate velocity in each stage.

143. Compound Turbines. This consideration led Parsons to develop the Compound Turbine, which became in his hands the most efficient means of utilizing the heat-drop of steam in the production of mechanical power. In the compound turbine each stage uses only a fraction of the whole heat-drop, leaving the remainder to be used in later stages. In each stage there is a conversion of part of the steam's heat-energy into work and there is frictional loss both in the nozzle and the blades. The heat produced by that loss augments the quantity of total heat which the steam carries on to the next stage; there is, therefore, in the subsequent stages a recovery of part of the loss. When the stages are very numerous, as they are in a Parsons' turbine, the steps in the resulting expansion process are so short that the process becomes approximately continuous and may be represented by a continuous curve on the pressure-volume diagram or on other diagrams. A diagram such as fig. 75 then exhibits the complete action; the outer

curve BC' is a continuous line drawn through points which represent the volume of the steam at the beginning of each stage, and the difference between it and the adiabatic curve BC shows how the volume is increased as a con-

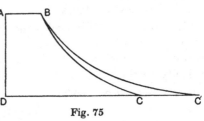

Fig. 75

sequence of all the internal losses that occur as the operation proceeds. The diagram differs from fig. 71 only in this, that the curve BC' is now to be understood as including all internal frictional losses instead of only nozzle friction. What was said in Art. 140 applies to the efficiency of the turbine as a whole, and so long as no heat is lost by conduction the equation holds good,

$$I_2' - I_2 = Q - \text{Area } BCC',$$

where Q is the heat generated within the turbine by fluid friction, I_2' is the total heat actually present in the steam at its exit from the turbine, and I_2 is, as before, the total heat which would be found in it after adiabatic expansion to the same final pressure.

144. Theoretical Efficiency-Ratio. Whether the stages are many or few, provided no heat escapes to the outside by conduction or by leakage of steam, and provided the kinetic energy of the current of steam is negligible on its exit from the turbine, the actual heat-drop $I_1 - I_2'$ is all represented by work done upon the rotor. Let η_t stand for the ratio of the actual heat-drop to the adiabatic heat-drop. By this definition,

$$\eta_t = \frac{I_1 - I_2'}{I_1 - I_2}.$$

Under the conditions stated above this fraction expresses the efficiency-ratio of the turbine as a whole, namely the ratio of the work done on the rotor to the work ideally obtainable by adiabatic expansion through the same range. The whole adiabatic heat-drop $I_1 - I_2$ would be converted into work only if the turbine were reversible and therefore thermodynamically perfect. Owing to internal irreversibility the heat converted into work is less, apart from any loss of heat by conduction to the outside.

We may call η_t the *theoretical efficiency-ratio*. It is what the efficiency-ratio would be if the whole *actual* heat-drop $I_1 - I_2'$ were converted into work.

145. Action in Successive Stages. The action of a compound turbine is most clearly shown by using the Mollier diagram of entropy and total heat to exhibit what happens in each step. Beginning with the initial pressure, let a series of constant-pressure

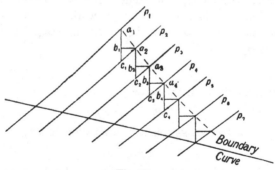

Fig. 76

lines be drawn, p_1, p_2, p_3, etc. (fig. 76), corresponding to the pressures at which the steam enters the successive stages. In the first stage the pressure drops from p_1 to p_2, in the second stage from p_2 to p_3, and so on. In the first stage, adiabatic expansion from p_1 to p_2 would be represented by a_1c_1, and the length of that line would be a measure of the adiabatic heat-drop; but the actual heat-drop is the smaller quantity a_1b_1. Still treating the action as adiathermal, a_1b_1 is the heat converted into work while the steam passes through the first stage. The condition of the steam at the end of the first stage, and beginning of the second, is represented by the point a_2, which is found by drawing a line of constant total heat through b_1 to meet the constant-pressure curve p_2. In the second stage adiabatic expansion would give the line a_2c_2. The actual heat-drop, which also measures the work done, is a_2b_2, and the condition of the steam as it passes on to the third stage is represented by a_3. Similarly in the third stage the work done is a_3b_3, the steam passes to the fourth stage in the condition a_4, and so on. The diagram shows the process, as carried down to the boundary curve, with steam initially superheated; it is readily extended into the wet region. In each stage the fraction ab/ac measures the ratio of the work done to the adiabatic heat-drop for that stage. The points a_1, a_2, a_3, etc. lie on what is called the "curve of condition," a curve showing what the condition of the steam would be as it passes from stage to stage on the assumption that no heat is lost to the outside. The curve of condition consequently corresponds to the outer curve BC' of fig. 75. The total work done on the rotor is the sum of the amounts of work done in the successive stages, namely, Σab.

146. **Stage Efficiency and Reheat Factor.** Taking any stage of a compound turbine, the ratio of the work done to the adiabatic heat-drop, in that stage, may be called the stage efficiency and denoted by η_s; thus

$$\eta_s = \frac{ab}{ac}.$$

The total work done on the rotor

$$\Sigma ab = \Sigma \eta_s (ac),$$

and if η_s can be treated as constant from stage to stage,

$$\Sigma ab = \eta_s \Sigma ac.$$

The quantity Σac is called by some writers the "cumulative heat-drop." This quantity is greater than the whole adiabatic heat-drop between the initial and final pressures, $I_1 - I_2$, to an extent that depends upon the stage efficiency. The ratio

$$R = \frac{\Sigma ac}{I_1 - I_2}$$

is called the *Reheat Factor*. The reheat factor is relatively high when the stage efficiency is low, or, in other words, when there is much loss through irreversible action within each stage.

Treating η_s as constant we have

$$\eta_s R = \frac{\eta_s \Sigma ac}{I_1 - I_2} = \frac{\text{work done on rotor}}{\text{adiabatic heat-drop}} = \eta_t,$$

under the conditions postulated, which make the actual heat-drop a measure of the work done on the rotor.

From the equation

$$\eta_t = \eta_s R$$

it will be seen that in a compound turbine η_t is greater than the stage efficiency η_s, since R is greater than unity.

We might have defined the reheat factor by reference to fig. 75 as

$$R = \frac{\text{area } ABC'D}{\text{area } ABCD},$$

for in a compound turbine of many stages the curve of condition is represented by BC' and the area $ABC'D$, which was called the "gross apparent work" in Art. 140, is the mechanical equivalent of the "cumulative heat-drop" Σac. The work done on the rotor is $\eta_s \times$ area $ABC'D$, and is less than the area $ABCD$, the efficiency-ratio being

$$\eta_t = \frac{\eta_s \times \text{area } ABC'D}{\text{area } ABCD}.$$

147. Real Efficiency-Ratio. The foregoing expressions involve the proviso that there is no leakage of heat. But when there is leakage of heat, or appreciable kinetic energy in the steam at its exit from the turbine, the actual heat-drop $I_1 - I_2'$ includes a quantity representing the loss due to these causes, in addition to the work done on the rotor. Let that loss be expressed as a fraction of the adiabatic heat-drop, namely,

$$x (I_1 - I_2).$$

Then
$$I_1 - I_2' - x (I_1 - I_2)$$

is that part of the actual heat-drop which is converted into work on the rotor.

Hence allowing for this loss, the net or real efficiency-ratio of the turbine becomes

$$\frac{I_1 - I_2' - x (I_1 - I_2)}{I_1 - I_2} = \eta_t - x,$$

since η_t is, by definition (Art. 144), the ratio of the actual heat-drop $I_1 - I_2'$ to the adiabatic heat-drop.

The amount of work obtained from the steam is therefore

$$(\eta_t - x)(I_1 - I_2).$$

Writing η_r for the real efficiency-ratio, its relation to the other quantities is given by the equation

$$\eta_r = \eta_t - x = \eta_s R - x.$$

In the process of designing a turbine a value is estimated by special methods for the stage efficiency η_s; then the curve of condition is deduced, which allows the reheat factor to be found and also the probable volume and velocity of the steam at each stage.

148. Types of Turbines. An "impulse" turbine is one in which the rotor is driven entirely by the impulse of a jet or jets against blades which are attached to it. In such turbines the expansion of the steam occurs in fixed nozzles, or passages which act as nozzles. The turbines of De Laval, Curtis, and Zölly or Rateau are examples of the impulse type. In De Laval's the whole expansion takes place in one step, and the extraction of energy from the jets also takes place in one step. As we have seen, this requires the ring of blades to have an extremely high velocity if it is to utilize a fairly large fraction of the kinetic energy of the jets. De Laval's turbine is used only for small powers; its efficiency is limited by the difficulty of making a wheel that will run safely at an enormous speed. The distinguishing characteristic of the Curtis turbine is that the kinetic energy of the jets is extracted in steps, by making the jets impinge successively on two or more rings of moving blades, with fixed guide-blades between to deflect the jets. This arrangement is often applied in the first stage of a compound turbine. It allows the first stage to utilize a considerable drop of pressure without making the blade-speed excessive. In the Rateau or Zölly type of turbine there are many stages, each involving a small drop of pressure and consequently a moderate velocity of jet; the jets in each stage give up their energy by impinging on a single ring of moving blades. Each ring runs in a separate chamber, and the jets are formed by nozzles or passages in the diaphragm which separates one chamber from the next.

A "reaction" turbine is one in which the nozzles or passages in which the steam expands are themselves the moving part, and are driven by the reaction which results from the fact that the steam is acquiring momentum as it passes through them. An ancient toy described by Hero of Alexandria, in which nozzles were caused to revolve backwards by discharging steam into the air, is an example of a pure reaction turbine. The type has not come into use; it would require an enormous speed of recoil to work efficiently. But a combination of reaction and impulse is applied in the most important turbine of all, that of Parsons, which was the first to be developed on economic lines. Parsons' is a compound turbine with many stages. Each stage comprises a ring of fixed blades, projecting inwards from the case and making up convergent passages which act as nozzles, and a ring of moving blades projecting outwards from the rotor. The rings of fixed blades and moving blades alternate from end to end of the turbine and are alike in shape. The moving blades, like the fixed ones, make up convergent passages which are completely filled by the steam as it passes through. In each set of passages, moving as well as fixed, there is some expansion; consequently any ring of moving blades is urged to move not only by the impulse of the jets which strike it, but by the reaction that arises from the expansion of steam within it, since that expansion gives the steam new velocity. The general direction in which the steam flows through the turbine is parallel to the axis.

In an early form of the Parsons turbine the general direction of flow was radial, the fixed blades being attached to a fixed disc, and the moving blades to a parallel disc which revolved about an axis through the centre of the fixed disc. An interesting modification of this arrangement was devised by Ljungström, who let both discs revolve, but in opposite directions. In the Ljungström turbine (which is also compounded of many stages) there are, therefore, no fixed blades; both sets are urged forward by reaction, and a high relative velocity, on which the stage efficiency depends, is obtained with a lower frequency of revolution.

Other turbines are made up by combining the various types which have been named.

It may be added that in compound turbines with many stages the drop of pressure in each stage is so small that the nozzles, or blade passages which act like nozzles, are not of the convergent-divergent kind described in Art. 132. They are only convergent, for the drop of pressure in each stage does not involve expansion beyond the "throat." In each stage the passages must be made sufficiently larger than those of the preceding stage to allow for the increase of volume that has taken place; in the final stages, when the pressure is approaching that of the condenser, the passages are relatively very large.

149. Performance of a Steam Turbine. In practice the steam turbine, especially in large sizes with high initial pressure, considerable superheat, and a good vacuum in the condenser, is much more efficient than the piston engine. For this reason, as well as for its greater mechanical simplicity, and the facility with which it may be used in units of great power, it has quickly come to be the chief means of converting heat into work on a large scale, in power-stations and in the propulsion of ships.

It is difficult to compare the thermodynamic efficiency of a turbine and an internal-combustion engine, for each has its own sphere. The turbine is commercially more economical for large power units, the internal-combustion engine where small units are needed. A good turbine will convert over 35 per cent. of the thermal energy of the fuel, whereas an internal-combustion engine may approach 40 per cent. The internal-combustion engine works through a very different temperature range. Though it rejects heat at a much higher level, its upper limit is relatively so high that the realized efficiency may be greater. But it requires a rather more expensive kind of fuel; its mechanism includes many reciprocating parts; and it cannot be built to give anything like the same concentration of power. However, where weight per unit of power is the primary consideration, as for aeronautical purposes, the internal-combustion engine easily wins, and even in the propulsion of ships of moderate size the internal-combustion engine of the Diesel type has proved a successful competitor of the steam turbine, but in the largest ships, as well as in central-station working on land, the turbine has no rival.

150. Utilization of Steam. Low Pressure and High Pressure. As we briefly pointed out in Arts. 94 and 95, the reason why the steam turbine came to be a more efficient means of converting heat into work than the piston engine, was the greater facility with which it could be adapted to make effective use of the energy in low-pressure steam. When the average steam pressures were about 200 pounds per square inch, the main advantage the turbine had over the piston engine was that in the later stages of expansion it was a far better agent of conversion, for it continued to be efficient down to the lowest pressure that is practically attainable in a condenser. In a piston engine, on the other hand, it would be useless to carry expansion so far, for not only would the bulk of the cylinder become impracticable, but the increased waste of power through friction between the piston and the cylinder would become greater than the gain of indicated work. Hence, with a piston engine, expansion in the cylinder is seldom in practice carried beyond an absolute pressure of 7 pounds or even 10 pounds per square inch. With a turbine the expansion is continued effectively almost down to the condenser pressure, and it is a matter of the utmost consequence to make that as low as the temperature of the condensing water will allow.

This point will be apparent if we use the entropy-temperature diagram and compare the work obtainable (under ideal adiabatic

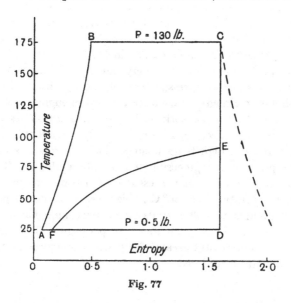

Fig. 77

conditions) when expansion is complete down to a low condenser pressure, with the work obtainable when release takes place at a pressure of say 10 pounds absolute. In the diagram (fig. 77) the area *ABCD* represents the work obtainable in the complete adiabatic expansion of initially saturated steam from a pressure of 130 pounds to a condenser pressure of 0·5 pound; and the area *ABCEF* represents the work obtainable when release takes place after expansion down to 10 pounds absolute, *EF* being a line of constant volume (Art. 95). The same condenser pressure is assumed in both cases. The area *FED* represents what is lost by incomplete expansion, such as necessarily occurs in a piston engine. The figure applies to an ideal performance in each case, with adiabatic expansion, but in the conditions of actual work the steam turbine would save most of the area *FED*. It is to be noticed that any reduction of vacuum will diminish the output of work from the turbine much more than it will diminish the output from the piston engine; for when the line *AD* is raised it affects the turbine area *ABCD* along the whole length of *AD*, whereas it affects the piston-engine area *ABCEF* only along the short distance *AF*.

The importance of high vacuum in a steam turbine is best realized by working out values of the adiabatic heat-drop with various back pressures. As an example take steam with an initial pressure of 250 pounds per square inch and moderately superheated: it will be found that the adiabatic heat-drop is increased by more than 12 per cent. when the vacuum is improved from $28\frac{1}{2}$ inches of mercury to $29\frac{1}{2}$ inches, with the barometer at 30 inches.

Nowadays the position has very materially changed: while formerly it was the low-pressure end of the expansion that gave the turbine its main advantage over the piston engine as a means of converting heat into work. it is now in addition its greater power of making effective use of the energy available with the high pressures and high temperatures which are made possible as a result of progress in engineering design and metallurgy. It is now common practice to employ pressures of 600 pounds per square inch and temperatures of 450° C.; this increase of initial pressure from 200 pounds to 600 pounds per square inch has meant a gain of thermal efficiency of about 10 per cent. These remarks therefore apply with still greater force when pressures of 1500 to 2000 pounds are considered.

CHAPTER VI

INTERNAL-COMBUSTION ENGINES AND
PROPERTIES OF GASES

151. Internal Combustion. In an internal-combustion engine the fuel which is to supply heat-energy for conversion into work forms part of the working substance, and its combustion takes place within the vessel or system of vessels in which the working substance does work by expanding. The working substance, therefore, undergoes a chemical change during its operation and the thermodynamic process is not cyclic. In the early stages, before combustion, the substance is a mixture of fuel with air, generally in excess of what is required to provide enough oxygen for complete combustion. In the later stages, after combustion, it is a mixture of the products of combustion with nitrogen and with any surplus of air. The fuel commonly enters as a gas or vapour, drawn in along with a suitable proportion of air; but it may be injected as a liquid, becoming vaporized after admission or directly burnt on entry. As a rule the only chemically active constituents of the fuel are hydrogen, hydrocarbons and carbonic oxide. In their combustion they unite with oxygen to form water-vapour and carbonic acid. The nitrogen of the air takes no part in the chemical process beyond acting as a diluent.

Typical examples of internal-combustion engines are the ordinary gas-engine or the petrol motor, in which a "charge" of air mixed with combustible gas or vaporized liquid fuel is drawn in by the piston, then compressed into a clearance space, and there ignited by an electric spark or other means, so that explosive combustion takes place while the volume of the charge is nearly constant. The heat thus internally developed gives the working substance a high temperature and pressure: it then expands, doing work as the piston advances. In all modern engines of this class the charge is brought to a fairly high pressure before being ignited. It will be shown later that this compression secures thermal efficiency; with increased compression a larger fraction of the heat of combustion of the fuel is converted into effective work.

From the thermodynamic point of view internal-combustion engines have this advantage over the steam-engine, that their working substance "takes in" heat (by its own combustion) at a much higher temperature. In the combustion of the charge a temperature of 2800° C. or so is possible. The average temperature at which the heat is developed is far above that at which heat is received by the working substance of a steam-engine. On the other hand it is not practicable to discharge heat at nearly so low a lower limit: the temperature of discharge usually well exceeds 500° C. and may be even higher than 1000°. But the actual working range of temperature is so wide that a gas-engine or oil-engine can in fact convert into work a larger fraction of the heat-energy of the fuel than is converted by any engine which burns its fuel to raise steam in a boiler, and uses the steam, however efficiently, as working substance.

152. The Four-Stroke Cycle. In the most usual type of internal-combustion engine the mechanical cycle is completed in four strokes or two revolutions. During the first forward stroke, gas and air are drawn in, so that the whole cylinder is filled with explosive mixture, at practically atmospheric pressure. During the first back-stroke this mixture is compressed into a clearance space at the end of the cylinder. The mixture is then ignited, while the piston is at or close to the "dead-point" or extreme of its travel*. The pressure consequently rises to a much higher value than was reached by compression. During the second forward stroke the fired mixture expands, doing work and falling in pressure. During the second backstroke it is discharged through an exhaust-valve into the atmosphere. A small quantity of the burnt mixture remains in the clearance space, and is mixed with the next charge unless special means are taken to remove it, by what is called "scavenging." As a rule there is no scavenging.

The four-stroke cycle was first described by Beau de Rochas in 1862; it was brought into use by Otto in 1876, and is often called by his name. It is still the most usual mode of action, notwithstanding the practical drawback of having only one working stroke out of four, a drawback which arises from the fact that the working cylinder serves also as inhaler and compressing pump.

153. The Clerk or Two-Stroke Cycle. To escape this defect of the Otto cycle, Sir Dugald Clerk introduced in 1881 an engine which completes its action in two strokes. Clerk's Engine has a

---

* The expressions "top dead centre" and "bottom dead centre" apply to the two extreme positions of the piston when the cylinder is vertical.

separate pump or displacer which inhales the charge and delivers it to the working cylinder just after the piston has completed its working stroke. The fresh charge drives the products of combustion before it, expelling them through exhaust ports, and filling the working cylinder. It is then compressed into the clearance space before ignition, just as in the four-stroke cycle. There is accordingly a compression stroke before firing and an expansion stroke after firing, and these two strokes complete the cycle, the displacer enabling the other two strokes of the four-stroke cycle to be dispensed with. Clerk's device is used in many of the largest gas-engines. For petrol-engines the greater power output possible from a given size of engine when operating with a two-stroke cycle rather than with a four, gives the former a considerable advantage where a low weight-power ratio is essential as in the case of aero engines, or when the engine is taxed according to its cylinder capacity. The disadvantages, which hitherto have prevented any extensive application of this method of operation, are the necessarily increased mechanical complication and the high fuel consumption which results from some of the fresh charge passing directly out of the exhaust valve during scavenging. This latter objection is obviated in the case of fuel-injection engines, where the fuel is not introduced into the cylinder with the air but is injected independently at the end of the compression stroke. The thermodynamic action now to be described is essentially the same in both.

154. **Ideal Action.** In any real engine the action is complicated by exchanges of heat, through conduction and radiation, between the working gas and the walls of the containing vessel, and also by the fact that the process of explosive combustion of the charge is not instantaneous, but takes an appreciable time to be completed. It is convenient, however, to consider an ideal action in which (1) there is no exchange of heat between the gas and the walls, and (2) all the heat of combustion is generated at a particular instant, namely when the volume is constant at the end of the compression stroke, before expansion begins. Such an ideal action affords a useful standard for comparison with the performance of a real engine.

Consider then an ideal engine in which there is no transfer of heat to or from the cylinder walls, and in which combustion occurs only while the piston is at the dead-point. The indicator diagram of this ideal engine, working on the Otto or four-stroke cycle, would take the form shown in fig. 78. *OM* is the volume of the clearance space into which the charge is compressed, and this is the constant volume which the charge occupies during its combustion. *MN* is the volume swept through by the piston in each stroke. *AB* represents the process of admitting the charge at atmospheric pressure; *BC* represents the compression, which by assumption is adiabatic; *CD* is the rise of pressure caused by the explosion; *DE* is the expansion, also adiabatic, which constitutes the effective working stroke; at *E* the exhaust-valve opens, with the result that part of the gas at once

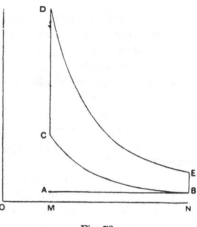

Fig. 78

escapes and the pressure falls to that of the atmosphere, giving the line *EB*; *BA* represents the exhaust stroke by which the cylinder is emptied of gas preparatory to receiving a fresh charge in the first stroke of the next cycle. If the ideal engine were of the two-stroke cycle type, the lines *AB* and *BA* would be omitted from the indicator diagram for the working cylinder, which would then consist simply of the figure *BCDEB*.

From *C* to *D* the whole heat-energy developed by the combustion of the charge goes to heat the working substance, since by hypothesis none is lost by conduction or radiation to the walls. The heat of combustion can be calculated when the composition of the charge is known, or may be measured directly by burning a sample of the gas in a calorimeter. In all cases one of the products of the combustion is water-vapour, and as any water-vapour formed in the cylinder of an internal-combustion engine remains uncondensed throughout the action it is proper to take, in calculating the heat developed by combustion, what is called the "lower" value, that is to say, a value which does not include the latent heat of the water-vapour.

Between *C* and *D* the mixture undergoes a chemical change which may or may not affect its specific volume: that is to say, the burnt products when brought to the same pressure and temperature as the unburnt mixture may not fill exactly the same volume. In general the specific volume after combustion is a little less, but with such mixtures as are used in gas-engines or petrol-engines the effect of this "chemical contraction," as it is called, on the specific volume is so small as to be unimportant. With mixtures of coal-gas and air it amounts to between two and three per cent. in ordinary cases. With some explosive vapours the specific volume is slightly increased (see Art. 158). The changes being in any case small, it is convenient in considering an ideal engine to ignore them, and to treat the working substance as if it were a gas whose specific volume does not alter. Further, the largest constituent of the unburnt charge is air, and that of the burnt charge is nitrogen, and the specific heat of nitrogen is, for equal volumes, the same as that of air. Hence for the purpose of obtaining a simple standard with which real engines may be compared, a practice has sometimes been adopted of treating the working substance as if it were air, to which between C and D there is imparted a definite quantity of heat, namely the heat of combustion of the charge.

155. Air Standard. It was on this basis that a Committee of the Institution of Civil Engineers* devised what is known as the "Air Standard" as a measure of the ideal efficiency of an internal-combustion engine.

Besides assuming (as in Art. 154) for the purposes of their ideal standard:

(1) No transfer of heat between the working substance and the metal;

(2) Instantaneous complete combustion;

and (3) No change of specific volume;

they made the further assumption:

(4) That the specific heat might be treated as constant (independent of temperature as well as pressure).

It is now recognized that this last assumption is by no means true even of air, and is still more untrue of the mixed gases in the cylinder. The specific heat is nearly independent of pressure, but it increases with rise of temperature to an extent which greatly affects the action of the engine. This point will be considered later: but it should be said here that because the specific heat of the work-

---

* Report of a Committee on the Efficiency of Internal-Combustion Engines, *Min. Proc. Inst. Civ. Eng.* vols. CLXII and CLXIII (1905 and 1906). Reference should also be made in this connection to Sir Dugald Clerk's book on *The Gas, Petrol and Oil Engine*, vol. I, and to D. R. Pye's book on *The Internal Combustion Engine.* 1931.

ing gas is much greater at high temperatures than at low temperatures, the "air standard," as defined by the Committee, is an unreasonably high criterion to apply to any actual performance. The efficiency of a real engine must fall far short of that standard, not only because of such more or less avoidable losses as occur through radiation and conduction of heat to the cylinder walls, but because the standard postulates, on the part of the working substance, a characteristic which it does not and cannot possess. Even if there were no loss of heat, the limit of temperature which the gases reach after explosion must be much lower than that which would be reached if the specific heat were constant. However much the heat losses are minimized, the hypothesis of constant specific heat makes the air standard an impossible ideal.

It is nevertheless instructive to use the air standard as a means of examining some of the effects that follow from varying the conditions of working. We may apply it for instance to show how the efficiency of the gas-engine cycle is improved by increasing the compression.

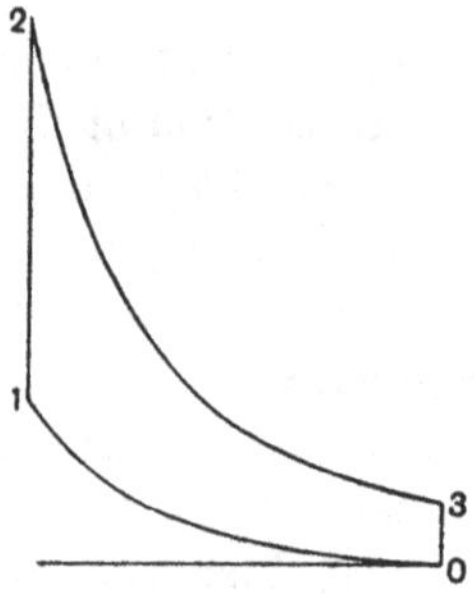

Let T_0 and T_1 be the absolute temperatures of the charge before and after compression, and let T_2 and T_3 be the temperatures before and after expansion. Fig. 79 shows the cycle with its stages numbered to correspond with these suffixes. Write r for the ratio in which the charge is compressed before ignition, which is also the ratio in which it is afterwards expanded during its working stroke. Then by Art. 26, since the compression and expansion are assumed to be adiabatic,

Fig. 79

$$\frac{T_0}{T_1} = \left(\frac{1}{r}\right)^{\gamma-1} \text{ and } \frac{T_3}{T_2} = \left(\frac{1}{r}\right)^{\gamma-1},$$

from which also

$$\frac{T_3 - T_0}{T_2 - T_1} = \frac{T_0}{T_1} = \left(\frac{1}{r}\right)^{\gamma-1}.$$

Here γ is the ratio of C_p the specific heat at constant pressure to C_v the specific heat at constant volume, and is treated as a constant because the specific heats are assumed to be constant in the "air cycle" whose efficiency we are now finding.

The heat supplied, namely the heat generated in the explosion, is $C_v(T_2 - T_1)$. The heat rejected is $C_v(T_3 - T_0)$, for it makes no difference whether the products of combustion are cooled on release to the atmosphere, or kept in the cylinder and cooled there to atmospheric temperature, at constant volume, before being released. Hence the thermal equivalent of the work done in the air cycle is

$$C_v(T_2 - T_1) - C_v(T_3 - T_0)$$

and the "air-standard" efficiency is

$$\frac{C_v(T_2 - T_1) - C_v(T_3 - T_0)}{C_v(T_2 - T_1)} \text{ or } 1 - \frac{T_3 - T_0}{T_2 - T_1},$$

which is equal to

$$1 - \left(\frac{1}{r}\right)^{\gamma - 1}.$$

This expression is important as showing how greatly the efficiency is raised by increasing the compression.

From it, taking γ to be $1\cdot4$, we have

Ratio of Compression	Air-Standard Efficiency
2	0·242
3	0·356
4	0·426
5	0·475
6	0·512
7	0·541
8	0·565
10	0·602
15	0·661
20	0·698

It will be seen from these figures and from the curve (fig. 80) that there is at first a very rapid gain of efficiency with increased compression, but that when the compression is high the thermodynamic advantage of increasing it becomes slight. When account is taken of variation in specific heat, figures are obtained for the theoretical limit of efficiency which fall short of the air standard by about 20 per cent. but preserve a nearly constant ratio to it throughout the usual range.

In favourable cases the measured thermal efficiency is as high as 0·37, corresponding to about 68 per cent. of the air standard, or to about 83 per cent. of the theoretical standard that is obtained when account is taken of variations in specific heat. This is for gas-engines and petrol-engines of the ordinary type in which combustion occurs at approximately constant volume, after the compression of a mixed charge.

In all such engines there is a practical limit to the amount of compression: it must not be so great as to cause pre-ignition by unduly raising the temperature before the end of the compression stroke, nor so great as to give to the explosion, when it does occur, the peculiarly violent character known as "knock*." The highest useful compression ratio differs with different kinds of fuel. With coal-gas the ratio used in practice is about 6½ or 7; in petrol-engines it seldom exceeds 5½. The limit arises from the compression of air and fuel together. When air alone is compressed and the fuel is injected only when combustion is intended to occur, compression may with advantage be carried much farther: there is then no restriction due to pre-ignition or "knock."

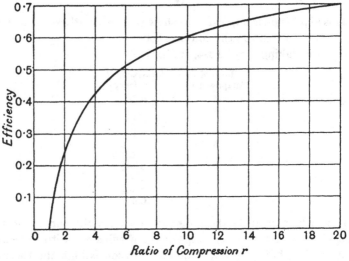

Fig. 80. Efficiency of "air standard."

156. Constant-Pressure Type. Besides the constant-volume type of internal-combustion engine, to which ordinary gas-engines and petrol-engines approximately conform, we may imagine a type in which the pressure of the working substance does not change while combustion is taking place. Suppose that the air is separately compressed into the clearance space before any fuel is admitted and that fuel is then forced in, burning as it enters, while the piston begins its forward movement. By suitably regulating the rate of admission of the fuel the pressure may be kept constant till the combustion is completed.

In this imaginary cycle the heat is supplied at constant pressure. We may further imagine the rejection of heat to occur at constant

* See Ricardo on *The Internal-Combustion Engine*, vol. II, Chapter II, also Pye, *loc. cit.* Chapter IV.

pressure, if we suppose that the products of combustion are expanded adiabatically down to atmospheric pressure before they are discharged. The ideal indicator diagram would then take the form sketched in fig. 81. Under these conditions (which are not realized in practice) we should have an engine of constant-pressure type, rejecting as well as receiving heat at constant pressure. Its air-standard efficiency is readily expressed in a form corresponding to that found for an engine of constant-volume type. We are concerned here with the specific heat at constant pressure, C_p. Treating it as

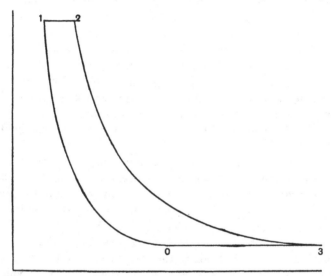

Fig. 81. Constant-pressure type.

constant, the heat taken in is $C_p (T_2 - T_1)$, the heat rejected is $C_p (T_3 - T_0)$, and the efficiency is

$$\frac{C_p (T_2 - T_1) - C_p (T_3 - T_0)}{C_p (T_2 - T_1)} \text{ or } 1 - \frac{T_3 - T_0}{T_2 - T_1}.$$

The ratio of adiabatic expansion is equal to the ratio r of adiabatic compression, $T_0/T_1 = T_3/T_2$. Hence the air-standard efficiency is given by the same expression as before*, namely

$$1 - \left(\frac{1}{r}\right)^{\gamma - 1}$$

* It may be noted that this same expression applies to *three* ideal types of engine:
    (1) The constant-volume type, in which heat is received and rejected only at constant volume.

It follows that for equal ratios of compression there would be no thermodynamic advantage in substituting a constant-pressure type of engine for the constant-volume type. But by avoiding any admixture of the fuel with the air before compression it becomes practicable to use a higher ratio of compression, and consequently to obtain a higher efficiency.

157. Diesel Engine. This advantage is secured in engines of the Diesel type, which compress the air separately to a pressure of 500 pounds per square inch or more, before the fuel is admitted. The air is compressed by the backward stroke of the piston. The fuel is oil and is introduced into the cylinder by means of a separate pump. The injection of the fuel commences a short time before the piston reaches top dead centre and combustion occurs spontaneously because of the high temperature to which the air has been brought by compression. No electric spark is employed to initiate combustion as in the case of petrol engines, and for this reason Diesel engines are often referred to as compression ignition engines.

During the first part of the fuel injection, while the piston is practically stationary at top dead centre, combustion takes place under approximately constant-volume conditions, but as the piston moves on its outward stroke the remainder of the combustion assumes a constant-pressure characteristic. When the injection is complete the combustion products expand as in the other internal-combustion engines, but not to the extent shown in the imaginary engine of fig. 81, for expansion is continued only to a volume equal to that of the air before compression, so that when release takes place, the pressure is much higher than that of the atmosphere. The heat rejection operation is therefore a constant-volume one.

In practice the injection process is carefully timed so that the maximum pressure which is consistent with the mechanical reliability of the engine is attained. By this means the constant-

---

(2) The constant-pressure type, in which heat is received and rejected only at constant pressure.

(3) The constant-temperature type (Carnot's engine of fig. 4, Art. 36) in which heat is received and rejected only at constant temperature.

For its efficiency is $1 - \dfrac{T_2}{T_1}$, and $\dfrac{T_2}{T_1} = \left(\dfrac{1}{r}\right)^{\gamma-1}$, using r to denote the

ratio of *adiabatic* compression (not isothermal, as in Art. 36). Since these can all be operated in a reversible manner they inevitably must have the same efficiency.

pressure portion of the cycle is reduced to a minimum. This condition is necessary if maximum thermal efficiency is to be attained, as will be seen from a study of the Diesel engine given in fig. 81 *a*. For, if the diagram is split up into a large number of small strips bounded by adiabatics, the several strips provide either the ideally maximum amount of work for the volume range through which the particular expansion occurs, or a close approximation to it for those strips bounded at the top by the constant-pressure line. These latter strips, however, provide varying amounts of work since the expansion ratios decrease towards the right of the diagram. In other words the total work available in a Diesel engine for a given heat input will be less than would be given by

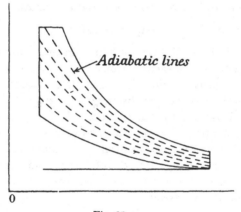

Fig. 81 *a*

a constant-volume cycle working over the whole expansion range. Thus, the greater the constant-volume portion and the shorter the constant-pressure portion of the combustion, the greater the efficiency. The high compression ratio of the Diesel engine, however, enables it to convert more of the thermal energy of the fuel into indicated work than is converted in other internal-combustion engines. It can also utilize much less volatile, and therefore safer, kinds of oil fuel.

158. Combustion of Gases. Molecular Weights and Volumes.
In calculations that relate to the combustion of gases the quantities involved are most conveniently reckoned per unit of volume, at a standard condition as to temperature and pressure. A chief reason for this is that the densities of gases are proportional to their com-

bining weights, and consequently the volumes in which they unite have a very simple ratio.

The combining weights for the substances with which we are now concerned are (in round numbers*):

$$
\begin{aligned}
\text{Hydrogen} \quad & H = 1 \\
\text{Oxygen} \quad & O = 16 \\
\text{Nitrogen} \quad & N = 14 \\
\text{Carbon} \quad & C = 12
\end{aligned}
$$

Hydrogen, oxygen and nitrogen are diatomic gases; that is to say their molecules, H_2, O_2, N_2, each comprise two atoms, and their *molecular* weights are accordingly 2, 32 and 28 respectively. The volumes represented by these weights are the same for all three, when brought to the same pressure and temperature.

The equation $\qquad 2H_2 + O_2 = 2H_2O$

means that in the combustion of hydrogen two molecules of hydrogen unite with one molecule of oxygen to form two molecules of water. As regards weights, it means that 4 parts by weight of hydrogen unite with 32 parts by weight of oxygen to form 36 parts by weight of water, and that the molecular weight of water is 18. As regards volumes, it means that two volumes of hydrogen unite with one volume of oxygen to form two volumes of water-vapour, assuming that the comparison of volumes is made under such conditions of temperature and pressure that the water-vapour may be treated as perfectly gaseous.

Again, the equation $\qquad 2CO + O_2 = 2CO_2$

means that two molecules of carbon monoxide (56 parts by weight) unite with one molecule of oxygen (32 parts by weight) to form two molecules of carbon dioxide (88 parts by weight). It also means that two volumes of carbon monoxide unite with one volume of oxygen to form two volumes of carbon dioxide.

In the combination of any gases the proportion by volume is given directly by the relative numbers of molecules. The principle involved—known as Avogadro's Law—is that equal volumes of all gases (in the perfectly gaseous state and under the same conditions as to pressure and temperature) contain the same number of molecules. The weight contained in unit volume—in other words the density—is therefore proportional to the molecular weight.

* More exactly, taking O as 16, H is 1·0078 and N is 14·008.

A few further illustrations may be useful:

Marsh gas (Methane) CH_4.

$$CH_4 + 2O_2 = CO_2 + 2H_2O$$

Weights	$16 + 64 = 44 + 36$
Volumes	$1 + 2$ form $1 + 2$

Ethylene C_2H_4.

$$C_2H_4 + 3O_2 = 2CO_2 + 2H_2O$$

Weights	$28 + 96 = 88 + 36$
Volumes	$1 + 3$ form $2 + 2$

Butylene C_4H_8.

$$C_4H_8 + 6O_2 = 4CO_2 + 4H_2O$$

Weights	$56 + 192 = 176 + 72$
Volumes	$1 + 6$ form $4 + 4$

Alcohol C_2H_6O.

$$C_2H_6O + 3O_2 = 2CO_2 + 3H_2O$$

Weights	$46 + 96 = 88 + 54$
Volumes	$1 + 3$ form $2 + 3$

It will be observed that with alcohol and with the heavy hydro-carbons, of which C_4H_8 is one, the specific volume is increased by combustion; with marsh gas and ethylene it undergoes no change; and with hydrogen and carbon monoxide it is reduced. The change of specific volume which any given gas mixture will undergo on complete combustion is readily predicted by applying this method of calculation to each of the constituents of the fuel, when the composition of the mixture is known. Another obvious application is to calculate the volume of oxygen, and by inference the volume of air, required for the complete combustion of a given gaseous fuel. For the purpose of such calculations, dry air may be taken as a mixture of 20·9 per cent. by volume of oxygen with 79·1 per cent. by volume of nitrogen and other inactive gases.

The following example will serve to show how the air required for the complete combustion of a gas of known composition is calculated, and also the change of specific volume, or the "chemical contraction," which will take place on combustion. The fuel is coal-gas, of the composition shown in the first column.

Composition of the gas by volume		Vol. of oxygen required for complete combustion	Volume of products		
			H_2O	CO_2	N_2
H_2	42·2	21·1	42·2	—	—
CH_4	34·0	68·0	68·0	34·0	—
C_2H_4	4·8	14·4	9·6	9·6	—
C_4H_8	2·1	12·6	8·4	8·4	—
CO	6·4	3·2	—	6·4	—
N_2	6·8	—	—	—	6·8
CO_2	3·7	—	—	3·7	—
	100·0	119·3	128·2	62·1	6·8

219·3   197·1

The chemical contraction is $219·3 - 197·1 = 22·2$ volumes.

Since 119·8 volumes of oxygen are required for complete combustion, the air required is $\dfrac{119·3 \times 100}{20·9}$ or 571 volumes. Hence if this gas is exploded in the richest possible mixture, with no surplus air or other diluent, the contraction amounts to 22·2 on a total volume of 671, or 8·8 per cent. In a gas-engine where the mixture is diluted by excess air, and by residual products from a previous charge, the contraction will of course be a smaller proportion of the whole volume.

159. **Avogadro's Principle.** The molecular theory of gases, which will be considered in the next chapter, is based on the idea that any gas is a group of moving particles called molecules each of which is indivisible so long as the gas preserves its chemical character. Each particle moves freely except when it collides with another or with the wall of the containing vessel. The weight of a gas, for a given volume, is the sum of the weights of the molecules. According to the principle of Avogadro mentioned in the last article equal volumes of all gases, under the same conditions as to pressure and temperature, contain the same number of molecules. The number has been determined by various experimental methods with results that are in fairly good agreement. It appears that $2·705 \times 10^{19}$ may be taken as the number of molecules in one cubic centimetre of any gas at 0° C. and the standard pressure of one atmosphere.

It follows that if we divide the density of a gas (at standard pressure and temperature) by $2·705 \times 10^{19}$ we shall find the weight of one molecule. Thus in hydrogen, whose density is 0·00008987 gramme per cubic centimetre, the weight of the molecule is

3·32 × 10⁻²⁴ gramme, and in oxygen it is about 52·7 × 10⁻²⁴ gramme.

160. The Gramme-Molecule and Avogadro's Number. Suppose that for the purpose of comparing gases we take a different quantity of each, namely a quantity equal to w grammes, where w is the molecular weight. The quantity so specified is called the Gramme-Molecule or "Mol." Thus for oxygen the gramme-molecule is 32 grammes, for hydrogen it is 2·016 grammes, for CO it is 28 grammes, for CO_2 44 grammes, and so on. By Avogadro's principle the gramme-molecule has the same volume for all gases and contains the same number of molecules, subject to small differences which are due to the gases not being strictly perfect in the sense of Art. 18. The volume of the gramme-molecule or Mol for the nearly perfect gases is 22,412 cubic centimetres at 0° C. and a pressure of one standard atmosphere. The number of molecules it contains is $6·062 \times 10^{23}$. This is called *Avogadro's Number*. Properties of gases such as the specific heats or the heat of combustion are often stated per mol.

A quantity of heat stated in gramme-calories per mol may be converted into foot-pounds per cubic foot by multiplying by

$$\frac{0·002205 \times 1400}{(0·032808)^3 \times 22,412} \text{ or } 3·90,$$

since 1 gr.-calory = 0·002205 lb.-calory, 1 lb.-calory = 1400 foot-pounds, and 1 cm. = 0·032808 ft.

As an alternative to the gramme-molecule English engineers sometimes use a lb.-molecule, which means a number of lbs. equal to the molecular weight of the gas. The lb.-molecule is 453·6 times the gramme-molecule: it has a volume of 359 cubic feet; and it contains $2·75 \times 10^{26}$ molecules.

161. The Universal Gas-Constant. The gas equation

$$PV = RT$$

is strictly applicable only to ideal gases which are "perfect" in the sense of obeying Boyle's Law and also Joule's Law (Art. 19), T being the absolute temperature on the thermodynamic scale. It is approximately true of all gases at low or moderate pressures, provided the conditions as to pressure and temperature are not such that the gas approaches liquefaction. At any given temperature a real gas is more and more nearly "perfect" the more the pressure is reduced. In the equation $PV = RT$ let V stand for the volume of w grammes instead of one gramme—a volume which, as we have just seen, does not differ for different gases. Then R takes the same numerical value for all gases that satisfy the equation: it becomes the Universal Gas-Constant of Art. 18. The dimensions of the universal gas-constant are those of work, expressed per gramme-molecule of gas. It may be interpreted as the work that is done by expansion under constant pressure when one mol of a gas is heated through one degree. To obtain its numerical value we have $V = 22,412$ cubic centimetres when T is 273·1 and P is one standard atmosphere or 1033·2 grammes per square centimetre. Hence the gas-constant is

$$\frac{1033\cdot2 \times 22,412}{273\cdot1} \text{ or } 84,790 \text{ centimetre-grammes.}$$

We may also express it in heat units. Since the gramme-calory is equivalent to 42,699 gramme-centimetres, the gas-constant is

$$\frac{84,790}{42,699} = 1\cdot986 \text{ gramme-calories.}$$

Or it may be expressed in ergs. The standard atmosphere being $1\cdot01325 \times 10^6$ dynes per centimetre (Art. 12) the gas-constant is

$$\frac{1\cdot01325 \times 10^6 \times 22,412}{273\cdot1} = 83\cdot15 \times 10^6 \text{ ergs.}$$

Again, if the lb.-molecule be used instead of the gramme-molecule the gas-constant is 1·986 lb.-calories, equivalent to 2782 foot-pounds. This figure refers to the pound where gravity has the standard value. The equivalent figure for use in London would be 2780 foot-pounds as given in Art. 14.

Knowing the gas-constant we can readily calculate the value of R in the equation $PV = RT$ for any gas to which that equation applies, by dividing the constant by w. Values of R calculated in this manner are given below.

Values of R calculated from the molecular weights

	in gramme-calories per gramme or lb.-calories per lb.	in foot-pounds per lb. (London)
Oxygen	0·0621	86·9
Nitrogen	0·0709	99·2
Air	0·0686	96·0
Hydrogen	0·985	·1380
Helium	0·497	695
Carbon monoxide	0·0709	99·3
Carbon dioxide	0·0451	63·2

It should be recalled that the value of R is equal to the difference between the specific heats at constant pressure and at constant volume, C_p and C_v (Art. 20).

Another quantity may be deduced from the universal gas-constant which we shall find useful later. The universal constant calculated for a gramme-molecule is a quantity of work that relates to an aggregate of N molecules, where N is Avogadro's number, namely $6 \cdot 062 \times 10^{23}$. If therefore we divide it by N we get the gas-constant for a single molecule of any perfect gas, a quantity which is sometimes called Boltzmann's constant and is usually denoted by the letter k. Taking the universal constant as $83 \cdot 15 \times 10^6$ ergs and dividing by N we find $1 \cdot 372 \times 10^{-16}$ erg for k.

162. **Specific Heats of Gases in Relation to their Molecular Weights. Volumetric Specific Heats.** In Art. 20 we reckoned the specific heats C_p and C_v per unit of mass. For many purposes it is more convenient to reckon the specific heats of gases per unit of volume: when so reckoned they are sometimes called volumetric specific heats. Most convenient of all is to reckon them per gramme-molecule or mol. This is in effect a volumetric method, for the volume of the mol is the same in all gases that satisfy the equation $PV = RT$. When the specific heats of such gases are reckoned per mol their difference is equal to the gas-constant. Thus

$$C_p = C_v + 1 \cdot 986,$$

when C_p and C_v are reckoned in gramme-calories per mol.

It follows that in all such gases the ratio γ of C_p to C_v is

$$\gamma = 1 + \frac{1 \cdot 986}{C_v}.$$

The volumetric method of reckoning specific heat has this further advantage that when so reckoned the specific heat (C_p or C_v) of the simpler gases is nearly the same, provided the gases have the same number of atoms in the molecule.

The kinetic theory (see Chapter VII) shows that in an ideal* diatomic gas $C_v = \frac{5}{2}R$. Therefore in any such gas $C_p = \frac{7}{2}R$ and $\gamma = \frac{7}{5} = 1 \cdot 40$. This is found to agree well with the values of γ got by direct measurement in air, oxygen, nitrogen and other diatomic permanent gases. It follows that what may be called the theoretical values of C_p and C_v, in calories per mol, are

$$C_p = \tfrac{7}{2} \times 1 \cdot 986 = 6 \cdot 951; \quad C_v = \tfrac{5}{2} \times 1 \cdot 986 = 4 \cdot 965.$$

* Ideal in the sense that the gas satisfies the equation $PV = RT$ and also that its molecules have no sensible energy of vibration (Art. 179).

The corresponding figures in calories per gramme are found for any gas by dividing by w. Thus for air we have 0·2401 and 0·1715 as the theoretical values of C_p and C_v respectively, in calories per gramme.

163. Summary of Methods of expressing the Specific Heats. A short summary of methods of stating C_p and C_v in gases may help to avoid confusion. Either of these quantities may be stated as follows:

(a) In gramme-calories per gramme-molecule or mol, the gramme-molecule or mol being a mass equal to w grammes, where w is the number which expresses the molecular weight.

(b) In gramme-calories per gramme.

(c) In lb.-calories per lb.

(d) In foot-pounds per cubic foot.

(e) In foot-pounds per lb.

To convert from (a) to (b) or to (c) divide by w. The numbers in (b) and in (c) are the same. To convert from (a) to (d) multiply by 3·90. To convert from (c) to (e) multiply by 1400.

The difference between C_p and C_v, which is nearly constant in all gases, has the following values:

In (a), 1·986 calories.

In (b) and (c), $\dfrac{1\cdot986}{w}$ calories.

164. Measured Values of Specific Heats. It is to be expected that the actual specific heats of gases should slightly exceed the values calculated from the kinetic theory, owing to the departure of real gases from the ideal conditions assumed in the theory.

Measurements of C_p by Regnault for a number of gases gave values which are somewhat *less* than the theoretical values, but the method used by him is now believed to have been affected by a systematic error, the effect of which was to make the measured values too small, apparently by about 2 per cent.[*]

A more modern measurement of C_p for air by Swann[†], by means of electric heating under constant pressure, gives (when reduced to the mean calory used in this book) $C_p = 0\cdot2418$ calory per unit of mass, at 0° C. and one atmosphere, which is, as we should expect, slightly greater than the theoretical number.

---

[*] Swann, *Phil. Trans. R.S.* A, vol. ccx, p. 281. Also Report of the British Association Committee on Gaseous Explosions, B.A. Rep. 1908.

[†] *Loc. cit.*

The corresponding value of C_v would be 0·1727, taking R to be 0·0686 as in Art. 161.

C_v has been directly measured by Joly for several gases, by the device of applying steam externally to heat a copper globe containing the gas, and comparing the amount of steam thereby condensed on the surface with the amount condensed on another exactly similar but empty globe*. His observed value of C_v for air, when corrected for the revised value of the latent heat of steam and for the mean calory, is 0·1729 in calories per unit of mass. This is in good agreement with the value of C_v inferred from Swann's measurement of C_p.

There is conclusive evidence that the specific heat of most gases rises with the temperature, so that substantially higher values apply when a gaseous mixture is fired. This point, which is important in relation to internal-combustion engines, will be considered in the next article.

165. **Variation of Specific Heat with Temperature.** It was pointed out in Art. 21 that a gas might be perfect in the sense of conforming to Boyle's Law and to Joule's Law, so that the equation $PV = RT$ is strictly applicable, and still have its specific heat vary with the temperature, though there would be no variation with the pressure.

Any variation of specific heat with pressure is due to imperfection of the gas. In the permanent gases, there is but little departure from the equation $PV = RT$ except at pressures much higher than those that are found in gas-engines. Hence their specific heat is nearly independent of the pressure. Even the mixture produced by a gas-engine explosion, comprising some water-vapour and carbon dioxide along with much nitrogen, conforms to the equation $PV = RT$ nearly enough to allow that equation to be applied in calculating the temperature from the observed pressure. Although the specific heat of such a mixture is undoubtedly somewhat greater at high pressures than at low pressures, the difference is not so considerable as to be taken into account in gas-engine calculations. On the other hand, the specific heat of such a mixture, and of most gases, varies largely with the temperature, becoming greater as the temperature rises, within the range that occurs in an engine cylinder.

In monatomic gases, such as argon or helium, there is little, if

* Joly, *Phil. Trans. R.S.* A, vol. CLXXXII, p. 73, 1891.

any, increase of specific heat within that range; in diatomic gases such as oxygen or nitrogen the increase is considerable; in gases of more complex constitution, such as the triatomic gases H_2O and CO_2, it is larger still. The presence of these constituents in a gas-engine mixture makes its rate of change of specific heat with temperature greater than that of air. The specific heat of a gas-engine mixture at 2000° C. is about 1·8 times what it is at 0° C.

An obvious result of the increase of specific heat with temperature is that when a definite quantity of heat is given to a gas or a mixture of gases—as, for instance, by an explosion completed at constant volume—the rise of temperature is less than it would be were the specific heat to keep constant, for as the gas gets hotter each degree of rise absorbs more and more of the available heat. When the experiment is made of exploding a charge in the cylinder of an engine or in any closed vessel, it is found that the temperature actually reached is far short of that calculated on the basis of constant specific heat, after making full allowance for loss of heat to the walls of the vessel. When this fact was first observed it was put down to imperfect or rather delayed combustion of the charge; the suggestion was that a large part of the heat of combustion was developed gradually, in a comparatively slow process called "after-burning," which was supposed to continue after the explosion had spread through the whole vessel and after the temperature and pressure had risen quickly to the observed maximum. The notion that there is any considerable effect due to "after-burning" in this sense is now abandoned, though it is recognized that after a high temperature has been reached in an explosion there is some small continued evolution of heat as the temperature falls, owing to the recombination of dissociated molecules among the products of combustion (see Art. 172).

Measurements of the specific heats of gases have been made in various ways*: by direct heating, up to high temperatures, under constant pressure; by observing the rise of temperature in explosions; and also by a method due to Clerk†, in which the gas in an engine cylinder is successively expanded and compressed several times while the valves are kept closed. In that process, the work done by or upon the gas between any two points of the stroke is

---

* Particulars of these, and a valuable discussion of the results, will be found in the Reports of the British Association Committee on Gaseous Explosions, 1908–16. See also Sir D. Clerk's book on *The Gas, Petrol and Oil Engine*, vol. i; also Partington and Shilling, *The Specific Heats of Gases*, Ernest Benn Ltd.

† D. Clerk, *Proc. R.S. A*, vol. LXXVII.

determined by measuring the area under the indicator curve, and is used as a basis for reckoning the change of internal energy, while the change of temperature is inferred from the change in the product of pressure and volume. The method can be applied either to imprisoned air or to an exploded charge. It is subject to some uncertainty in the estimate that has to be made of the heat which is given to, or taken from, the gas by the cylinder walls.

The results of these various methods of experiment are not very accordant, but they agree in showing that there is an important rise in specific heat with temperature, greater in triatomic gases such as water-vapour or carbon dioxide than in nitrogen or air. The rate of increase is not uniform but becomes greater at high temperatures.

Within the usual range of temperature the specific heat may be expressed by a formula of the type

$$C_v = (C_v)_0 + \alpha t + \beta t^2,$$

where $(C_v)_0$ is the specific heat at $0°$ C., t is the excess of temperature above $0°$ C., and α and β are empirical constants.

We are dealing here with gases which are far removed from the conditions of liquefaction. It will be recalled (Art. 71) that when water-vapour is formed, especially at high pressure, the initial stages of superheating are marked by specially high specific heats which become quickly less as superheating proceeds. In fact they pass a minimum, and after that they slowly rise at higher temperatures as happens in gases generally. The high specific heat that is found in the initial stages may be ascribed to the presence of molecular aggregations when the vapour is still near the state of saturation. From this point of view it is only when the molecules have become completely separated that the gaseous condition is attained, and until that happens some heat is absorbed in the process of breaking up the aggregates.

166. Internal Energy of a Gas. What we are practically concerned with in the gas-engine is not so much the specific heat as a quantity closely related to it, namely the internal energy E. When the charge is exploded at constant volume its internal energy increases by the amount of heat developed, less what is lost to the cylinder walls. What is wanted is a curve showing the relation of E to the temperature in the exploded charge.

The relation between the internal energy E and the specific heat is that

$$dE = C_v dT.$$

Hence, at any temperature, the slope of the curve of E and T, namely $\frac{dE}{dT}$, measures C_v, and $E = \int C_v dT$.

If C_v were constant the curve of internal energy would be a straight line and we should have

$$E = C_v t.$$

Here t is the temperature on the centigrade scale, and the constant of integration is zero if the usual convention be adopted of reckoning the energy of the gas from an arbitrary starting point at $0°$ C. This of course does not mean that a gas at $0°$ C. has no internal energy, but only that the stated value at any temperature is the excess above the value at $0°$ (compare Art. 66).

Taking $C_v = (C_v)_0 + \alpha t + \beta t^2$,

we have $E = (C_v)_0 t + \dfrac{\alpha t^2}{2} + \dfrac{\beta t^3}{3}.$

We may accordingly construct a curve of E and t when an expression for C_v is given, or conversely find an expression for C_v from a given curve of E and t.

Further, when the curve of E and t for a gas or mixture of gases is drawn, the value of C_v at any temperature is readily found by measuring the slope of the curve there. From that C_p may be deduced by adding the gas-constant R to C_v, namely $1·986$ if C_v is expressed in gramme-calories per gramme-molecule (Art. 162). In this way the ratio γ of C_p to C_v may be determined for any temperature.

167. Curve of Internal Energy for a Typical Gas-Engine Mixture. The British Association Committee on Gaseous Explosions give in their first Report (1908) a curve of internal energy and temperature for a typical gas-engine mixture. This mixture was the product of combustion of a charge of one part by volume of coal-gas to about nine parts of air, together with the burnt gases in the clearance space: it contained 5 per cent. by volume of carbon dioxide and 12 per cent. of water-vapour, the remaining 83 per cent. being made up of nitrogen and surplus oxygen. The curve is reproduced in fig. 82.

Examination of the curve shows that it is well represented by the formula $E = 5·2t + 0·00043t^2 + 0·0000002t^3$,

which corresponds (Art. 166) to

$$C_v = 5·2 + 0·00086t + 0·0000006t^2.$$

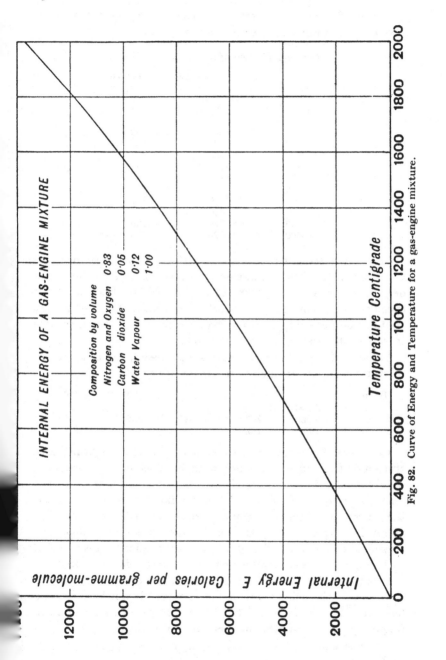

INTERNAL ENERGY OF A GAS-ENGINE MIXTURE

Composition by volume
Nitrogen and Oxygen 0·83
Carbon dioxide 0·05
Water Vapour 0·12
1·00

Fig. 82. Curve of Energy and Temperature for a gas-engine mixture.

Values of E measured from the curve and calculated from the above formula are compared in the table below:

Internal Energy of Gas-Engine Mixture.

E in calories per gramme-molecule

Temp. °C.	measured from the curve	calculated from the formula
0	0	0
200	1060	1059
400	2160	2162
600	3320	3318
800	4540	4538
1000	5840	5830
1200	7200	7205
1400	8680	8672
1600	10,240	10,240
1800	11,920	11,920
2000	13,720	13,720

As was pointed out in Art. 166, the slope of the curve measures C_v at any point. The initial slope, for $t=0°$ C., is 5·2 calories per degree. The slope for $t=2000°$ C. is 9·82. Within that range the specific heat has increased by nearly 80 per cent. At 2000° the gas contains 30 per cent. more internal energy than it would contain if the specific heat were constant at 5·2.

Values of C_v, C_p (taken as equal to $C_v + 1·986$), and of γ, for this mixture at various temperatures are given below:

Temp. °C.	0°	500°	1000°	1500°	2000°
C_v	5·200	5·780	6·660	7·840	9·320
C_p	7·186	7·766	8·646	9·826	11·306
γ	1·38	1·35	1·30	1·25	1·21

168. The Process of Explosive Combustion. Much light has been thrown on this process by experiments in which gas mixtures have been exploded in closed vessels of constant volume, with devices for registering the rise of pressure in relation to the time, and also the progressive changes of temperature at various points within the vessel. The experiments of Hopkinson on explosions of coal-gas and air should be specially referred to in this connection*. It is essential to distinguish between a mixture that is at rest before it is fired and one that has turbulent motion. Turbulence causes ignition to spread farther from the point of origin and may greatly quicken the combustion. It is in fact essential to the proper working of high-frequency petrol and other motors; without it the charge could not

* B. Hopkinson, *Proc. R.S.* A, vol. LXXVII, p. 387, 1906; also vol. LXXXIV, p. 155, 1910.

be burnt in the available time. The importance of turbulence is clear when we study what happens in a non-turbulent mixture.

Let an explosive mixture, homogeneous and at rest to begin with, be ignited at any point. A flame spreads in all directions from the point of ignition, travelling at a rate which depends on the pressure, so that each portion of the mixture ignites in turn, the most distant portions last. When the initial pressure is that of the atmosphere, the flame may travel at the rate of only about five feet per second to begin with, even in a rich mixture such as one of gas to nine of air. The rate depends on the richness of the mixture as well as on the pressure: in a weak mixture it takes much longer for the ignition to spread through the whole volume. When the initial pressure of the mixture is high the ignition flame travels much faster.

The portion which is first ignited, close to the ignition plug, burns at nearly constant pressure, being surrounded by a large elastic cushion of unignited gas. Its combustion is practically completed before the pressure has risen. Then the spread of the flame brings more of the gas into action; the pressure rises, and the portion which was first burnt is compressed. This compression is nearly adiabatic: its effect is to raise the temperature of that portion much above the temperature to which it was brought by combustion, and above the temperature which is reached in combustion by the outlying parts of the gas, which are compressed before they become ignited.

In Hopkinson's experiments a mixture of nine parts of gas to one of air was fired at atmospheric pressure in a cylindrical vessel with a capacity of about 6 cubic feet. It was ignited by an electric spark at the centre, and developed a maximum pressure of about 80 pounds per sq. inch, which was reached a quarter of a second after firing. The temperature was observed near the centre and at other points. On ignition the temperature at the centre rose very rapidly to 1225° C., while the pressure remained nearly constant. In the later stages of the explosion, when the burnt gas at the centre was being adiabatically compressed, its temperature rose above the melting point of platinum, probably to 1900°. This is a higher temperature than was reached in the outlying portions, which were first compressed adiabatically and then heated by combustion. When the maximum pressure was reached, the mean temperature inferred from it was 1600°. Hopkinson concluded that even in a vessel impervious to heat, the portion of the mixture first fired would be hotter than the outlying portions by about 500°, when the combustion of the whole was practically complete.

These experiments go to show that there is no substantial amount of "after-burning," and that the effects formerly ascribed to after-burning are due to the increase of specific heat with temperature. Practically the full evolution of heat in each portion of the gas takes place at once when the flame reaches that portion, but there is some delay in completing the ignition of those portions which are in proximity to the cold walls, especially when the mixture is weak.

In explosion experiments with weak mixtures the spread of the flame is much slower, so slow indeed that it is largely affected by convection currents set up by the ignition of the gas nearest to the spark. The gas in the upper part of the vessel may be completely ignited while the lower part of the vessel is still full of unburnt gas. By stirring the contents of the vessel, so that the gases are in motion when the spark passes, a much more rapid combustion of the whole can be secured. This was demonstrated in some of Hopkinson's experiments by the use of a fan to stir the mixture before it was fired.

169. Effect of Turbulence. This effect of turbulence in promoting rapid ignition of the whole contents is felt, though to a less degree, in strong mixtures as well as weak mixtures. When a fresh charge is drawn in and compressed, the gases are still in more or less violent motion at the moment of ignition. This has the great advantage that combustion is rapidly propagated throughout the charge and the maximum pressure occurs early in the expansion stroke. Clerk observed that when the explosive charge in a gas-engine was not fired after the first compression, but was fired after three successive compressions, so that the turbulence set up on its entry had time in part to subside, the process of combustion was generally prolonged, with the result of giving a flat diagram and a wasteful action. The action of a high-speed petrol-engine depends very much on the effect of turbulence in helping the flame to spread quickly throughout the clearance space in which the gases are compressed. In a compression ignition engine the effect of directed turbulence (swirl) is still more important. Turbulence in the petrol-engine has the further advantage that it reduces the tendency of highly compressed mixtures to *detonate*, or explode violently in the manner that is described in the next Article.

170. "Knocking." Detonation in an explosive gas mixture is the nearly simultaneous explosion of all parts of an explosive substance by means of a compression wave, as distinguished from

the normal process in which ignition is carried from point to point by a flame. In the explosion of certain mixtures, especially when the compression ratio is high, it may happen that although the process has begun in the normal manner a part of the gas may undergo a peculiar and violent type of ignition whereby a compression wave is caused to travel back through the burnt charge and gives rise to a sharp knocking sound. The part of the gas in which this occurs is generally the last part to be burnt, and has been longest exposed to the conditions in which the gases find themselves at the end of the compression stroke. "Knock" has to be distinguished from "detonation" in gaseous mixtures because in the former the detonating kind of explosion is confined to only a very small region of the gas, whereas in a true detonation the explosion is propagated throughout the whole mixture simultaneously with the compression wave. It is a familiar trouble in the running of petrol motors, especially when the constituents of the fuel are mainly of the "paraffin" or chain-molecule class: it is less liable to happen when the petrol contains a good proportion of "aromatics" such as benzene. It is objectionable not only because it may produce damaging stresses, but because it is apt to cause local heating which may result in the pre-ignition of a subsequent charge. The tendency to detonate, more than anything else, determines the highest useful compression ratio for any particular kind of petrol or other vapour mixture. In a spirit composed wholly of normal paraffins that ratio would be very low: by adding a substantial proportion of benzol it can be greatly improved.

It is found that the tendency to detonate may be much reduced by adding to the fuel small quantities of certain substances called "dopes" which have a definitely anti-knock effect. Of these the most notable is tetra-ethyl-lead, $Pb\,(C_2H_5)_4$, which is commercially used for the purpose. The reason why "dopes" operate in this manner has been shown to be due to their inhibiting effect on the chemical reactions which occur in the gases in the cylinder prior to inflammation.

170 a. **Fuels for Petrol-Engines.** The fuel for a petrol-engine has to satisfy a number of conditions. The volatility of the lowest boiling fraction should be high to obtain good starting; its mid-boiling range should be low to provide good acceleration, and good fuel distribution with a minimum of pre-heating in the induction pipe; its latent heat should be high so as to give a cool charge, thereby

increasing the charge density and hence the power; its calorific value should be high; its density should be not too low otherwise it occupies too great a volume; and its "knockrating" should be high, so that the engine can be utilized to the best advantage at as high a compression ratio as possible.

Petrol is a mixture mainly of paraffinic, naphthenic and aromatic hydrocarbons. The straight chain paraffinic constituents knock worse than the naphthenic and they in turn more than the aromatic constituents. On the other hand the more hydrogenated the fuel the greater in general its calorific value, hence the saturated paraffins have advantage in this respect. Hydrogen itself, however, even if it were practical to use it in the liquid condition, would be unsuitable on account of its very low density and therefore large bulk.

Alcohols are excellent fuels, both because they do not knock and because of their high latent heat, but their calorific value is lower than for the paraffinic fuels, and so larger weights of fuel would have to be carried for the same power available. The knocking tendency of fuels is obtained by determining what mixture of n. heptane and iso-octane will show the same tendency to knock as the sample in a standard engine constructed for the purpose. n. Heptane knocks badly and is at one end of the scale and iso-octane at the other. For instance a fuel which behaves similarly as regards knock to a mixture of 75 per cent. iso-octane and 25 per cent. n. heptane would have an "octane number" 75.

171. **Radiation in Explosions.** In closed-vessel experiments, the maximum of pressure is reached a little before the combustion is complete, for it occurs when the rate of loss of heat by radiation and conduction to the walls just balances the rate at which heat is being generated in the gas. Hopkinson investigated the effect of radiation by comparing the rate of cooling in an explosion vessel which was lined with highly polished silver, with the rate in the same vessel when its inner surface was blackened*. The rate of cooling after explosion was notably greater when the walls were blackened, and the maximum pressure was less for charges of the same composition. The rate at which heat was lost to the polished walls was on the average about two-thirds of the rate of loss to blackened walls. Hopkinson concluded that of the heat given by the gas to the

* *Proc. R.S.* A, vol. LXXXIV, p. 155, 1910. See also W. T. David, *Phil. Trans. R.S.* A, vol. CCXI, p. 375; *Phil. Mag.* Feb. 1918 and Jan. 1920.

walls of a blackened enclosure during the first quarter-second after maximum pressure at least 30 per cent. is radiant heat, and possibly a good deal more, for the reflecting quality of the polished walls may have been impaired by the deposit of a film of moisture at an early stage of the cooling. Further experiments, in which the vessel was fitted with a fluorite window to allow the radiation to fall on an absorbing screen outside, confirmed this view. The effects of radiation continue to be perceptible as the temperature of the gas falls and have been detected below 700° C.

The energy of the radiation from an exploded gas-engine mixture is due almost wholly to two or three bands of rays of definite wavelength, corresponding to much slower vibrations than those which produce the visible spectrum. The existence of these bands may be demonstrated by examining the heat which is radiated from a gas flame when it is made non-luminous by using a Bunsen burner.

Experiments with such flames show that when hydrogen is burnt to form water-vapour most of the radiant heat that is given off is in a band with a wave-length of about $2\cdot8\mu$*, but some has a longer wave-length: also that when carbon monoxide is burnt most of the radiant heat is in a band with a wave-length of about $4\cdot4\mu$, but some is in two bands whose wave-lengths are about $2\cdot7\mu$ and between 14μ and 15μ. In one case the radiation comes from vibrating molecules of H_2O, in the other from vibrating molecules of CO_2.

It is also found that cold CO_2 absorbs strongly the radiation from a CO flame, and water-vapour absorbs strongly the radiation from a hydrogen flame. It may be concluded that the modes of free vibration of a molecule of cold CO_2 or water-vapour have periods corresponding to the chief wave-lengths which the gas gives out when it is so violently agitated as to become a source of radiation. This happens when the molecules are formed by the coming together of their constituent atoms.

It is further found† that a mixed or compound gas burning to form CO_2 and H_2O gives out both wave-lengths ($4\cdot4\mu$ and $2\cdot8\mu$), and that the whole energy it radiates is equal to the sum of the energies separately computed for the molecules of H_2O and CO_2 that are formed by its combustion. For equal volumes of H_2O and CO_2, at the same flame temperature, the radiation from CO_2 appears to be about $2\frac{1}{2}$ times that from H_2O.

* μ stands for millionths of a metre. The wave-lengths in the visible spectrum range from about $0\cdot39$ to $0\cdot77\,\mu$.

† R. von Helmholtz: see the Third Report of the British Association Committee.

These results point to the conclusion that when a gas-engine mixture is fired, the energy that is radiated comes mainly from molecules of CO_2 and H_2O in the burnt gases: very little of it comes from the nitrogen or the surplus oxygen.

172. Dissociation. In any gas the molecules suffer mutual encounters and are moving with various speeds, such that their average kinetic energy is proportional to the absolute temperature. The velocities of individual molecules are very unequal. Some of the encounters may be so violent as to break up compound molecules, separating them for a time into parts which after a time find fresh partners and recombine. The probability of such disruptive encounters is increased by heating the gas, because that raises the average velocity. In a hot gas in equilibrium, a process of dissociation and recombination goes on continually, to an extent depending on the temperature, with the result that at any instant a certain proportion of the gas is in the dissociated state. Hence in the normal combustion of an explosive mixture a stage is reached at which the temperature ceases to rise because a balance is established between the amount of dissociation and the amount of recombination. Thus when a petrol-vapour charge is fired and the temperature has risen to its maximum there is still a proportion of unburnt material. Molecules of hydrogen and of carbon monoxide are present although there may be an excess of oxygen waiting to combine with them. The combustion becomes complete only when the temperature has fallen sufficiently to make the effect of dissociation negligible. A result of this is that the maximum temperature is substantially lower than that which would be found if there were no dissociation, and further that the maximum temperature is somewhat higher with a rich mixture than with one in which the ratio of air to fuel is so adjusted as to be sufficient for complete combustion*.

* For examples see Pye on *The Internal Combustion Engine*, Chapters III and v.

CHAPTER VII

MOLECULAR THEORY OF GASES

173. Pressure due to Molecular Impacts. According to the molecular theory, a gas consists of a very large number of particles called molecules moving with great velocity. Each molecule moves freely, with uniform velocity in a straight line, except when it encounters another molecule or the wall of the containing vessel. In an encounter the velocity changes in direction, and generally in amount, but there is no dissipation of energy; the molecules behave like perfectly elastic bodies. As a result of many encounters, a stable distribution of speed among the molecules is established, but the speed of any one molecule is being constantly changed, by its encounters, within very wide limits. The length of the free path, which it traverses between one encounter and the next, is also quite irregular. The average of that length, or what is called the "mean free path," is very long compared with the dimensions of the molecule itself. This characteristic distinguishes a gas from a liquid. In a gas the average time during which a molecule is moving in its free path is very large compared with the time of an encounter. By the time of an encounter is meant the time during which the molecule is either in contact with another, or so near it that there is a sensible force acting between them. When a gas is compressed, the mean free path is reduced, and the encounters become more frequent between one molecule and another and also between the molecules and the walls of the vessel. When a gas is heated the speed with which the molecules move is increased; we shall see immediately that their average kinetic energy is proportional to the temperature. The molecular theory is now well established: there is conclusive evidence that actual gases do consist of particles moving in the manner which the theory prescribes.

The pressure of the gas, that is to say, the pressure which the gas exerts on every unit of surface of the containing vessel, is due entirely to the blows of the molecules upon the surface: the momentum given to the surface by their blows, per unit of area and per unit of time, measures the pressure in kinetic units.

In the early days of the molecular theory it was supposed that all the atoms of a chemical element, and consequently all the molecules of a chemically homogeneous gas, had the same mass. We now know that this is incorrect. Most gases, including oxygen and hydrogen, are mixtures of isotopes in which the atoms though chemically alike have different weights. But the proportion of the isotopes is constant so long as the gas retains its chemical character-istics, and from the dynamical point of view the behaviour of the gas is the same as if all the molecules had the same mass, namely a mass equal to the average mass per molecule. It will simplify the theory without affecting the general conclusions if we treat the molecules as having the same mass. Call that mass m. Let N be the number of molecules present in unit of volume of the gas in any actual state as to pressure and temperature. Then mN represents the density, namely, the whole mass per unit of volume, and V, the volume per unit of mass, is equal to $1/mN$.

Before proceeding to consider the pressure caused by molecular blows, we shall make the following postulates:

(1) That the molecules are perfectly free except during en-counters, and therefore move in straight lines with uniform velocity, from one encounter to the next;

(2) That the time during which an encounter lasts is negligibly small in comparison with the time during which the molecule is free;

(3) That the dimensions of a molecule are negligibly small in comparison with the free path.

These three postulates are equivalent to assuming that the gas is perfect in the sense of Art. 18. They are not strictly true of any real gas; but we shall assume them to be true in what immediately follows, and shall thereby deduce from the molecular theory a result which corresponds to the ideal formula $PV = RT$.

Suppose the gas to be in equilibrium in a vessel at rest, and let the velocity v of any molecule be resolved into rectangular com-ponents v_x, v_y and v_z, along three fixed axes.

Consider the pressure due to molecular blows upon a containing wall, of area S, forming a plane surface at right angles to the direction of x. The contribution which any molecule makes to the pressure on that wall is due entirely to the component velocity v_x: nothing is contributed by the components v_y or v_z. Any molecule which strikes the wall has the normal component of its velocity reversed by the collision. Hence the momentum due to the blow is $2mv_x$, where v_x is the normal component of the velocity and m is the mass of the molecule.

Consider next how to express the sum of the effects of such blows in a given time. For this purpose we may think of the molecules as divided into groups according to their velocities at any instant. Let n be the number, in unit volume of the gas, whose x-component of velocity, v_x, has the same numerical value or does not differ from it by more than some assigned very small quantity. Since the number of molecules is very great, we may take the number to be the same in one cubic inch (say) as in another. There will of course be very many such groups, each with a different value of v_x. Think, in the first place, only of those in the group n. Half of the whole number of molecules in the group are moving towards S; the other half are moving away from it. At any instant of time there will therefore be within a small distance δx of the surface S, and moving towards it with component velocity v_x, a number of molecules of that group equal to $\frac{1}{2}nS\delta x$. A molecule distant δx from S, and having a component velocity v_x towards S, would reach S in a time $\delta t = \dfrac{\delta x}{v_x}$, provided it did not encounter any other molecule on its way. Hence the number of blows delivered to S by molecules of that group, in the time δt, would (on the same proviso) be equal to the number of such molecules as originally lay within a distance δx, namely the number $\frac{1}{2}nS\delta x$.

Hence also the momentum due to the blows on the area S in the time δt would be equal to $\frac{1}{2}nS\delta x \times 2mv_x$, which becomes, per unit of area and per unit of time,

$$nmv_x \frac{\delta x}{\delta t} = nmv_x{}^2,$$

since

$$v_x = \frac{\delta x}{\delta t}.$$

This is the momentum contributed by one group only. The pressure P is made up of the sum of the quantities of momentum contributed by all the groups; hence

$$P = \Sigma nmv_x{}^2 = m\Sigma nv_x{}^2,$$

or

$$P = mN\overline{v_x{}^2},$$

where N is as before the whole number of molecules per unit of volume, and $\overline{v_x{}^2}$ is the average of $v_x{}^2$ for all the molecules.

Now the velocity v of any molecule is related to its components by the equation

$$v^2 = v_x{}^2 + v_y{}^2 + v_z{}^2.$$

Hence, if we write $\overline{v^2}$ for the average value of v^2 for all the molecules,

$$\overline{v^2} = \overline{v_x{}^2} + \overline{v_y{}^2} + \overline{v_z{}^2} = 3\overline{v_x{}^2},$$

since the motions take place equally in all directions.

The square root of $\overline{v^2}$ is called the "velocity of mean square." It is not the same thing as the average velocity, but is the velocity a molecule would have whose kinetic energy is equal to the average kinetic energy of all the molecules.

The expression for P may therefore be written

$$P = \tfrac{1}{3} m N \overline{v^2}.$$

Further, since mN is the quantity of gas in unit volume, or $1/V$, where V is (as usual) the volume of unit mass, this gives

$$PV = \tfrac{1}{3} \overline{v^2}.$$

In obtaining this result we made (in order to simplify the argument) a proviso that each molecule of a particular group, lying initially within the distance δx of the wall, struck the wall without encountering other molecules on the way. This is not true, but any encounter on the way does not affect the final result in a gas to which the three postulates apply. For in any encounter, some momentum, perpendicular to the wall, is simply transferred to another molecule, and reaches the wall without loss. The molecule which takes it up has to travel the full remainder of the distance in the direction of x, neither more nor less, since the dimensions of the molecules are negligibly small (Postulate 3), and no time is lost in the encounter (Postulate 2). Hence the general result of the encounters is not to alter the amount of momentum which reaches the wall in any given time, and the conclusion remains valid that

$$PV = \tfrac{1}{3} \overline{v^2}.$$

Comparing this with the perfect-gas equation

$$PV = RT$$

we see that $\overline{v^2}$ is proportional to the absolute temperature; and consequently the average kinetic energy which the molecules possess in virtue of their velocity of translation is proportional to the absolute temperature. We shall call their energy of translation E'; they may, in addition, have energy of other kinds.

The energy of translation of the molecules E' is equal to $\tfrac{1}{2} \overline{v^2}$ per unit mass of the gas. Hence by the molecular theory

$$PV = \tfrac{2}{3} E',$$

and the pressure is equal to two-thirds of the energy of translation, per unit volume of the gas.

It may be noted in passing that the molecular theory explains why a gas is heated by compression. Think of the gas as contained in a cylinder, and being compressed by the pushing in of a piston. Then any molecule which strikes the piston recoils with an increased velocity because it has struck a body that is advancing towards it. The component velocity v_x normal to the piston is not simply reversed by the blow, but is increased by an amount $2v'$, where v' is the velocity with which the piston is moving when the molecule strikes it, for the quantity which is reversed is the relative velocity $v_x + v'$. The result is that the motion of the piston in compressing the gas augments the average velocity of the molecules, and consequently increases $\overline{v^2}$, on which the temperature depends.

174. Boyle's, Avogadro's, and Dalton's Laws. These laws follow from the molecular theory, for gases that obey the three postulates. Keeping $\overline{v^2}$ constant, we have the law of Boyle, $PV = \text{constant}$, since $PV = \frac{1}{3}\overline{v^2}$.

If there are two gases at the same pressure, since $P = \frac{1}{3}mN\overline{v^2}$ in each,

$$m_1 N_1 \overline{v_1^2} = m_2 N_2 \overline{v_2^2}.$$

Maxwell has shown that if two gases are at the same temperature, the average kinetic energy of a molecule is the same in both, or

$$m_1 \overline{v_1^2} = m_2 \overline{v_2^2}.$$

Hence if they are at the same pressure and the same temperature

$$N_1 = N_2,$$

that is to say, the number of molecules in unit volume is the same for both, which is Avogadro's Law. It follows that the density, or mass of unit volume, differs in the two gases in the ratio of the masses of their molecules; or, in other words, the density is proportional to the molecular weight.

Again, the molecular theory shows that in a mixture of two or more gases, each of which obeys the three postulates,

$$P = \frac{1}{3}m_1 N_1 \overline{v_1^2} + \frac{1}{3}m_2 N_2 \overline{v_2^2} + \text{etc.}$$

In other words, the partial pressure due to each constituent of the mixture is the same as it would be if the other constituents were not there. This is in agreement with Dalton's Law (Art. 62).

175. Perfect and Imperfect Gases. Thus the molecular theory, for gases which satisfy the three postulates, gives results identical with those we already know as laws of ideal perfect gases.

In a real gas the postulates do not strictly hold. The size of the molecules is not negligible, and in any encounter there is an appreciable time during which the molecules concerned exert forces on one another. There may even be temporary pairing or co-aggregation on the part of some molecules. It is interesting to enquire, in a general way, how these departures from the ideal conditions affect the calculation of the pressure.

For this purpose, consider the simple case in which one of a group of molecules, advancing towards the wall, meets a molecule, initially at rest, to which it passes on the whole of its momentum, and the other molecule then completes the journey and delivers the blow. If there were no loss of time in the encounter, and if the second molecule could be regarded as travelling over exactly the remainder of the distance, the rate at which the wall receives momentum would be exactly the same as if the encounter had not taken place. But if there were loss of time in any encounter, such, for example, as would occur if the two colliding molecules moved together for any appreciable time, with their velocity reduced below that of the molecule which was originally moving, then the rate at which the wall receives momentum would be reduced, with the result of reducing P. On the other hand, if the molecules have a finite size, so that the one which was initially at rest had less distance to travel in completing the journey, the rate at which the impacts succeed one another on the wall would be increased, with the result of increasing P.

This indicates that the pressure in a real gas will differ from the ideal pressure, which is given by the equation $PV = \frac{1}{3}\overline{v^2}$, by two small terms, one positive, depending on the size of the molecules, and one negative, depending on their cohesion. Such, in effect, is the kind of modification which finds expression in characteristic equations like those of Van der Waals, to be discussed later.

176. Calculation of the Velocity of Mean Square. Taking, for any gas that may be treated as sensibly perfect, the equation

$$P = \frac{1}{3}mN\overline{v^2},$$

it is easy to calculate the value of the velocity of mean square $\bar{v}$ when we know the density of the gas at a given pressure. The product mN is the density, and we do not need to know m or N separately to find $\bar{v}$. In oxygen, for example, at 0° C., the density is 0·001429 gramme per c.c., when the pressure is one atmosphere, or $1\cdot0182 \times 10^6$ dynes per sq. cm. (Art. 12). Hence in oxygen at standard temperature and pressure, $\bar{v}$ is $\sqrt{\dfrac{3 \times 1\cdot0182 \times 10^6}{0\cdot001429}}$, equal to 461 metres per second. Similarly in nitrogen it is 493 metres per second, and in hydrogen 1888 metres per second.

177. Average Distance Apart. As we saw in Art. 159, the number N of molecules in a cubic centimetre at $0°$ C. and a pressure of one atmosphere is $2\cdot705 \times 10^{19}$. Hence their average distance apart, which is $\dfrac{1}{\sqrt[3]{N}}$, is one three millionth of a centimetre. In a "vacuum" of one millionth of an atmosphere there are still $2\cdot705 \times 10^{13}$ molecules of residual gas per cubic centimetre and their average distance apart is $0\cdot00033$ millimetre.

178. Internal Energy and Specific Heat. Consider next the bearing of the molecular theory on the internal energy and specific heats of a gas. We have seen that, in an ideal gas,

$$PV = \tfrac{2}{3}E',$$

where E' is the energy of translation of the molecules, or $\tfrac{1}{2}mN\bar{v^2}$. This may be written

$$RT = \tfrac{2}{3}E' \quad \text{or} \quad E' = \tfrac{3}{2}RT.$$

E' is therefore proportional to the temperature. Now E' may or may not be the whole internal energy, E, which the gas acquires when it is heated. It would be the whole if, when the gas was heated, the molecules could only take up energy of translation, and could not also be set rotating or vibrating. Suppose, for instance, that each molecule behaved like a perfectly smooth rigid billiard ball, or like a massive point with no appreciable moment of inertia about any line passing through it. In that case, it could not have any energy of vibration, nor acquire any energy of rotation in the course of its encounters with other molecules, and the only kind of communicable kinetic energy would be energy of translation. We should then find $E = E'$, and consequently $E = \tfrac{3}{2}RT$.

When a gas of this kind is heated, we should therefore have

$$dE = \tfrac{3}{2}RdT.$$

But in any gas (regarded as perfect)

$$dE = C_v dT \quad \text{and} \quad C_p = C_v + R.$$

Hence for a gas whose molecules have energy of translation only

$$C_v = \tfrac{3}{2}R, \quad C_p = \tfrac{5}{2}R,$$

and

$$\gamma \text{ or } \frac{C_p}{C_v} = \frac{5}{3} \text{ or } 1\cdot667.$$

This value of γ would not apply if E' were only a part of E. But it is found that in a monatomic gas, such as argon, or helium, or the vapour of mercury, the value of γ is in fact equal to 1·667 or very near it. The inference is that in a monatomic gas, the structure of the molecule is such that substantially all its communicable energy consists of energy of translation.

In any gas each molecule possesses three degrees of freedom of translation, namely, freedom to move along each of three independent axes. Since $E' = \frac{3}{2}RT$, each degree of freedom of translation accounts for a quantity of kinetic energy equal to $\frac{1}{2}RT$. This is true whatever be the number of atoms in the molecule, and whether or no the molecules have other energy besides energy of translation.

Consider next a diatomic gas, each molecule of which consists of two atoms. According to modern views an atom is a complex system made up of a minute positively charged central nucleus in which the mass of the atom is almost all concentrated, with electrically negative particles called electrons distributed around it, at distances which are large compared with the dimensions of the nucleus*. The structure of the atom and the nature of the forces cannot be fully represented by any mechanical model, but for our present purpose it will suffice to picture an atom as a massive point, surrounded by a massless quasi-elastic fender due to forces which keep other atoms at a distance. Under normal conditions a diatomic molecule is equivalent, as regards inertia, to two masses held some distance apart: dynamically it may be compared to a dumb-bell; a more exact comparison would be to a light stick capable of some elastic extension and carrying a heavy ball at each end. Considered as a rigid body it has five effective degrees of freedom—effective as regards the storing and communication of kinetic energy—namely, three of translation and two of rotation†. The two effective degrees of freedom of rotation are about axes in a plane perpendicular to the line joining the two atoms: about that line itself, the system has no effective moment of inertia. Under these conditions it can

* In an electrically neutral atom the positive electricity in the nucleus is equal to the negative electricity in the electrons. Removal of one or more of the electrons would therefore leave the atom as a whole positively charged: this happens when a gas is "ionized."

† A free rigid body has six degrees of freedom: it can move parallel to itself along three independent axes, and it can rotate about these axes. Any possible movement is made up of these six components. In a diatomic molecule one of the degrees of freedom of rotation is ineffective, for it does not appear that energy of rotation about the line joining the two atoms is communicated in any encounter.

be shown that the ultimate result of collisions is that the kinetic energy becomes equally shared by each of the five degrees of freedom. The energy of translation E' is equal to $\frac{3}{2}RT$, and each degree of freedom of translation accounts for an amount of energy equal to $\frac{1}{2}RT$. It follows that each of the two degrees of freedom of rotation accounts in addition for $\frac{1}{2}RT$, and that the energy of translation and rotation together amounts to $\frac{5}{2}RT$. Hence if there were no sensible energy of vibration as well, we should have the whole energy $E = \frac{5}{2}RT$ and

$$C_v = \tfrac{5}{2}R, \quad C_p = \tfrac{7}{2}R, \quad \text{and} \quad \gamma = \tfrac{7}{5} \text{ or } 1\cdot4.$$

Now in most diatomic gases, such as oxygen (O_2), nitrogen (N_2), air, hydrogen (H_2), nitric oxide (NO), or carbon monoxide (CO), it is in fact found that γ is equal, very nearly, to $1\cdot4$ at ordinary temperatures, and the inference is that the structure of their molecules is such as to give five effective degrees of freedom, namely the five that have just been described, and that their molecules do not, at ordinary temperatures, hold any considerable amount of communicable energy in any other form than as energy of translation and energy of rotation. But when such gases are strongly heated we know that the specific heat increases and γ is reduced. This means that energy of vibration is then developed, which at high temperatures becomes an important part of the whole energy.

In triatomic gases it may be conjectured that the three atoms of any molecule group themselves not in one straight line—which would be an unstable arrangement—but so that the massive centres lie at the corners of a triangle. Similarly when there are more than three atoms in the molecule, they will place themselves with their massive centres at the corners of a polyhedron. In any such triangular or polyhedral structure, considered as a rigid system, there are six effective degrees of freedom, namely three of rotation as well as three of translation, for there is a finite moment of inertia about any axis, and the structure is such that the molecule can be set spinning about any axis by encounters with other molecules. As an ultimate result of many such encounters, it may be shown that each of the three degrees of freedom of rotation takes up a share of the kinetic energy equal to that of each of the three degrees of freedom of translation, namely, $\frac{1}{2}RT'$, and consequently that the six degrees together account for a total of $3RT$. That is the energy which the molecules possess in virtue of their movements as rigid structures. If there were no other way in which they could take up energy when the gas is heated, we should consequently find, in a triatomic or polyatomic gas,

$$C_v = 3R, \ C_p = 4R, \ \text{and} \ \gamma = \tfrac{4}{3} \ \text{or} \ 1 \cdot 333.$$

The actual value of γ, as experimentally measured, in the triatomic gases CO_2 and H_2O, is rather less than this, and in gases of more complex constitution it is generally a good deal less. It is also found that the specific heats are greater than $3R$ and $4R$. The inference is that in such gases even at moderate temperatures, the molecule takes up some energy of vibration in addition to its energy of translation and rotation. It appears that a complex molecule can absorb energy not only by moving as a rigid body but by internal vibrations which come about through elastic deformation of its own structure. The energy so absorbed is half potential and half kinetic. In any vibrating body the energy of the vibration undergoes a periodic alternation between the potential and kinetic forms. Hence when a large group of bodies is vibrating independently the whole vibratory energy at any instant will be equally divided between the two.

The main part of this energy of vibration probably results from to and fro movements on the part of the massive centres of the linked atoms. It is obvious that such a motion might occur in any molecule that is made up of more than one atom. The effect in a complex molecule is such as would occur if the lines joining the massive centres of the constituent atoms behaved like stiff springs. Thus in a diatomic molecule we might think of the "dumb-bell" as having an elastic shank which allowed the distance between the two masses to vary. The fact that in a diatomic gas at ordinary temperatures the observed specific heats are approximately $\tfrac{5}{2}R$ and $\tfrac{7}{5}R$, and γ is approximately $1 \cdot 4$, shows, however, that the diatomic molecule then behaves like a dumb-bell with a nearly inextensible shank. But when the temperature is high, the vibratory motion becomes relatively more important, and it accounts for an appreciable part of the whole energy, even in a diatomic molecule, and still more in a triatomic or polyatomic molecule. To this we must ascribe the progressive increase in specific heat, and the fall in γ, which are observed when any gas is heated that has two or more atoms in the molecule.

In a monatomic gas there is no possibility of this kind of vibratory motion, and there is no experimental evidence of any change of specific heat with temperature. The energy depends only on motion of translation, and when the gas is heated its energy increases in simple proportion to the temperature. But when diatomic, tri-atomic, or polyatomic gases are strongly heated, the energy increases in a more rapid ratio than the temperature. This means that the ratio of the total energy E to the energy of translation E' is not constant.

In any gas that satisfies the equation $PV = RT$,

$$\gamma = 1 + \frac{R}{C_v}.$$

If the total energy E preserved a constant ratio to E', the specific heat would be constant, and in that case we should have γ constant and equal to $1 + \frac{2}{3}E/E'$, since E, reckoned from the absolute zero, is C_vT, and E' is $\frac{3}{2}RT$. The fact, however, that γ falls with rising temperature shows that the total energy does not preserve a constant ratio to the energy of translation, and hence that there is not equipartition of the energy among the possible modes of motion.

In any gas we may write

$$E = E' + E'' + E'''.$$

The energy of translation E' varies as T, being equal to $\frac{3}{2}RT$. The energy of rotation E'' bears, in any given type of molecule, a constant ratio to E', and therefore also varies as T. If the energy of vibration E''' also bore a constant ratio to E', the whole energy would vary as T, which is inconsistent with the results of experiment.

179. Energy of Vibration. The term E''' includes not only energy due to vibrations of the constituent atoms relatively to one another within the molecule (E_m'''), but energy due to vibrations (movements of electrons) within the constituent atoms themselves (E_a'''). It is known that E_a''' is a very small part of the whole energy, even at temperatures as high as 2000° C. The vibrations that make up E_a''' have much higher frequencies than those that make up E_m'''. It is to vibrations within the constituent atoms that one attributes the bright lines which make up the visible spectrum of an incandescent gas, and the corresponding dark lines due to absorption in the visible spectrum of light transmitted through a cold gas. The longer-period vibrations that make up E_m''' emit or absorb rays which lie in the infra-red region, beyond the range of the visible spectrum. It is these longer-period vibrations that constitute the main part of the vibratory energy when a gas is strongly heated, as in a flame or an explosion, and give rise to most of its radiant energy.

From the theory that has been outlined above, of the constitution of a diatomic molecule, we should expect it to have one well-marked period of free vibration, and therefore to show a strong emission band when heated, or when excited by electric discharge in a vacuum tube, and also a corresponding strong absorption band when cold. A good example is furnished by carbon monoxide (CO), whose infra-red spectrum is found to consist almost entirely of one characteristic band, the wave-length of which is about $4\cdot7\mu$.

Again, in the infra-red spectrum of the triatomic gas CO_2 we should expect to find three prominent bands corresponding to the three modes of vibration that can be set up within a CO_2 molecule by relative movements of the carbon and oxygen atoms. This is in agreement with what is observed. There are, both in absorption and emission, three distinct infra-red bands, namely a strong band whose wave-length is about $4\cdot4\mu$, a weak band with a wave-length of $2\cdot7\mu$, and another with a much longer wave-length, between 14μ and 15μ (Art. 171). This long-period vibration accounts for the fact that even at ordinary temperatures the specific heat of CO_2 exceeds the value it would have if there were no vibratory energy, making γ distinctly less than $1\cdot333$. For the principle holds that vibrations of long period require no more than a comparatively low temperature to excite them into taking up some considerable share of the energy, so that they then contribute substantially to the specific heat, whereas those of short period do not begin to take up an appreciable share until the gas is strongly heated.

180. Departure from the Principle of Equipartition. It follows that the vibratory energy of the molecules is not governed by the principle of equipartition which applies under normal conditions to the energies of translation and rotation. Equipartition would require that each degree of freedom should account for the same quantity of energy, after a state of equilibrium has been reached as a result of many encounters. In considering the translations and rotations of the molecules of an ideal gas, apart from any effects of vibration, we inferred from the kinetic theory that a quantity of energy equal to $\frac{1}{2}RT$ is taken up in respect of each of the three freedoms of translation and in respect of each effective freedom of rotation. Hence in a polyatomic molecule, when all the six degrees of freedom are effective the whole energy due to translations and rotations amounts to $3RT$ and is equally distributed among the six degrees. So far as these items of the total energy are concerned there is equipartition. But the case is different when we attempt to include items that are due to vibration. The molecule may be capable of vibrating in various modes, each contributing an item of energy to the total. But the energy that is due to any particular mode of vibration bears no constant ratio to the other items. At low temperatures it is insignificantly small; at high temperatures it may become a substantial part of the whole. As the temperature rises each mode of vibration in turn becomes important, beginning with that mode which has the longest period. When the vibration in any particular mode is fully excited its energy tends, in the limit, to become equal to RT, which may be regarded as the sum of two terms, namely $\frac{1}{2}RT$ for the kinetic energy and $\frac{1}{2}RT$ for the potential energy of vibrations in that particular mode.

Thus at very high temperatures each mode of vibration natural to the molecule approaches a state in which the energy it absorbs is equivalent to the quota of two degrees of freedom—a fact which abundantly accounts for the rise in specific heat that is observed when the gas is strongly heated.

It will be convenient if at this stage we adopt the notation (given in Art. 161) where k represents the average value of R *per molecule*. Then $\frac{1}{2}kT$ is the average energy that each molecule acquires in respect of each degree of freedom of translation or effective rotation, under the principle of equipartition, and kT is the limiting amount of energy of vibration that would on the average be acquired by each molecule for any assigned mode of vibration if that mode were fully excited. What actually happens when a gas is heated is that only a fraction of the limiting quantity kT is excited, the fraction depending on the temperature reached and the consequent intensity of vibration that is developed.

181. **The Quantum Theory. Planck's Formula.** The step-by-step development of vibrational energy in gas molecules, with rising temperature, is one of many physical facts which have been clarified if not fully rationalized by the Quantum Theory of Max Planck. That theory modifies and supplements the older conceptions of "classical" mechanics. It supplies rules which although they conflict with those conceptions are accepted as valid because they yield the results established by experiment. We are concerned only with

one aspect of the Quantum Theory here, namely its bearing on the specific heats of gases.

According to Planck's theory, when an atom or a molecule can vibrate with a frequency ν it will take up or give up vibrational energy only in amounts equal to $h\nu$ or multiples of $h\nu$, where h is a universal constant of nature. The energy passes, so to speak, only in packets: the whole of the packet must be taken, or none. But the amount in each packet, being equal to $h\nu$, depends on the frequency: it is relatively large when the frequency is high. The constant h, which is the basis of the theory, is called Planck's Constant of Action. Its physical dimensions are $\dfrac{\text{mass} \times (\text{length})^2}{\text{time}}$, which are also those of angular momentum, or of the product of work into time. It enters into so many physical actions that its numerical value has been found by several methods: they agree in making it about $6\cdot55 \times 10^{-27}$ erg-second.

As a result of the Quantum Theory the average vibratory energy of a gas molecule, corresponding to any particular frequency ν, depends on the statistical probability that some of the encounters are sufficiently violent to communicate the quantum of energy $h\nu$. That probability becomes rapidly greater at a certain stage in the heating process when the temperature of the gas is raised. This consideration leads to the formula

$$E_\nu = \frac{h\nu}{e^{h\nu/kT} - 1},$$

where E_ν represents the average vibratory energy in ergs per molecule, in respect of vibration with the frequency ν; h is Planck's constant, and k has the value already stated of $1\cdot372 \times 10^{-16}$ erg. To find the corresponding quantity per mol we should multiply by Avogadro's constant N.

The frequency ν is often expressed, with reference to the wavelength of light, as c/λ, where λ is the wave-length in centimetres and c is the velocity of light or 3×10^{10} centimetres per second.

In a gas whose molecules are capable of more than one mode of vibration the whole vibrational energy would be the sum of as many terms, in the above form, as are required to express the various modes. Thus in carbon dioxide, for example, there would be three terms for frequencies of vibration corresponding to the three main infra-red bands.

At any one frequency ν let the quantity $h\nu/kT$ be represented by x. Then Planck's formula becomes

$$E_\nu = \frac{x}{e^x - 1}\, kT.$$

The factor $\dfrac{x}{e^x - 1}$ depends on both ν and T. For any given ν it tends to an upper limit of 1 when T is indefinitely increased and to a lower limit of zero when $T = 0$. It follows that when the molecules of a gas are free to vibrate in a particular mode they will take up, in respect of that freedom, an average quantity of energy which approaches the limit kT per molecule when the gas is strongly heated. This will also be true of any other mode of free vibration which the molecules possess. When the gas is heated to a given temperature the fraction of kT which is taken up will be different for different modes, for it is smaller when the frequency is high. This is why the high-frequency modes of vibration which are revealed by the visible spectrum, and those that are known to occur beyond it, do not contribute substantially to the whole energy of a gas, even at temperatures such as are reached in an ordinary flame or in a gas-engine explosion, and why, in the reckoning of energy and of specific heat it is only vibrations of infra-red frequency that need be taken into account. For the same reason a gas whose molecules have one or more long-period types of vibration may, at ordinary temperatures, hold a considerable quantity of energy in the vibratory form, and have a specific heat markedly greater than the ideal (vibrationless) value.

The amount by which any one mode of vibration will augment the specific heat is found by differentiating (with respect to T) the expression for the extra internal energy that is due to that mode. We may write it

$$(C_v)_\nu = \frac{dE_\nu}{dT} = \frac{e^x x^2}{(e^x - 1)^2}\, k \text{ per molecule, or } \frac{e^x x^2}{(e^x - 1)^2}\, R \text{ per mol.}$$

Here $\dfrac{e^x x^2}{(e^x - 1)^2}$ is a factor, depending on the wave-length and the temperature, which ranges from zero to unity as the quantity $1/x$ is increased from zero to infinity*. Fig. 83 exhibits the manner in which this factor increases relatively to $1/x$. It shows that there is a very rapid rise in the factor, and therefore in the specific heat,

* The quantity $1/x$ is proportional to the temperature, being equal to $kT/h\nu$, or $0\cdot698\lambda T$ when we substitute c/λ for ν and assign the stated numerical values to k, h, and c.

after $1/x$ has reached a value of about 0·1, but up to that point the effect of vibration on the specific heat is quite insignificant. At 0° C. the value of λ which corresponds to $1/x = 0·1$ is 0·00052; hence it is only those modes of vibration whose wave-lengths are greater than say 5μ that sensibly affect the specific heat of a gas at normal temperature.

As an example, take the diatomic gas CO with its characteristic vibration for which λ is about $4·7\mu$ or 0·00047 centimetre. For that wave-length the value of $1/x$, at 0° C., is 0·09; and at 2000° C., it is 0·74. The factor $e^x x^2/(e^x - 1)^2$ is therefore insignificantly small at 0° C., but becomes about 0·86 at 2000° C. Hence the calculated

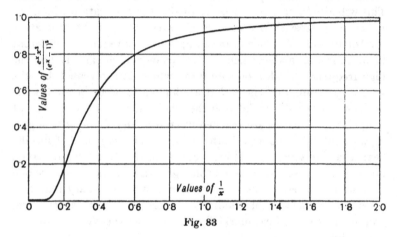

Fig. 83

specific heat C_v, which is $\frac{5}{2}R$ at 0° C., rises, as a consequence of this vibrational energy, to $(\frac{5}{2} + 0·86) R$ at 2000° C.; and the corresponding value of γ falls from 1·4 to barely 1·3.

Again, take the triatomic gas CO_2, one of whose characteristic vibrations has a wave-length of nearly 15μ. So slow a vibration contributes substantially to the specific heat even when the gas is cold. At 0° C. a wave-length of 15μ makes $1/x = 0·24$ and

$$e^x x^2/(e^x - 1)^2 = 0·28.$$

Hence a single mode of vibration with that frequency should bring the specific heat of the cold gas up to about $(3 + 0·28) R$, and reduce γ from 1·333 to 1·305. When the gas is strongly heated, account has to be taken of three modes of vibration whose wave-lengths are long enough to be important. In respect of the three together, C_v obviously tends, at very high temperatures, to increase towards a limit of $6R$, and γ to fall to $\frac{7}{6}$, apart from anything that other vibrations may contribute, and apart from effects of dissociation.

The fact that in polyatomic gases generally the specific heats, at normal temperature, are greater than the ideal (vibrationless) values, and γ is notably less than 1·333, is to be ascribed to their possessing long-period modes of vibration which are responsive to low-temperature encounters. A complex polyatomic molecule may have many such modes, each producing a substantial augmentation of the specific heat.

Similarly the characteristic mode of vibration in a diatomic gas may be so slow as to affect the specific heat at normal or comparatively low temperature, making C_v greater than $\frac{5}{2}R$, and C_p greater than $\frac{7}{2}R$, and γ less than 1·4. This is notably the case with the vapours of the halogen elements Cl_2, Br_2, I_2. These elements have high atomic weights, and it would seem that in each of them the pair of heavy atoms in the molecule, perhaps rather loosely held together, have a slow type of vibration, which explains the observed high specific heats and low value of the ratio γ. When a hydrogen atom is substituted for one of the pair, this characteristic disappears, for the gases HCl, HBr and HI, when cold, are found to have specific heats that approximate to the normal values, with the ratio 1·4.

182. Molecular Rotations: Limits imposed by the Quantum Theory. When the rules of the quantum theory are applied to the dynamics of rotation it appears that the least amount of rotational energy that can be communicated is $\frac{1}{2}h\nu$, where ν is the frequency with which the body receiving the energy rotates. This imposes a lower limit on the frequency of rotation which a gas molecule may acquire in the course of its encounters. To determine the limit in question, let us assume a molecule to be rotating with frequency ν and that its moment of inertia about the axis of rotation is I. Then the angular velocity is $2\pi\nu$ and the energy of the rotation is $\frac{1}{2}I(2\pi\nu)^2$. Under quantum rules the lowest permissible value of this quantity will correspond to $\frac{1}{2}h\nu$.

Hence to find the least possible frequency we have

$$\tfrac{1}{2}I(2\pi\nu)^2 = \tfrac{1}{2}h\nu,$$

from which $$\nu = \frac{h}{4\pi^2 I}.$$

The physical meaning of this is that unless an encounter is violent enough to produce a frequency of rotation equal to $h/4\pi^2 I$ it will not communicate any rotational energy at all. The value of ν so found is a limit below which the molecules cannot rotate: they may however rotate, not only at this frequency, but at a series of higher frequencies which are multiples of it.

This curious deduction from the quantum theory finds a remarkable confirmation in the observed behaviour of hydrogen and other light diatomic gases when they are cooled to a very low temperature. The violence of the encounters is then so much reduced that the molecules cease to take up any rotational energy and the only energy that is transferred is energy of translation. At ordinary temperatures these gases show a specific heat C_v of $\frac{5}{2}R$ which, as we have seen, implies five effective degrees of freedom, namely three translations and two rotations. But it has been found experimentally that when hydrogen is cooled to 70° K. or lower its specific heat falls to a value not much greater than that of a monatomic gas and γ approaches the corresponding value of 1·667. This remarkable result, first observed for C_v in hydrogen by Eucken, has been confirmed by independent measurements of C_p and of γ, and has been extended to nitrogen and other diatomic gases*.

At first these experiments seemed to suggest that the structure of the molecule was modified by extreme cold, but it is now recognized that they are to be interpreted as examples of the limits imposed by quantum rules. At very low temperatures the encounters are insufficiently violent to cause molecular rotation because they fail to communicate the limiting value of ν stated above. Consequently the two freedoms of rotation then become ineffective and the energy depends only on motions of translation. When ordinary conditions of temperature are restored the violence of the encounters is increased; more and more of them become capable of communicating rotational energy, and the normal condition is recovered under which all five degrees of freedom share alike. Numerical estimates of the moment of inertia of hydrogen and other light diatomic gases are given by Jeans† which show that at ordinary temperatures, say from 0° C. upwards, the slowest rotation permitted by the quantum theory has an energy that is small compared with the average energy of translation and consequently we may expect the two freedoms of rotation to be effective.

* Eucken, *Sitzungsberichte d. k. Preuss. Akad.* Berlin, Feb. 1912; Scheel and Heuse, *ibid.* Jan. 1913, also *Ann. Physik*, vol. XL, p. 473, 1913; M. C. Shields, *Phys. Review*, Nov. 1917; Schreiner, *Zeits. Phys. Chem.* vol. CXII, p. 1, 1924.

† Jeans, *Dynamical Theory of Gases*, 4th edition, 1925, p. 393.

On the other hand, when the temperature approaches the absolute zero the average energy of translation becomes so much less that rotations with even one quantum of energy must be very rare*.

The same quantum considerations obviously account for the fact that the single atoms of a monatomic gas do not acquire rotation. Single atoms have a finite but very small moment of inertia about any central axis. Hence the least frequency of rotation permitted by the quantum theory, namely $h/4\pi^2 I$, is very large, and its energy, which is $\frac{1}{2}h\nu$, is so considerable that the encounters fail to supply it even when the gas is warm.

Again, the quantum rules explain why a diatomic molecule acquires no spin about the axis joining the two atoms. Here also, though I is finite it is very small, the least permissible frequency is very large and the encounters are unable to provide the quantum of energy it would acquire. We are therefore justified in treating rotation about that axis as ineffective and negligible in the reckoning of specific heats.

* It was discovered also from a study of the shape of the C_p/T curve for hydrogen that there were two forms, called ortho and para, which differ in their nuclear rotations and so have very different specific heats.

CHAPTER VIII

GENERAL THERMODYNAMIC RELATIONS

183. Introduction. In the earlier chapters but little use was made of formal mathematics in introducing the reader to the fundamental ideas of thermodynamics. To most students there is an advantage in having these ideas so presented: their physical significance is more likely to be appreciated. Once that is grasped, the student may proceed to a more mathematical treatment with less risk that the real meaning of the symbols will be obscured in the analysis. But a mathematical treatment must be resorted to if we wish to express with anything like completeness the relations that hold between the various properties of a fluid.

One of the uses to which these relations can be put is in framing tables or charts of the properties of the fluid. By their aid such tables can be compiled from a small number of experimental data, and the experimental data themselves, as well as the numbers computed from them, can be tested for thermodynamic consistency.

In this chapter it will be shown how the methods of the differential calculus may be applied to obtain, by inference from the First and Second Laws of Thermodynamics, certain general relations between the properties of any fluid. With some of these results the reader of the earlier chapters is already acquainted.

In the next chapter some applications of these general relations to particular substances will be considered, including imperfect gases, or real fluids in the state of vapour, along with "characteristic equations" which have been devised to connect the properties of fluids in the gaseous and liquid states.

184. Functions of the State of a Fluid. Assume that we are dealing with unit mass of a homogeneous fluid. As was pointed out in Art. 75, the six quantities named there, T, P, V, E, I, and ϕ, are all functions of the state of the fluid, that is to say their values depend only on the actual state. When the fluid passes in any manner from one state to another, each of these quantities changes by a definite amount which does not depend on the nature of the operation by which the change is effected, but only on what the state was before and what it is after the operation has taken place.

This fact is expressed in mathematical language by saying that the differential of any of these quantities is a "perfect" differential. Other quantities might be added to the list, which are also functions of the state of the fluid, such as the quantities G (or ζ, which is $-G$) and ψ mentioned in Art. 90.

In what follows it is to be understood that T means (as usual) the absolute temperature on the thermodynamic scale (Art. 42).

We defined the entropy ϕ in Art. 44 by the equation $d\phi = \dfrac{uQ}{T}$ in a reversible operation; and the fact that ϕ is a function of the state was proved there as a consequence of the result that $\displaystyle\int \dfrac{dQ}{T} = 0$ for a reversible cycle, a result which follows from the Second Law of Thermodynamics. The Second Law is therefore involved in treating ϕ as a function of the state. Hence the fact that $d\phi$ is a perfect differential is sometimes spoken of as a mathematical expression of the Second Law. It is important to notice that while $\dfrac{dQ}{T}$, which is $a\phi$, is a perfect differential, dQ itself is not a perfect differential, for the amount of heat involved in a change is not a function of the state alone. When a substance changes from one state to another, the amount of heat taken in depends not simply on what the two states are, but also on the nature of the operation by which the change occurs. For the same reason, if W represents the work done during a change of state, dW is not a perfect differential.

Since E, P, and V are all functions of the state, it follows that the total heat I, which is equal to $E + PV$, is also a function of the state. And since T and ϕ are also functions of the state, it follows that this is also true of ζ, which is $I - T\phi$, and of ψ, which is $E - T\phi$. Hence dI, $d\zeta$ and $d\psi$, as well as $d\phi$, dE, dP, dV and dT, are perfect differentials.

185. Relation of any one Function of the State to two others. The state of the fluid (assumed to be homogeneous) is completely specified when any two of the functions of the state are known. Any third function is then determinate; that is to say, it can have only one value in any particular substance. Thus if any two functions (such for example as the pressure and the volume) be selected as "independent variables," by reference to which the state is to be specified, then any third function (such for example as the temperature, or the total heat) may be represented in relation to them by the familiar device of drawing a figure in which the two functions selected as independent variables are represented by rectangular coordinates X and Y, and the third function is represented by a third coordinate Z, perpendicular to the plane of X and Y. This gives a solid figure, the height of which shows, for any given state of the substance, the value of the function Z in relation to the values of the functions X and Y which serve to specify that state. The surface of such a figure may be called a thermodynamic surface.

Suppose now that the substance undergoes an infinitesimal change of state, so that the independent variables change by dX and dY respectively. That is to say, we suppose X to change to $X + dX$ and Y to change to $Y + dY$. Then the third function changes from Z to $Z + dZ$, by an amount dZ which may be expressed thus:

$$dZ = MdX + NdY \quad \dots\dots\dots\dots\dots\dots(1),$$

where M and N are quantities depending on the relations of the functions to one another, and are therefore also functions of the state.

This expression applies whether both functions X and Y vary, or only one of them. If X varies but not Y, then $dY = 0$ and $dZ = MdX$: similarly if Y varies but not X, $dX = 0$ and $dZ = NdY$. Hence

$$M = \left(\frac{dZ}{dX}\right)_Y \text{ and } N = \left(\frac{dZ}{dY}\right)_X \quad \dots\dots\dots\dots(2).$$

In this notation, $\left(\frac{dZ}{dX}\right)_Y$ means the rate of variation of Z with respect to X when Y is constant. In the language of the calculus, $\left(\frac{dZ}{dX}\right)_Y$ is the partial differential coefficient of Z with respect to X when Y is constant, and $\left(\frac{dZ}{dY}\right)_X$ is the partial differential coefficient of Z with respect to Y when X is constant.

We might regard the change of Z as occurring in two steps. In the first step suppose X to change and Y to keep constant. The corresponding part of the change of Z is MdX, and M is the slope of the thermodynamic surface in a section-plane ZX. In the second step X is constant and Y changes. The corresponding part of the change of Z is NdY, and N is the slope of the thermodynamic surface in a section-plane ZY. The whole change of Z is the sum of these two parts, as expressed in equation (1). The slopes along the two section-planes are expressed in equation (2).

Combining these equations we have

$$dZ = \left(\frac{dZ}{dX}\right)_Y dX + \left(\frac{dZ}{dY}\right)_X dY \dots\dots\dots\dots(3).$$

These equations apply when X, Y, and Z are interpreted as any three functions of the state of a fluid. Thus, for instance, if we think of a small change of state in which the temperature changes from T to $T + dT$, and the pressure from P to $P + dP$, the consequent change of volume will be

$$dV = \left(\frac{dV}{dT}\right)_P dT + \left(\frac{dV}{dP}\right)_T dP.$$

Similarly, if the volume and pressure change, the consequent change of temperature is

$$dT = \left(\frac{dT}{dV}\right)_P dV + \left(\frac{dT}{dP}\right)_V dP.$$

Or again, the change of entropy, consequent on a change of temperature and pressure, is

$$d\phi = \left(\frac{d\phi}{dT}\right)_P dT + \left(\frac{d\phi}{dP}\right)_T dP,$$

and so on. It will be obvious that a very large number of similar equations might be written out, each using one pair of functions of the state as independent variables, and expressing in terms of their variation the variation of some third function of the state. These are merely forms of the general equation (3).

Returning now to the general form in X, Y, and Z, suppose a small change of state to occur of such a character that the function Z undergoes no change. In that special case $dZ = 0$; the steps MdX and NdY cancel one another. Consequently

$$\left(\frac{dZ}{dX}\right)_Y dX = -\left(\frac{dZ}{dY}\right)_X dY,$$

when dX and dY are so related that there is no variation of Z.

Hence the general conclusion follows that

$$\left(\frac{dZ}{dX}\right)_Y = -\left(\frac{dZ}{dY}\right)_X \left(\frac{dY}{dX}\right)_Z \dots\dots\dots\dots(4).$$

This relation between the three partial differential coefficients holds, in all circumstances, for any three functions of the state of any fluid. It may be expressed in the form

$$\left(\frac{dX}{dY}\right)_Z \left(\frac{dY}{dZ}\right)_X \left(\frac{dZ}{dX}\right)_Y = -1 \dots\dots\dots\dots(4a).$$

Further, since X, Y, and Z are all functions of the state the differential of each is a perfect differential and each may be expressed as a function of the other two. We may write $Z = F(XY)$, where F means some function of X and Y. Hence the perfect differential

$$dZ = \left(\frac{dF}{dX}\right)_Y dX + \left(\frac{dF}{dY}\right)_X dY.$$

Thus in equation (1)

$$M = \left(\frac{dF}{dX}\right)_Y \text{ and } N = \left(\frac{dF}{dY}\right)_X,$$

and

$$\left(\frac{dM}{dY}\right)_X = \frac{d^2F}{dYdX} = \left(\frac{dN}{dX}\right)_Y \quad \dots\dots\dots\dots(5).$$

In other words, $\left(\frac{dM}{dY}\right)_X$ and $\left(\frac{dN}{dX}\right)_Y$ are equal when dZ is a perfect differential, a condition which is satisfied in dealing with functions of the state of a fluid.

186. Energy Equations and Relations deduced from them. Consider now the heat taken in when a small change of state occurs in any fluid. Calling the heat dQ we have, by the First Law,

$$dQ = dE + dW \quad \dots\dots\dots\dots\dots\dots(6),$$

where dE is the gain of internal energy and dW is the work which the fluid does through increase of its volume. Since $dW = PdV$ the equation may be written

$$dE = dQ - PdV \quad \dots\dots\dots\dots\dots(7).$$

Here and in what follows we shall assume that quantities of heat are expressed in work units. This simplifies the equations by allowing the factor J or A to be omitted.

We are concerned for the present only with reversible operations. In any such operation $dQ = Td\phi$; hence

$$dE = Td\phi - PdV \quad \dots\dots\dots\dots\dots(8).$$

Again, $I = E + PV$, by definition of I.

Hence $dI = dE + d(PV)$

$$= Td\phi - PdV + PdV + VdP$$

$$= Td\phi + VdP \quad \dots\dots\dots\dots\dots(9).$$

It will be noted that when a fluid takes in heat dQ at constant volume, so that no work is done, $dE = dQ$, and when it takes in heat at constant pressure $dI = dQ$. When it expands adiabatically $dQ = 0$

and $d\phi = 0$; hence $dI = VdP$, and the work done in a finite process, namely $\int_b^a VdP$, is equal to the heat-drop $I_a - I_b$ (compare Art. 80).

Again, $\qquad \zeta = I - T\phi$, by definition of ζ*.

Hence $\qquad d\zeta = dI - d(T\phi)$

$$= Td\phi + VdP - (Td\phi + \phi dT)$$

$$= VdP - \phi dT \quad\dotfill(10).$$

Again, $\qquad \psi = E - T\phi$, by definition of ψ.

Hence $\qquad d\psi = dE - d(T\phi)$

$$= Td\phi - PdV - (Td\phi + \phi dT)$$

$$= -PdV - \phi dT \quad\dotfill(11).$$

But dE, dI, $d\zeta$, and $d\psi$ are all perfect differentials. Hence, applying equation (5) in turn to equations (8), (9), (10), and (11) we obtain at once the following four relations between partial differential coefficients:

From (8), $\qquad \left(\dfrac{dT}{dV}\right)_\phi = -\left(\dfrac{dP}{d\phi}\right)_V \dotfill(12).$

From (9), $\qquad \left(\dfrac{dT}{dP}\right)_\phi = \left(\dfrac{dV}{d\phi}\right)_P \dotfill(13).$

From (10), $\qquad \left(\dfrac{dV}{dT}\right)_P = -\left(\dfrac{d\phi}{dP}\right)_T \dotfill(14).$

From (11), $\qquad \left(\dfrac{dP}{dT}\right)_V = \left(\dfrac{d\phi}{dV}\right)_T \dotfill(15).$

These are known as Maxwell's four thermodynamic relations.

The following further relations are immediately deducible from equations (8) to (11). Taking equation (8), imagine the fluid to be heated at constant volume. Then $dV = 0$ and $dE = Td\phi$; hence

$$\left(\frac{dE}{d\phi}\right)_V = T.$$

Again, imagine the fluid to expand adiabatically. Then $d\phi = 0$ and $dE = -PdV$; hence

$$\left(\frac{dE}{dV}\right)_\phi = -P.$$

---

* For the sake of symmetry ζ, which is $-G$, is used here rather than G. The Committee of the International Union of Physics has advised that the function ζ should be represented by G, where $G = H - TS$; which is $-G$ as used in this book.

Similarly from equation (9) we obtain

$$\left(\frac{dI}{d\phi}\right)_P = T, \text{ and } \left(\frac{dI}{dP}\right)_\phi = V;$$

from equation (10) $\quad \left(\frac{d\zeta}{dP}\right)_T = V, \text{ and } \left(\frac{d\zeta}{dT}\right)_P = -\phi;$

from equation (11) $\quad \left(\frac{d\psi}{dV}\right)_T = -P, \text{ and } \left(\frac{d\psi}{dT}\right)_V = -\phi.$

Collecting these results,

$$\left(\frac{dI}{d\phi}\right)_P = T = \left(\frac{dE}{d\phi}\right)_V \quad\ldots\ldots\ldots\ldots\ldots(16),$$

$$\left(\frac{dI}{dP}\right)_\phi = V = \left(\frac{d\zeta}{dP}\right)_T \quad\ldots\ldots\ldots\ldots\ldots(17),$$

$$\left(\frac{d\psi}{dV}\right)_T = -P = \left(\frac{dE}{dV}\right)_\phi \quad\ldots\ldots\ldots\ldots\ldots(18),$$

$$\left(\frac{d\psi}{dT}\right)_V = -\phi = \left(\frac{d\zeta}{dT}\right)_P \quad\ldots\ldots\ldots\ldots\ldots(19).$$

187. Expressions for the Specific Heats C_p and C_v. In general the specific heats of a fluid are not constant; they are functions of the state of the fluid. We shall proceed to find differential expressions connecting them with the temperature, volume and pressure. Such expressions enable other properties to be calculated when the relation between T, V, and P is known.

Consider, as before, a small change of state during which the fluid takes in an amount of heat dQ while it expands in a reversible manner. Its entropy accordingly increases by an amount $d\phi$ such that $Td\phi = dQ$. Its temperature changes from T to $T + dT$ and its volume from V to $V + dV$. Take, in the first place, the temperature and volume as the two independent variables by means of which the state of the fluid is specified. The change in any third quantity may be stated with reference to the changes in T and in V. Thus the heat taken in may be written

$$dQ = C_v dT + l dV \quad\ldots\ldots\ldots\ldots\ldots(20).$$

Here C_v, which is the specific heat at constant volume, is $\left(\frac{dQ}{dT}\right)_V$ and l is a symbol for $\left(\frac{dQ}{dV}\right)_T$.

Since
$$dQ = T d\phi, \quad l = T\left(\frac{d\phi}{dV}\right)_T.$$

But by equation (15)
$$\left(\frac{d\phi}{dV}\right)_T = \left(\frac{dP}{dT}\right)_V.$$

Hence
$$l = T\left(\frac{dP}{dT}\right)_V \quad\dots\dots\dots\dots\dots\dots\dots(21),$$

and
$$dQ = C_v dT + T\left(\frac{dP}{dT}\right)_V dV.$$

Dividing both sides by T, we have

$$d\phi = \frac{C_v}{T} dT + \left(\frac{dP}{dT}\right)_V dV \quad\dots\dots\dots\dots(22).$$

This is a perfect differential, and therefore, by equation (5),

$$\left(\frac{d}{dV}\right)_T \frac{C_v}{T} = \left(\frac{d}{dT}\right)_V \left(\frac{dP}{dT}\right)_V.$$

Hence
$$\frac{1}{T}\left(\frac{dC_v}{dV}\right)_T = \left(\frac{d^2P}{dT^2}\right)_V,$$

or
$$\left(\frac{dC_v}{dV}\right)_T = T\left(\frac{d^2P}{dT^2}\right)_V \quad\dots\dots\dots\dots\dots(23).$$

This is an important property of C_v.

To obtain a corresponding property of C_p, take the temperature and pressure as the two independent variables and express the heat taken in with reference to them. The heat taken in, dQ, is the same as before, being still equal to $T d\phi$. We may write

$$dQ = C_p dT + l' dP \quad\dots\dots\dots\dots\dots\dots(24).$$

Here C_p, which is the specific heat at constant pressure, is $\left(\frac{dQ}{dT}\right)_P$ and l' is a symbol for $\left(\frac{dQ}{dP}\right)_T$.

Since
$$dQ = T d\phi, \quad l' = T\left(\frac{d\phi}{dP}\right)_T.$$

But by equation (14)
$$\left(\frac{d\phi}{dP}\right)_T = -\left(\frac{dV}{dT}\right)_P.$$

Hence
$$l' = -T\left(\frac{dV}{dT}\right)_P \quad\dots\dots\dots\dots\dots\dots(25),$$

and
$$dQ = C_p dT - T\left(\frac{dV}{dT}\right)_p dP.$$

Dividing both sides by T, we have

$$d\phi = \frac{C_p}{T} dT - \left(\frac{dV}{dT}\right)_P dP \quad \ldots\ldots\ldots\ldots\ldots(26).$$

And by equation (5), since this is a perfect differential,

$$\left(\frac{d}{dP}\right)_T \frac{C_p}{T} = -\left(\frac{d}{dT}\right)_P \left(\frac{dV}{dT}\right)_P .$$

Hence

$$\frac{1}{T} \left(\frac{dC_p}{dP}\right)_T = -\left(\frac{d^2V}{dT^2}\right)_P,$$

or

$$\left(\frac{dC_p}{dP}\right)_T = -T \left(\frac{d^2V}{dT^2}\right)_P \quad \ldots\ldots\ldots\ldots\ldots(27),$$

which is the property of C_p corresponding to that of C_v in equation (23).

Further, from equations (20) and (24),

$$C_p dT + l' dP = C_v dT + l dV,$$

$$(C_p - C_v) dT = l dV - l' dP.$$

By writing $dP = 0$ it follows that

$$C_p - C_v = l \left(\frac{dV}{dT}\right)_P .$$

Or by writing $dV = 0$,

$$C_p - C_v = -l' \left(\frac{dP}{dT}\right)_V .$$

By equation (21) or (25), either of these gives this important expression for the difference between the two specific heats,

$$C_p - C_v = T \left(\frac{dP}{dT}\right)_V \left(\frac{dV}{dT}\right)_P \ldots\ldots\ldots\ldots\ldots(28).$$

And since by equation (4)

$$\left(\frac{dV}{dT}\right)_P = -\left(\frac{dV}{dP}\right)_T \left(\frac{dP}{dT}\right)_V,$$

this result may be written

$$C_p - C_v = -T \left(\frac{dV}{dP}\right)_T \left(\frac{dP}{dT}\right)_V^2 \quad \ldots\ldots\ldots\ldots(28a).$$

From equation ($28a$) it will be seen that C_p can never be less than C_v, for $\left(\frac{dV}{dP}\right)_T$ is essentially negative, increase of pressure causing decrease of volume in any fluid, and therefore the whole expression on the right is positive. Accordingly C_p is always greater than C_v, except in the special case when one of the factors on the right-hand side is equal to zero, in which case C_p is equal to C_v. This is possible in a fluid which has a temperature of maximum density (as water has at about 4° C.). At the temperature of maximum density $\left(\frac{dV}{dT}\right)_P = 0$, and consequently at that point $C_p - C_v = 0$.

Return now to equations (22) and (26). In heating at constant volume $dV = 0$; hence by equation (22)

$$C_v = T\left(\frac{d\phi}{dT}\right)_V \quad \text{......................(29)}.$$

In heating at constant pressure $dP = 0$; hence by equation (26)

$$C_p = T\left(\frac{d\phi}{dT}\right)_P \quad \text{......................(30)}.$$

In an adiabatic operation $d\phi = 0$; hence by equation (22)

$$\frac{C_v}{T}\left(\frac{dT}{dV}\right)_\phi = -\left(\frac{dP}{dT}\right)_V \text{......................(31)},$$

and by equation (26) $\dfrac{C_p}{T}\left(\dfrac{dT}{dP}\right)_\phi = \left(\dfrac{dV}{dT}\right)_P$ ....(32).

Further, by equation (4)

$$\left(\frac{dV}{dP}\right)_T = -\frac{\left(\frac{dV}{dT}\right)_P}{\left(\frac{dP}{dT}\right)_V} = \frac{\frac{C_p}{T}\left(\frac{dT}{dP}\right)_\phi}{\frac{C_v}{T}\left(\frac{dT}{dV}\right)_\phi} = \frac{C_p}{C_v}\left(\frac{dV}{dP}\right)_\phi.$$

or $\dfrac{C_p}{C_v} = \left(\dfrac{dV}{dP}\right)_T\left(\dfrac{dP}{dV}\right)_\phi$ ....(33).

This is the ratio usually called γ.

Thus in the adiabatic expansion of any fluid the slope of the PV line is γ times its slope in isothermal expansion,

$$\left(\frac{dP}{dV}\right)_\phi = \gamma\left(\frac{dP}{dV}\right)_T \quad \text{......................(33a)}.$$

188. Further deductions from the Equations for E and I. By equation (7) $$dE = dQ - PdV.$$

Hence by equation (20) $dE = C_v dT + ldV - PdV$
$$= C_v dT + (l - P)\, dV.$$

In heating at constant volume $dV = 0$; hence

$$\left(\frac{dE}{dT}\right)_V = C_v \text{......................(34)}.$$

In isothermal expansion $dT = 0$; hence, using equation (21),

$$\left(\frac{dE}{dV}\right)_T = l - P = T\left(\frac{dP}{dT}\right)_V - P \dots\dots\dots\dots(35).$$

We may therefore write

$$dE = C_v dT + \left[T\left(\frac{dP}{dT}\right)_V - P\right] dV \dots\dots\dots(36).$$

Again, by equation (9) $dI = dQ + VdP.$

Hence by equation (24) $dI = C_p dT + l'dP + VdP$
$$= C_p dT + (l' + V) dP.$$

In heating at constant pressure $dP = 0$; hence

$$\left(\frac{dI}{dT}\right)_P = C_p \dots\dots\dots\dots\dots\dots(37).$$

In isothermal compression $dT = 0$; hence, using equation (25),

$$\left(\frac{dI}{dP}\right)_T = l' + V = V - T\left(\frac{dV}{dT}\right)_P \dots\dots\dots\dots(38).$$

We may therefore write

$$dI = C_p dT + \left[V - T\left(\frac{dV}{dT}\right)_P\right] dP \dots\dots\dots(39).$$

189. The Joule-Thomson Effect. In a throttling process $dI = 0$ (Art. 72); hence, from equation (39),

$$\left(\frac{dT}{dP}\right)_I = \frac{1}{C_p}\left[T\left(\frac{dV}{dT}\right)_P - V\right] \dots\dots\dots\dots(40).$$

This is the "cooling effect" in the Joule-Thomson porous plug experiment of Art. 19; the cooling effect which the working fluid of a refrigerating machine undergoes in passing the expansion-valve (Art. 110); the cooling effect used cumulatively by Linde for the liquefaction of gases (Art. 123). It expresses the fall of temperature per unit fall of pressure when a fluid suffers a throttling operation, during which it receives no heat from outside.

From equation (40) it follows that the cooling effect vanishes when

$$\left(\frac{dV}{dT}\right)_P = \frac{V}{T}.$$

This occurs in any ideal "perfect" gas under all conditions, that is to say in a gas which exactly satisfies the equation $PV = RT$. But it also occurs in real gases under particular conditions of temperature and pressure. A gas tested for the Joule-Thomson effect at moderate pressure, and at various temperatures, will be found to become warmer instead of colder on passing the plug if the temperature exceeds a certain value. At that temperature, which is called the temperature of inversion of the Joule-Thomson effect, throttling produces no change of temperature. Above the temperature of inversion the effect of passing the plug is to heat the gas; $\left(\dfrac{dV}{dT}\right)_P$ is then less than $\dfrac{V}{T}$ and the expression for the "cooling effect" is negative. Below the temperature of inversion the cooling effect is positive. The temperature of inversion depends to some extent on the pressure, in any one gas. It differs widely in different gases. In air, oxygen, carbon dioxide, steam and most other gases it is so high that the normal effect of throttling is to make the gas colder; in hydrogen, on the other hand, the normal effect of throttling is to make the gas warmer, for the temperature of inversion is exceptionally low, about $-80°$ C. In the Linde process it is essential that the gas to be liquefied should have, before throttling, a temperature lower than the temperature of inversion: the process can be applied to hydrogen only by cooling the gas beforehand to a suitably low temperature.

Taking equations (38) and (40) together we have

$$C_p \left(\frac{dT}{dP}\right)_I = T \left(\frac{dV}{dT}\right)_P - V = -\left(\frac{dI}{dP}\right)_T \quad \ldots\ldots\ldots(41).$$

This product, $C_p \left(\dfrac{dT}{dP}\right)_I$, is the quantity of heat that would just suffice to neutralize the Joule-Thomson cooling effect per unit drop in pressure, if it were supplied to the fluid in the process of throttling. It may conveniently be represented by the single symbol ρ. It measures the cooling effect, per unit drop in pressure by throttling, as a quantity of heat (expressed in work units), while $\left(\dfrac{dT}{dP}\right)_I$ measures that effect as a change in temperature.

It follows that if the range through which the pressure falls in a throttling process is from P_A to P_B, the whole quantity of heat that would have to be supplied to neutralize the cooling effect is

$$\int_{P_B}^{P_A} \rho \, dP = \int_{P_B}^{P_A} \left[T \left(\frac{dV}{dT}\right)_P - V \right] dP,$$

as was stated in a footnote to Art. 124*.

* Cf. E. Buckingham, *Bulletin of the Bureau of Standards (Washington)*, vol. VI, p. 125, 1909.

Since $I = E + PV$ we may write equation (41) in the form

$$\rho = -\left(\frac{dE}{dP}\right)_T - \left(\frac{d(PV)}{dP}\right)_T \quad \ldots\ldots\ldots\ldots(41a).$$

This is instructive as showing the analysis of the Joule-Thomson effect into two parts. When an imperfect gas or vapour is throttled, that part of the effect which is measured by the first term arises from the fact that the internal energy is not constant at any one temperature but depends to some extent on the pressure. In other words, the first term is due to departure from Joule's Law. There is in general an additional part of the effect, measured by the second term. It is due to departure from Boyle's Law, according to which PV should be constant for constant temperature. A gas may conform to Boyle's Law at a particular temperature and still be imperfect: in that case it will show a cooling effect due to the first term alone. It is only when both terms vanish that the gas is perfect.

Experiments show that in an imperfect gas the term $\left(\frac{d(PV)}{dP}\right)_T$ may be either negative or positive according to the conditions of pressure and temperature. Hence that part of the Joule-Thomson effect which is due to deviation from Boyle's Law will under some conditions assist, and under other conditions oppose, that part of the effect which is due to deviation from Joule's Law. The latter part is always a cooling effect; the former may be either a cooling or a heating effect. At the temperature of inversion the two parts cancel one another.

It may help the student to understand equation (41a) if we put the physical interpretation of that equation in another way. Suppose unit quantity of any fluid to undergo unit drop of pressure in passing a porous plug or other throttling device. We may then put $dP = -1$. Suppose also a quantity of heat ρ to be supplied to it from outside which just prevents any change of temperature. Then equation (41a) takes the form

$$\rho = dE + d(PV),$$

which is equivalent to saying that in the complete process,

Heat supplied = Increase of internal energy + Work done by the fluid.

Here $d(PV)$ is the net amount of work done by the fluid, because it is the excess of $P_2 V_2$, which is the work done by the fluid as it leaves the apparatus, over P_1V_1, which is the work spent upon the fluid as it enters the apparatus.

190. **Unresisted Expansion.** In the Joule-Thomson porous plug experiment the fluid, in expanding from a region of constant high pressure to a region of constant lower pressure, does some work on things external to itself, the net amount of which is

$$P_2V_2 - P_1V_1.$$

This quantity is not zero except in special cases.

But in the original Joule experiment with two closed vessels (Art. 19) the fluid did not work on anything external to itself. The expansion there may therefore be described as strictly *unresisted*. This distinction between it and the Joule-Thomson mode of expansion is important.

Imagine the two closed vessels of the Joule experiment to be completely impervious to heat, so that no heat passes out of, or into, the fluid as a whole during the process. Imagine also that heat may pass freely from the fluid in one vessel to the fluid in the other through the opening between them, so that after expansion T becomes the same in both as well as P. Under these conditions the internal energy E of the fluid as a whole is not altered by the expansion; for no heat is taken in or given out, and no work is done. This is true of any fluid. The characteristic, therefore, of such expansion is that E is unchanged, just as the characteristic of the Joule-Thomson expansion is that I is unchanged.

In the unresisted Joule expansion each vessel may of course be of any size. Think of the second vessel, into which the fluid expands, as consisting of a group of very small chambers which are successively opened, so that the volume of the fluid increases by steps, each dV. We still suppose the temperature of the fluid to attain equilibrium at each step, and no heat to come in from outside. Then for each step $dE = 0$. With infinitesimal steps the process becomes continuous. The cooling effect in this imaginary process is not identical with the cooling effect in the Joule-Thomson experiment. In this process it is $-\left(\dfrac{dT}{dV}\right)_E$, namely the rate at which the temperature falls with increase of volume, under the condition that E is constant.

By equation (36), writing $dE = 0$,

$$-\left(\frac{dT}{dV}\right)_E = \frac{1}{C_v}\left[T\left(\frac{dP}{dT}\right)_V - P \right] \quad\ldots\ldots\ldots\ldots(42),$$

and this, along with equation (35), gives

$$-C_v\left(\frac{dT}{dV}\right)_E = T\left(\frac{dP}{dT}\right)_V - P = \left(\frac{dE}{dV}\right)_T \quad \text{.........(43)}.$$

Equation (42) expresses the cooling effect in this imaginary process as a fall of temperature, per unit increase of volume; equation (43) expresses it as a quantity of heat, per unit increase of volume, namely the quantity that would have to be supplied from outside to neutralize the change of temperature caused by the expansion. We may call this quantity of heat σ.

Hence in unresisted expansion from any volume V_A to any volume V_B, under adiathermal conditions (Joule's expansion with vessels made perfectly impervious to heat), the whole quantity of heat that would have to be supplied to neutralize the cooling effect is, for any fluid,

$$\int_{V_A}^{V_B}\sigma dV = \int_{V_A}^{V_B}\left[T\left(\frac{dP}{dT}\right)_V - P\right]dV.$$

A further interesting relation follows. By equation (28), we had

$$C_p - C_v = T\left(\frac{dP}{dT}\right)_V\left(\frac{dV}{dT}\right)_P.$$

But by equation (35)

$$T\left(\frac{dP}{dT}\right)_V = P + \left(\frac{dE}{dV}\right)_T = P + \sigma \text{ (by equation (43)).}$$

Also, by equation (41), $\quad \left(\frac{dV}{dT}\right)_P = \frac{V+\rho}{T}.$

On substituting these values, equation (28) takes the new form

$$C_p - C_v = \frac{(P+\sigma)(V+\rho)}{T} \quad \text{...................(44)}.$$

This, like all the relations given in the present chapter, is true of any fluid. We shall return to it later in connection with imperfect gases (Art. 202).

191. **Slopes of Lines in the** $I\phi$, $T\phi$, **and** IP **charts, for any Fluid.** The slope of any constant-pressure line in the $I\phi$ chart is equal to the absolute temperature, for, by equation (16),

$$\left(\frac{dI}{d\phi}\right)_P = T.$$

It follows that all constant-pressure lines in that chart have the same slope at points where they cross any one line of constant temperature.

To find an expression for $\left(\dfrac{dI}{d\phi}\right)_T$, which is the slope of a constant-temperature line in the $I\phi$ chart, we shall proceed by a process of substitution which may be followed in finding other partial differential coefficients. It will serve as an example of a general method.

Starting with equation (9),

$$dI = Td\phi + VdP,$$

we shall eliminate dP by substituting for it an expression in terms of $d\phi$ and dT, got by applying the general relation of equation (4), namely,

$$dP = \left(\frac{dP}{d\phi}\right)_T d\phi + \left(\frac{dP}{dT}\right)_\phi dT.$$

This substitution gives

$$dI = \left[T + V\left(\frac{dP}{d\phi}\right)_T\right]d\phi + V\left(\frac{dP}{dT}\right)_\phi dT.$$

Hence, writing $dT = 0$,

$$\left(\frac{dI}{d\phi}\right)_T = T + V\left(\frac{dP}{d\phi}\right)_T \quad \dots\dots\dots\dots\dots(45)$$

$$= T - V\left(\frac{dT}{dV}\right)_P \quad \dots\dots\dots\dots(45\,a),$$

since, by equation (14), $\left(\dfrac{dP}{d\phi}\right)_T = -\left(\dfrac{dT}{dV}\right)_P$.

Similarly, to find an expression for $\left(\dfrac{dI}{d\phi}\right)_V$, which is the slope of a constant-volume line in the $I\phi$ chart, we start from the same equation for dI, but eliminate dP by substituting an expression for it in terms of $d\phi$ and dV, namely

$$dP = \left(\frac{dP}{d\phi}\right)_V d\phi + \left(\frac{dP}{dV}\right)_\phi dV.$$

This substitution gives

$$dI = \left[T + V\left(\frac{dP}{d\phi}\right)_V\right]d\phi + V\left(\frac{dP}{dV}\right)_\phi dV.$$

Hence, writing $dV = 0$,

$$\left(\frac{dI}{d\phi}\right)_V = T + V\left(\frac{dP}{d\phi}\right)_V \quad \dots\dots\dots\dots\dots(46),$$

$$= T - V\left(\frac{dT}{dV}\right)_\phi \quad \dots\dots\dots\dots(46\,a),$$

since, by equation (12), $\left(\dfrac{dP}{d\phi}\right)_V = -\left(\dfrac{dT}{dV}\right)_\phi$.

Turning next to the $T\phi$ chart, the slope of a constant-volume line is given by equation (29),

$$\left(\frac{dT}{d\phi}\right)_V = \frac{T}{C_v},$$

and the slope of a constant-pressure line by equation (30),

$$\left(\frac{dT}{d\phi}\right)_P = \frac{T}{C_p}.$$

To find the slope of a line of constant total heat $\left(\frac{dT}{d\phi}\right)_I$ we may again apply the method of substitution. Starting with the equation

$$Td\phi = dI - VdP,$$

substitute for dP an expression in dT and dI (equation (4)),

$$dP = \left(\frac{dP}{dT}\right)_I dT + \left(\frac{dP}{dI}\right)_T dI.$$

This gives $\quad Td\phi = \left[1 - V\left(\frac{dP}{dI}\right)_T\right] dI - V\left(\frac{dP}{dT}\right)_I dT,$

from which, writing $dI = 0$,

$$\left(\frac{dT}{d\phi}\right)_I = -\frac{T}{V}\left(\frac{dT}{dP}\right)_I \quad\dots\dots\dots\dots\dots(47).$$

But by equation (40)

$$\left(\frac{dT}{dP}\right)_I = \frac{1}{C_p}\left[T\left(\frac{dV}{dT}\right)_P - V\right].$$

Hence $\quad\quad \left(\frac{dT}{d\phi}\right)_I = \frac{T}{C_p}\left[1 - \frac{T}{V}\left(\frac{dV}{dT}\right)_P\right] \quad\dots\dots\dots\dots(47a).$

Also, since $\quad \left(\frac{dV}{dT}\right)_P = \left(\frac{dV}{dI}\right)_P\left(\frac{dI}{dT}\right)_P$ and $\left(\frac{dI}{dT}\right)_P = C_p,$

we may put this result in the form

$$\left(\frac{dT}{d\phi}\right)_I = \frac{T}{C_p} - \frac{T^2}{V}\left(\frac{dV}{dI}\right)_P \dots\dots\dots\dots\dots(47b).$$

In the IP chart the slope of an adiabatic, or line of constant entropy, is given by equation (17),

$$\left(\frac{dI}{dP}\right)_\phi = V,$$

from which it follows that all adiabatics have the same slope at points where they cross any one line of constant volume.

The slope of a line of constant temperature is given by equation (38),

$$\left(\frac{dI}{dP}\right)_T = V - T\left(\frac{dV}{dT}\right)_P.$$

To find expressions for the slope of a line of constant volume, $\left(\frac{dI}{dP}\right)_V$, we may proceed thus:

$$dI = dE + d(PV) = dE + VdP + PdV.$$

Hence

$$\left(\frac{dI}{dP}\right)_V = \left(\frac{dE}{dP}\right)_V + V,$$

and since

$$\left(\frac{dE}{dP}\right)_V = \left(\frac{dE}{dT}\right)_V \left(\frac{dT}{dP}\right)_V = C_v \left(\frac{dT}{dP}\right)_V,$$

it follows that

$$\left(\frac{dI}{dP}\right)_V = V + C_v \left(\frac{dT}{dP}\right)_V \quad \dots\dots\dots\dots\dots(48).$$

By equation (31) this may be written

$$\left(\frac{dI}{dP}\right)_V = V - T\left(\frac{dV}{dT}\right)_\phi \quad \dots\dots\dots\dots\dots(48\,a).$$

Two other expressions which are sometimes useful may conveniently be given here, one for $\left(\frac{dE}{dP}\right)_T$ and one for $\left(\frac{dI}{dV}\right)_P$:

$$\left(\frac{dE}{dP}\right)_T = \left(\frac{dE}{dV}\right)_T \left(\frac{dV}{dP}\right)_T = \left[T\left(\frac{dP}{dT}\right)_V - P\right]\left(\frac{dV}{dP}\right)_T, \text{ by equation (35).}$$

But

$$\left(\frac{dP}{dT}\right)_V \left(\frac{dV}{dP}\right)_T = -\left(\frac{dV}{dT}\right)_P, \text{ by equation (4).}$$

Hence

$$\left(\frac{dE}{dP}\right)_T = -T\left(\frac{dV}{dT}\right)_P - P\left(\frac{dV}{dP}\right)_T \quad \dots\dots\dots(49).$$

$$\left(\frac{dI}{dV}\right)_P = \left(\frac{dI}{d\phi}\right)_P \left(\frac{d\phi}{dV}\right)_P = T\left(\frac{d\phi}{dV}\right)_P, \text{ by equation (16).}$$

Hence, by equation (13),

$$\left(\frac{dI}{dV}\right)_P = T\left(\frac{dP}{dT}\right)_\phi \quad \dots\dots\dots\dots\dots(50).$$

192. Application to a Mixture of Liquid and Vapour in Equilibrium: Clapeyron's Equation. Change of Phase. Equation (50) is applicable not only to homogeneous fluids, but to a mixture of two phases of the same substance, in equilibrium with each other and therefore both at the same pressure and the same temperature. I and V are then to be reckoned for unit mass of the mixture. Say for instance that the substance is a mixture, part liquid and part saturated vapour. Suppose the proportion of liquid to vapour to be changed by vaporizing some of the liquid part at constant pressure, and therefore also at constant temperature. During that process $\left(\dfrac{dI}{dV}\right)_P$ is constant, for the volume of the mixture as a whole increases in proportion to the heat taken in. Instead of $\left(\dfrac{dI}{dV}\right)_P$ in equation (50) we may therefore write

$$\frac{I_s - I_w}{V_s - V_w} \quad \text{or} \quad \frac{L}{V_s - V_w},$$

where the suffixes s and w relate to the two states, when all is vapour and all is liquid respectively. Further, the condition that ϕ is constant may be dropped in writing the coefficient $\dfrac{dP}{dT}$, which is no longer a *partial* differential coefficient. Since the vapour present in the mixture is always saturated, P is a function of T only; $\dfrac{dP}{dT}$ is simply the rate at which the pressure of saturation rises with the temperature. While the mixture is vaporizing or condensing under variable pressure it makes no difference in the relation of P to T whether the process is conducted with $\phi =$ constant, or with $V =$ constant, or in any other way: during that process $\left(\dfrac{dP}{dT}\right)_\phi$ or $\left(\dfrac{dP}{dT}\right)_V$ is the same as $\dfrac{dP}{dT}$. Hence when applied to an equilibrium mixture of liquid and vapour, or of any two phases, equation (50) may be written in the form

$$\frac{L}{V_s - V_w} = T \frac{dP}{dT},$$

or

$$V_s - V_w = \frac{L}{T} \frac{dT}{dP} \quad \dots\dots\dots\dots\dots\dots\dots(51).$$

This is Clapeyron's Equation, which was arrived at in Art. 98 in another way.

The same result may be got from equation (21):

$$\left(\frac{dQ}{dV}\right)_T = T \left(\frac{dP}{dT}\right)_V.$$

During vaporization at constant temperature $\left(\dfrac{dQ}{dV}\right)_T$ is constant and its value is $\dfrac{L}{V_s - V_w}$. Hence, dropping the suffix V for the reason just given, we have as before

$$V_s - V_w = \frac{L}{T} \frac{dT}{dP}.$$

This result may be extended to any reversible change of phase which a substance undergoes at constant pressure. During any such change the two phases of the substance are in equilibrium with one another, and the temperature is constant. Writing λ for the heat taken in during the change of phase, and V' and V'' for the volumes of the first and second phases respectively (as in Art. 99), we have

$$V'' - V' = \frac{\lambda}{T} \frac{dT}{dP} \qquad \dots\dots\dots\dots\dots(51\,a).$$

Similarly, the expression for $\left(\dfrac{dT}{d\phi}\right)_I$ in equation $(47\,b)$, namely

$$\left(\frac{dT}{d\phi}\right)_I = \frac{T}{C_p} - \frac{T^2}{V}\left(\frac{dV}{dI}\right)_P,$$

may be adapted to a mixture of liquid and vapour in equilibrium, during the change of phase which occurs in vaporization at constant pressure (and temperature). In this process C_p is infinite, for heat is taken in without rise of temperature; and also

$$\left(\frac{dV}{dI}\right)_P = \frac{V_s - V_w}{L}.$$

The equation therefore takes the form*

$$\left(\frac{dT}{d\phi}\right)_I = -\frac{T^2}{V} \cdot \frac{V_s - V_w}{L} \qquad \dots\dots\dots\dots(52).$$

This applies at any stage in the process of vaporization, V being the volume of the mixture at that stage, namely $qV_s + (1-q)\,V_w$, where q is the fraction that has been vaporized (Art. 74). It gives the slope of a line of constant total heat in the wet region (the region within the boundary curve) of the $T\phi$ chart.

A still more direct means of getting Clapeyron's Equation is to use the function G, which is $T\phi - I$ or $-\zeta$.

By equation (10)

$$dG = \phi dT - VdP.$$

In any change of phase which occurs at constant temperature and constant pressure, such as the conversion of water into steam at

---

* Used by Jenkin and Pye (*Phil. Trans. R.S.* A, vol. DXXXIV, p. 866) in correcting the $T\phi$ chart for carbon dioxide.

constant pressure, dT and dP are both zero. Hence in such a change G is constant, as was pointed out in Art. 90, where this property of G was turned to account.

Compare now the state of any substance at the beginning and end of a change of phase, during which G is constant. Use the suffix w for the first state (say water), and the suffix s for the second state (say steam):

$$G_s = G_w \quad \dots\dots\dots\dots\dots(53),$$

$$dG_s = dG_w,$$

$$\phi_s dT - V_s dP = \phi_w dT - V_w dP.$$

Therefore

$$V_s - V_w = (\phi_s - \phi_w)\frac{dT}{dP}.$$

But $\phi_s - \phi_w = \dfrac{L}{T}$. Hence this again gives Clapeyron's Equation,

$$V_s - V_w = \frac{L}{T}\frac{dT}{dP}.$$

193. Compressibility and Elasticity of a Fluid. Let a fluid be subjected to an increase of pressure dP, with the result that the volume is reduced from V to $V - dV$. Then $-\dfrac{dV}{V}$ measures the volume strain, and the ratio of this strain to dP measures the compressibility.

The reciprocal of the compressibility or $-V\left(\dfrac{dP}{dV}\right)$ measures what is called the elasticity of the fluid. Its value will obviously depend on the circumstances under which the compression takes place. We may for instance keep the temperature constant during the compression. In that case the expression for the elasticity becomes $-V\left(\dfrac{dP}{dV}\right)_T$. This is called the isothermal elasticity of a fluid, and will be denoted here by e_t. Or we may prevent any heat from leaving or entering the fluid during the compression. In that case the expression becomes $-V\left(\dfrac{dP}{dV}\right)_\phi$. This, which is called adiabatic elasticity of a fluid, will be denoted here by e_ϕ. We have accordingly the two elasticities

$$e_t = -V\left(\frac{dP}{dV}\right)_T \quad \dots\dots\dots\dots(54),$$

$$e_\phi = -V\left(\frac{dP}{dV}\right)_\phi \quad \dots\dots\dots\dots(55).$$

Hence

$$\frac{e_\phi}{e_t} = \frac{\left(\frac{dP}{dV}\right)_\phi}{\left(\frac{dP}{dV}\right)_T} = \frac{C_p}{C_v} \dots\dots\dots\dots\dots(56),$$

by equation (33). That is to say, the ratio of the adiabatic to the isothermal elasticity is equal to γ, the ratio of the specific heats. Since C_p is greater than C_v (Art. 187), e_ϕ is greater than e_t.

194. Collected Results. All the foregoing relations are true of any fluid. Before proceeding to apply them (in the next chapter) to particular fluids, it will be useful to collect them here for convenience of reference:

$$dE = Td\phi - PdV \quad \dots\dots\dots\dots\dots(8),$$

$$I = E + PV; \qquad dI = Td\phi + VdP \quad \dots\dots\dots(9),$$

$$\zeta = I - T\phi; \qquad d\zeta = VdP - \phi dT \quad \dots\dots\dots(10),$$

$$\psi = E - T\phi; \qquad d\psi = -PdV - \phi dT \quad \dots\dots(11),$$

$$\left(\frac{dT}{dV}\right)_\phi = -\left(\frac{dP}{d\phi}\right)_V \quad \dots\dots\dots\dots(12),$$

$$\left(\frac{dT}{dP}\right)_\phi = \left(\frac{dV}{d\phi}\right)_P \quad \dots\dots\dots\dots(13),$$

$$\left(\frac{dV}{dT}\right)_P = -\left(\frac{d\phi}{dP}\right)_T \quad \dots\dots\dots\dots(14),$$

$$\left(\frac{dP}{dT}\right)_V = \left(\frac{d\phi}{dV}\right)_T \quad \dots\dots\dots\dots(15),$$

$$\left(\frac{dI}{d\phi}\right)_P = T = \left(\frac{dE}{d\phi}\right)_V \quad \dots\dots\dots\dots(16),$$

$$\left(\frac{dI}{dP}\right)_\phi = V = \left(\frac{d\zeta}{dP}\right)_T \quad \dots\dots\dots\dots(17),$$

$$\left(\frac{d\psi}{dV}\right)_T = -P = \left(\frac{dE}{dV}\right)_\phi \quad \dots\dots\dots\dots(18),$$

$$\left(\frac{d\psi}{dT}\right)_V = -\phi = \left(\frac{d\zeta}{dT}\right)_P \quad \dots\dots\dots\dots(19),$$

$$d\phi = \frac{C_v}{T} dT + \left(\frac{dP}{dT}\right)_V dV \quad \dots\dots\dots\dots(22),$$

$$\left(\frac{dC_v}{dV}\right)_T = T \left(\frac{d^2P}{dT^2}\right)_V \dots\dots\dots\dots(23),$$

$$d\phi = \frac{C_p}{T} dT - \left(\frac{dV}{dT}\right)_P dP \quad \dots\dots\dots\dots(26),$$

$$\left(\frac{dC_p}{dP}\right)_T = -T\left(\frac{d^2V}{dT^2}\right)_P \quad \text{...............}(27),$$

$$C_p - C_v = T\left(\frac{dP}{dT}\right)_V\left(\frac{dV}{dT}\right)_P \text{...............}(28),$$

$$C_v = T\left(\frac{d\phi}{dT}\right)_V \quad \text{...................}(29),$$

$$C_p = T\left(\frac{d\phi}{dT}\right)_P \quad \text{...............}(30),$$

$$\frac{C_v}{T}\left(\frac{dT}{dV}\right)_\phi = -\left(\frac{dP}{dT}\right)_V \text{...............}(31),$$

$$\frac{C_p}{T}\left(\frac{dT}{dP}\right)_\phi = \left(\frac{dV}{dT}\right)_P \text{...............}(32),$$

$$\gamma = \frac{C_p}{C_v} = \left(\frac{dV}{dP}\right)_T\left(\frac{dP}{dV}\right)_\phi \quad \text{...............}(33),$$

$$\left(\frac{dP}{dV}\right)_\phi = \gamma\left(\frac{dP}{dV}\right)_T \text{...................}(33\,a),$$

$$\left(\frac{dE}{dT}\right)_V = C_v \text{.......................}(34),$$

$$\left(\frac{dE}{dV}\right)_T = T\left(\frac{dP}{dT}\right)_V - P \quad \text{...............}(35),$$

$$dE = C_v dT + \left[T\left(\frac{dP}{dT}\right)_V - P\right]dV \quad \text{...........}(36),$$

$$\left(\frac{dI}{dT}\right)_P = C_p \text{.......................}(37),$$

$$\left(\frac{dI}{dP}\right)_T = V - T\left(\frac{dV}{dT}\right)_P \quad \text{...............}(38),$$

$$dI = C_p dT + \left[V - T\left(\frac{dV}{dT}\right)_P\right]dP \quad \text{...........}(39),$$

$$\left(\frac{dT}{dP}\right)_I = \frac{1}{C_p}\left[T\left(\frac{dV}{dT}\right)_P - V\right] \text{...............}(40),$$

$$\rho = C_p\left(\frac{dT}{dP}\right)_I = T\left(\frac{dV}{dT}\right)_P - V = -\left(\frac{dI}{dP}\right)_T \text{.........}(41),$$

$$\rho = -\left(\frac{dE}{dP}\right)_T - \left(\frac{d(PV)}{dP}\right)_T \text{...............}(41\,a),$$

$$-\left(\frac{dT}{dV}\right)_E = \frac{1}{C_v}\left[T\left(\frac{dP}{dT}\right)_V - P\right]\text{...............}(42),$$

$$\sigma = -C_v \left(\frac{dT}{dV}\right)_E = T\left(\frac{dP}{dT}\right)_V - P = \left(\frac{dE}{dV}\right)_T \quad \text{......(43)},$$

$$C_p - C_v = \frac{(P+\sigma)(V+\rho)}{T} \quad \text{.................(44)},$$

$$\left(\frac{dI}{d\phi}\right)_T = T + V\left(\frac{dP}{d\phi}\right)_T = T - V\left(\frac{dT}{dV}\right)_P \quad \text{.........(45)},$$

$$\left(\frac{dI}{d\phi}\right)_V = T + V\left(\frac{dP}{d\phi}\right)_V = T - V\left(\frac{dT}{dV}\right)_\phi \quad \text{.........(46)},$$

$$\left(\frac{dT}{d\phi}\right)_I = -\frac{T}{V}\left(\frac{dT}{dP}\right)_I = \frac{T}{C_p}\left[1 - \frac{T}{V}\left(\frac{dV}{dT}\right)_P\right] = \frac{T}{C_p} - \frac{T^2}{V}\left(\frac{dV}{dI}\right)_P \text{...(47)},$$

$$\left(\frac{dI}{dP}\right)_V = V + C_v\left(\frac{dT}{dP}\right)_V = V - T\left(\frac{dV}{dT}\right)_\phi \quad \text{.........(48)},$$

$$\left(\frac{dE}{dP}\right)_T = -T\left(\frac{dV}{dT}\right)_P - P\left(\frac{dV}{dP}\right)_T \quad \text{...........(49)},$$

$$\left(\frac{dI}{dV}\right)_P = T\left(\frac{dP}{dT}\right)_\phi \quad \text{....................(50)}.$$

In a reversible change of phase at constant pressure

$$V_s - V_w = \frac{L}{T}\frac{dT}{dP} \quad \text{......................(51)},$$

$$\left(\frac{dT}{d\phi}\right)_I = -\frac{T^2}{V}\cdot\frac{V_s - V_w}{L} \quad \text{...............(52)},$$

and $\qquad G_s = G_w, \quad \text{or} \quad T\phi_s - I_s = T\phi_w - I_w \quad \text{...........(53)}.$

The isothermal and adiabatic elasticities:

$$e_t = -V\left(\frac{dP}{dV}\right)_T \quad \text{......................(54)},$$

$$e_\phi = -V\left(\frac{dP}{dV}\right)_\phi \quad \text{......................(55)},$$

$$\frac{e_\phi}{e_t} = \frac{C_p}{C_v} \quad \text{or} \quad e_\phi = \gamma e_t \quad \text{....................(56)}.$$

195. Heterogeneous Systems. Before applying these relations to particular fluids, a brief reference may be made to the general thermodynamics of systems containing more than one substance. We saw in Art. 43 that the entropy of any system of bodies might be treated as a whole by summing up the entropies of the various parts. This is also true of the total heat, the energy, and the functions ψ and ζ. In physical chemistry it is important to consider these functions not only for a single substance or a mixture of its phases, but for a heterogeneous system in which substances that are chemically different may be reacting on one another, as for instance the substances in a galvanic cell. It is essential to notice in dealing with a heterogeneous system that it may do work otherwise than by change of volume. Changes may occur within the system that do work against gravity, by altering the levels of its parts, or they may do work electrically by producing current in an external circuit. It will be obvious that in any such case dW must not be taken as equal to PdV. There may even be no change of volume. But in whatever way work is done the energy equation holds,

$$dW = dQ - dE,$$

where dW is the work done by the system, dQ is the heat it takes in, and $-dE$ is its loss of internal energy.

Since
$$\zeta = I - T\phi,$$
$$d\zeta = dI - Td\phi - \phi dT.$$

This is true of any change, reversible or not. Let the imposed conditions be such that the temperature is kept constant during the change, then $\phi dT = 0$. This condition would be realized if the system were surrounded by a capacious reservoir of heat, and the change were to take place very slowly. If the pressure also be kept constant dI is equal to dQ, the heat taken in. Hence under these conditions we have
$$d\zeta = dQ - Td\phi.$$

Now $d\phi$ is equal to dQ/T in any reversible change, but is greater than dQ/T in an irreversible change. It follows that in a change under the prescribed conditions of constant temperature and constant pressure $d\zeta$ vanishes only if the change be reversible, and is a negative quantity if the change be irreversible, $Td\phi$ being then greater than dQ. Hence if the system were to be a little disturbed from thermal equilibrium, with the result that for a time there is irreversible action, ζ is diminishing while the system adjusts itself so that stable equilibrium is restored. Accordingly the fact that $d\zeta$ is zero in the condition of equilibrium is to be interpreted as meaning that ζ is then a minimum. Thus under the stated conditions of constant temperature and constant pressure the criterion of stable equilibrium is that the function ζ for the system as a whole shall be a minimum.

Returning to the equation
$$d\zeta = dI - Td\phi - \phi dT,$$

in a reversible change at constant pressure $Td\phi = dI$ and, by equation (19), $(d\zeta/dT)_P = -\phi$. Substituting this in the expression for ζ, we have

$$\zeta = I + T\left(\frac{d\zeta}{dT}\right)_P \quad\text{.....................(57).}$$

Again, since

$$\psi = E - T\phi,$$

$$d\psi = dE - Td\phi - \phi dT.$$

Hence in a reversible change at constant temperature

$$d\psi = dE - dQ,$$

or

$$-d\psi = dW \quad\text{..........................(58).}$$

Therefore, if a system changes reversibly by a finite amount, at constant temperature, from state (a) to state (b),

$$\psi_a - \psi_b = W \quad\text{........................(58 a),}$$

where W is the work done (in any manner) during the change. In other words, the decrement of ψ measures the amount of energy actually converted into external work by a system, whether by expansion of volume, or by generating electricity, or otherwise, during any isothermal reversible process. If the process were not reversible the work done would be less. Thus the decrement of ψ measures how much of the energy of the system can, in the most favourable case, be converted into work while the system undergoes a change at constant temperature. It consequently also measures how much the energy of the system has lost of *availability* for further conversion into work under isothermal conditions. For this reason Helmholtz (*Wied. Ann.* 1879, VII, 337) called the function ψ the Free Energy of the system, regarding the whole energy E as made up of two parts, namely the "free" or available energy ψ and the "bound" energy $T\phi$. It should, however, be borne in mind that during the conversion some heat may be taken in from or given out to the isothermal envelope, in keeping the temperature of the system constant. Following Helmholtz, many writers on the thermodynamics of chemical processes speak of ψ as the "free energy."

From the above equation it also follows that in a system which is maintained at constant temperature $d\psi$ is zero for a change that does not involve the doing of any external work, when the change is reversible, but is negative when the change is irreversible. Hence if the system be such that work can be done only by expansion of volume, its criterion of stable equilibrium at constant temperature and constant volume is that the function ψ for the system as a whole shall be a minimum.

To these functions Willard Gibbs gave the name of "Potentials," from their analogy to the Potential function in statics. On account of the properties which have been indicated above, ζ is called the Thermodynamic Potential at constant pressure, and ψ is called the Thermodynamic Potential at constant volume.

Returning to the equation

$$d\psi = dE - Td\phi - \phi dT,$$

since in a reversible change

$$dE - Td\phi = dE - dQ = -dW,$$

$$d\psi = -dW - \phi dT.$$

Hence if the change is such that no external work (of any kind) is done,

$$d\psi = -\phi dT,$$

or

$$\left(\frac{d\psi}{dT}\right)_W = -\phi,$$

where the suffix implies that the partial differential coefficient $d\psi/dT$ is the rate at which ψ increases with the temperature when no work is being done. Substituting this in the expression $\psi = E - T\phi$, we have

$$\psi = E + T\left(\frac{d\psi}{dT}\right)_W \quad\ldots\ldots\ldots\ldots\ldots\ldots(59).$$

In applying this equation, or the foregoing one for ζ, to a system which undergoes any reversible transformation at constant temperature—as, for instance, to a galvanic cell (Chapter XII)—we are concerned only with changes in E and in ψ or ζ, not with the absolute values of these quantities. In such applications the symbols E, ψ and ζ may accordingly be taken as standing for corresponding finite changes in the quantities they name.

Equation (59) has many applications in chemical physics. It was first stated by W. Thomson (Lord Kelvin) in 1855[*] and is often called the Thomson Equation. As applied to a single substance, it may be expressed thus:

$$\psi = E + T\left(\frac{d\psi}{dT}\right)_V \quad\ldots\ldots\ldots\ldots(59a),$$

for the condition that no work is done is then equivalent to the condition that the volume does not change, since it is only by change of volume that a single substance can do work.

---

[*] *Collected Papers*, vol. I, p. 297.

CHAPTER IX

APPLICATIONS TO PARTICULAR FLUIDS

196. Characteristic Equation. The general thermodynamic relations considered in Chapter VIII can be applied to determine the properties of a particular fluid when an equation connecting one of its properties with two others is known. An equation of this kind is called the "Characteristic Equation" or "Equation of State" for the given fluid. It is based upon experimental knowledge of how the numerical values of some one property, such as the volume, depend upon those of two other properties, such as the pressure and the temperature, these two being used as independent variables for specifying the state. The most usual form of characteristic equation is one connecting V with P and T. Such an equation, when it can be established, is of fundamental importance in the calculation of other properties. But taken by itself it does not allow all the thermodynamic quantities to be determined: for that purpose it must be supplemented by data regarding the specific heat, or (what comes to the same thing) by data as to the relation of the internal energy to the temperature.

197. Characteristic Equation of a Perfect Gas. The simplest case to consider is that of an ideal gas conforming exactly to the equation
$$PV = RT,$$
where R is a constant and T is the absolute temperature on the thermodynamic scale. We discussed some of the properties of such a gas in Chapter I, but it will be instructive now, as a first example of the method, to show how certain results which were obtained there follow directly when this characteristic equation is interpreted by applying to it some of the general relations of Art. 194 which hold for all fluids.

By differentiating the characteristic equation of the ideal gas, we have
$$PdV + VdP = RdT.$$
Hence in such a gas
$$\left(\frac{dP}{dT}\right)_V = \frac{R}{V} = \frac{P}{T}; \quad \left(\frac{dV}{dT}\right)_P = \frac{R}{P} = \frac{V}{T}; \quad \left(\frac{dP}{dV}\right)_T = -\frac{P}{V};$$
$$\left(\frac{d^2P}{dT^2}\right)_V = 0; \quad \left(\frac{d^2V}{dT^2}\right)_P = 0.$$

By equations (23) and (27) of Chapter VIII, in any fluid

$$\left(\frac{dC_v}{dV}\right)_T = T\left(\frac{d^2P}{dT^2}\right)_V \text{ and } \left(\frac{dC_p}{dP}\right)_T = -T\left(\frac{d^2V}{dT^2}\right)_P.$$

Hence in the ideal gas

$$\left(\frac{dC_v}{dV}\right)_T = 0 \text{ and } \left(\frac{dC_p}{dP}\right)_T = 0.$$

Thus it follows from the characteristic equation that both C_v and C_p are constant at any one temperature; in other words they are independent of the pressure. They may however vary with temperature; the characteristic equation gives no information on that point.

By equation (28) of Chapter VIII, in any fluid

$$C_p - C_v = T\left(\frac{dP}{dT}\right)_V \left(\frac{dV}{dT}\right)_P.$$

Hence in the ideal gas

$$C_p - C_v = T \cdot \frac{R}{V} \cdot \frac{R}{P} = R.$$

This agrees with Art. 20. The factor A is omitted because quantities of heat are here expressed in work units (Art. 186).

By equation (40), Chapter VIII, in any fluid the cooling effect in the Joule-Thomson porous plug experiment is

$$\frac{1}{C_p}\left[T\left(\frac{dV}{dT}\right)_P - V\right].$$

In the ideal gas $\left(\frac{dV}{dT}\right)_P = \frac{V}{T}$; hence the quantity in square brackets vanishes and there is no cooling effect.

By equation (36), Chapter VIII, in any fluid

$$dE = C_v dT + \left[T\left(\frac{dP}{dT}\right)_V - P\right] dV.$$

In the ideal gas $T\left(\frac{dP}{dT}\right)_V = P$, hence

$$dE = C_v dT,$$

and since C_v is independent of the pressure it follows that the internal energy of the ideal gas depends upon the temperature alone.

By equation (39), Chapter VIII, in any fluid

$$dI = C_p dT + \left[V - T\left(\frac{dV}{dT}\right)_P\right] dP.$$

In the ideal gas $T\left(\dfrac{dV}{dT}\right)_P = V$, hence

$$dI = C_p dT,$$

and since C_p is independent of the pressure it follows that the total heat of the ideal gas also depends upon the temperature alone.

These results show that a gas which conforms exactly to the characteristic equation $PV = RT$ (T being the temperature on the thermodynamic scale) conforms exactly both to Boyle's Law (PV constant for any one temperature) and to Joule's Law (E a function of the temperature alone). It is therefore "perfect" in the sense of Art. 19.

When the equation $PV = RT$ was introduced in Art. 18 the symbol T denoted temperature on the scale of the gas thermometer, that is to say a scale defined by the expansion of the gas itself, and the gas was assumed to conform exactly to Boyle's Law. But if it also conforms exactly to Joule's Law, the scale of the gas thermometer coincides with the thermodynamic scale (Art. 42).

198. Isothermal and Adiabatic Expansion of an Ideal Gas. In the ideal gas, since E depends upon the temperature alone, it is constant during isothermal expansion, and therefore the work done by the gas is equal to the heat it takes in. The pressure varies inversely as the volume.

By equation (33 a), Chapter VIII, for the adiabatic expansion of any fluid

$$\left(\frac{dP}{dV}\right)_\phi = \gamma \left(\frac{dP}{dV}\right)_T.$$

Hence in the ideal gas $\left(\dfrac{dP}{dV}\right)_\phi = -\gamma\,\dfrac{P}{V}.$

So that in the adiabatic expansion of an ideal gas

$$\frac{dP}{P} + \gamma\,\frac{dV}{V} = 0.$$

If now we make the further assumption that γ is constant, which is equivalent to assuming that the specific heat does not vary with temperature, this gives on integration

$$\log_e P + \gamma \log_e V = \text{constant},$$

or $$PV^\gamma = \text{constant},$$

which is the adiabatic equation of a perfect gas with constant specific heat, arrived at otherwise in Art. 25.

199. Entropy, Energy and Total Heat of an Ideal Gas. By equations (8) and (9), Chapter VIII, in any fluid

$$d\phi = \frac{dE + P\,dV}{T} = \frac{dI - V\,dP}{T}.$$

In the ideal gas $\quad dE = C_v\,dT; \quad dI = C_p\,dT,$

and since $\qquad\qquad \dfrac{P}{T} = \dfrac{R}{V},$

$$d\phi = C_v \frac{dT}{T} + R \frac{dV}{V}$$

$$= C_p \frac{dT}{T} - R \frac{dP}{P}.$$

Hence if we again assume that the specific heat does not vary with the temperature,

$$E = C_v T + \text{constant},$$
$$I = C_p T + \text{constant},$$
$$\phi = C_v \log_e T + R \log_e V + \text{constant}$$
$$= C_p \log_e T - R \log_e P + \text{constant}.$$

The values of the constants depend on what initial state is chosen as the starting point of the reckoning. It is only changes in E, I and ϕ that can be determined by these formulas.

200. Ratio of Specific Heats. Method of inferring γ in Gases from the Observed Velocity of Sound. We saw (Art. 193) that in any fluid the ratio γ of the two specific heats, C_p/C_v, is equal to the ratio of the adiabatic elasticity e_ϕ to the isothermal elasticity e_t. Also that

$$e_t = - V \left(\frac{dP}{dV} \right)_T.$$

Hence, in a gas for which $PV = RT$,

$$e_t = V \left(\frac{P}{V} \right) = P, \text{ and } e_\phi = \gamma P.$$

This relation has been used as a means of finding γ experimentally in air and other gases which at ordinary temperatures and pressures very nearly conform to the equation $PV = RT$. The method is based on Newton's theory of the transmission of waves of sound. Newton showed that waves of compression and dilatation, such as those of sound, travel through any homogeneous fluid with a velocity which may be expressed as $\sqrt{eV}$, where V is as usual the volume of the fluid per unit mass (the reciprocal of the average density) and e is the elasticity, in kinetic units. It was afterwards pointed out by Laplace that in applying this result to the passage of sound through air or other gases e should be taken as the adiabatic elasticity e_ϕ, for the compressions and dilatations follow one another so fast as to leave no time for any substantial transfer of heat from the portions that are momentarily heated by compression to those that are momentarily cooled by expansion. Hence in air under atmospheric conditions, or in any other nearly perfect gas, sound travels at a rate equal to $\sqrt{\gamma P V}$. This fact is used as a means of determining γ by measuring the velocity of sound or (what comes to the same thing) by measuring the wave-length in sound of a known pitch.

In air at 0° C. and a pressure of one atmosphere the values given by various observers for the velocity of sound range from 33,060 to 33,240 centimetres per second*. Under these conditions the volume of one gramme of air is 773·5 cubic cms., and P is $1·0133 \times 10^6$ dynes per sq. cm. (Art. 12). Hence, taking an average of 33,150 for the velocity,

$$33,150 = \sqrt{\gamma \times 1·0133 \times 10^6 \times 773·5},$$

which gives $\gamma = 1·402.$

201. Measurement of γ by Adiabatic Expansion. Method of Clément and Desormes. Another method of determining the value of γ in a gas is by an experiment due originally to Clément and Desormes and improved on by Gay-Lussac and others. A quantity of the gas is contained in a large vessel at a pressure somewhat higher than that of the atmosphere, and at atmospheric temperature. There is a pressure-gauge attached, and a tap which may be opened to allow some of the gas to escape quickly. On opening the tap, the pressure falls suddenly to that of the atmosphere: when this happens the tap is at once closed. Then the pressure of the gas that remains in the vessel slowly rises, because the temperature, which had been reduced by the sudden expansion of the gas in the vessel while the tap was open, rises gradually to the value which it had at first, namely the temperature of the surrounding atmosphere. When this process is complete the final pressure is noted. Let the original pressure be P_1, the pressure of the atmosphere P_2 and the

* See Rayleigh's *Theory of Sound*, vol. ii. For recent measurements by the velocity of sound method and the correction for frequency, see Sherratt and Griffiths, *Proc. Roy. Soc.* vol. cxlvii, p. 292, 1934.

final pressure P_3. The change from P_1 to P_2 is approximately adiabatic on account of its suddenness: the change from P_2 to P_3 occurs at constant volume. Let V_1, V_2 and V_3 be the volumes of the gas *per unit mass*, at the three corresponding stages. Then $V_2 = V_3$. We have, in the adiabatic expansion,

$$P_1 V_1^\gamma = P_2 V_2^\gamma,$$

and since the initial and final temperatures are the same,

$$P_1 V_1 = P_3 V_3 = P_3 V_2.$$

Hence
$$\frac{P_1}{P_2} = \left(\frac{V_2}{V_1}\right)^\gamma = \left(\frac{P_1}{P_3}\right)^\gamma,$$

or
$$\gamma = \frac{\log P_1 - \log P_2}{\log P_1 - \log P_3}.$$

Values of γ are accordingly found by observing these three pressures. Experiments by Lummer and Pringsheim, using this method in an improved form, give 1·4025 as the value of γ for normal air*. An earlier application of the method by Röntgen gave 1·405†.

202. **Effect of Imperfection of the Gas on the Ratio of Specific Heats.** It has been already mentioned that in a perfect diatomic gas the ratio γ, as deduced from the molecular theory (see Chapter VII), should not exceed 1·4. In air the ratio, according to all the evidence, is, at ordinary temperatures and pressures, slightly greater. This is due partly to the presence of about one per cent. of (monatomic) argon, but mainly to the fact that air is an imperfect gas, deviating to a small extent both from Boyle's Law and from Joule's Law.

By equation (44), Chapter VIII, in any fluid,

$$C_p - C_v = \frac{(P+\sigma)(V+\rho)}{T},$$

where ρ is the cooling effect in the Joule-Thomson porous plug experiment (Art. 189), and σ is the cooling effect that would be found in unresisted expansion (Art. 190), without gain or loss of heat in either case. In a perfect gas ρ and σ are both nil, and the expression on the right becomes PV/T, as it should. With air (under usual conditions) both ρ and σ are small positive quantities: ρ was measured in the Joule-Thomson experiments, and σ, though it has not been directly measured, can be inferred from known experi-

* Cf. Partington, *Proc. Roy. Soc.* A, vol. c, 1921, p. 27.
† See Preston's *Theory of Heat*, Chapter IV.

mental data. Hence $C_p - C_v$ is a little greater than PV/T, which is the value it would have in a perfect gas.

The ratio γ is also a little greater in normal air than it would be in a perfect gas. In any fluid

$$\gamma = 1 + \frac{(P+\sigma)(V+\rho)}{C_v T}.$$

In air at ordinary temperatures the imperfection increases $(P+\sigma)(V+\rho)$ more than it increases C_v, and consequently makes γ slightly exceed the ideal value 1·4. But at high temperatures C_v is much increased (because the molecules then acquire energy of vibration) and γ is substantially reduced.

203. Relation of the Cooling Effects to the Coefficients of Expansion. The expressions for ρ and σ given in equations (41) and (43) of Chapter VIII may be put in another form which is convenient in dealing with imperfect gases.

By these equations, in any fluid

$$V + \rho = T\left(\frac{dV}{dT}\right)_P, \text{ and } P + \sigma = T\left(\frac{dP}{dT}\right)_V.$$

Here $\left(\frac{dV}{dT}\right)_P$ may be written αV, where α is the fractional increase of volume per degree, on the thermodynamic scale, when the fluid is heated at constant pressure. Measured at 0° C. α is the coefficient of expansion at constant pressure, or what is sometimes called the "volume-coefficient."

Similarly $\left(\frac{dP}{dT}\right)_V$ may be written βP, where β is the fractional increase of pressure per degree, on the thermodynamic scale, when the fluid is heated at constant volume. Measured at 0° C. β is what is called the "pressure-coefficient."

Hence at 0° C.

$$V_0 + \rho_0 = 273 \cdot 1 \alpha_0 V_0, \text{ and } P_0 + \sigma_0 = 273 \cdot 1 \beta_0 P_0,$$

the suffix being introduced to show that the quantities concerned are all to be taken as at 0° C.

The results of the Joule-Thomson porous plug experiments may be used to calculate ρ_0. They showed that with air the cooling effect of passing the plug was nearly proportional to the drop in pressure. It was different for different initial temperatures, becoming less when the initial temperature was raised. With air at 0° C. the cooling effect (according to the formula in Art. 123) was 0·275° for a pressure-drop of one atmosphere in passing the plug. Hence, using C.G.S. units, for air at 0° C.,

$$\left(\frac{dT}{dP}\right)_I = \frac{0\cdot275}{1\cdot0133\times10^6}.$$

C_p may be taken as 0·241 calory (Art. 14), equivalent in C.G.S. units of work to $0\cdot241\times4\cdot186\times10^7$. Multiplying the values of C_p and $\left(\dfrac{dT}{dP}\right)_I$ gives $\qquad\qquad \rho_0 = 2\cdot74,$

which is in cubic centimetres per gramme, the dimensions of ρ being the same as those of V, namely

$$\frac{\text{work}}{\text{pressure}\times\text{mass}} = \frac{\text{volume}}{\text{mass}}.$$

This result of the porous plug experiment may be applied to calculate the coefficient of expansion when air, at 0° C., is heated under a constant pressure of one atmosphere through one degree of the thermodynamic scale. We had

$$\alpha_0 = \frac{V_0 + \rho_0}{273\cdot1\,V_0}.$$

In air at 0° C. and a pressure of one atmosphere, the volume of one gramme is 773·5 cub. cms. Hence under these conditions we should have
$$\alpha_0 = \frac{773\cdot5 + 2\cdot74}{273\cdot1\times773\cdot5} = 0\cdot003675.$$

This is slightly larger than the mean coefficient that is found when the expansion of air at a constant pressure of one atmosphere is measured over a range of temperature from 0° C. to 100° C.

Again, taking the relation

$$P_0 + \sigma_0 = 273\cdot1\beta_0 P_0,$$

a value of σ_0 can be inferred when the pressure-coefficient is known. If β_0 for air be taken as about 0·003674, $P_0 + \sigma_0$ becomes $1\cdot0034P_0$, making $\sigma_0 = 0\cdot0034P_0$.

In a perfect gas both coefficients, α_0 and β_0, would be equal to $\dfrac{1}{273\cdot1}$ or 0·0036617. The scale of the perfect-gas thermometer, whether of the constant-volume or constant-pressure type, would coincide at all points with the thermodynamic scale*.

* Reference should be made to Callendar's paper "On the Thermodynamical correction of the Gas Thermometer" (*Phil. Mag.* January, 1903) for an account of how the absolute zero may be determined and intervals on gas and thermodynamic scales compared, by making use of the Joule-Thomson cooling effect and the measured coefficients of expansion.

204. **Forms of Isothermals. Diagrams of P and V, and of PV and P.** Taking any ideal gas, which satisfies the characteristic equation $PV = RT$, let us draw its isothermals on a diagram whose coordinates are the volume and the pressure. The characteristic equation shows that these curves are rectangular hyperbolas (fig. 84), for while any temperature remains constant P varies inversely as V. These isothermals for an ideal gas should be compared with those for a liquid and its vapour already illustrated in fig. 14 (Art. 76), to which we shall recur presently.

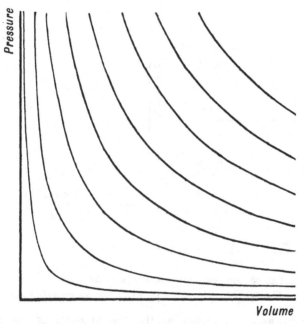

Fig. 84. Pressure-Volume isothermals for a perfect gas.

Another kind of isothermal curve, which Amagat showed to be useful in dealing with real gases, is drawn by taking as coordinates the product PV and the pressure. When this method is applied to an ideal gas the isothermals are simply horizontal straight lines (fig. 85), since at any temperature PV is constant. This is an obvious test of whether a gas obeys Boyle's Law. If it does, then the isothermals of PV in relation to P will be horizontal straight lines whether the gas also obeys Joule's Law or no. Any curvature in these lines, or any deviation from the horizontal, means a departure from Boyle's Law.

205. Imperfect Gases. Amagat's Isothermals of PV and P. No real gas conforms strictly to Boyle's Law. The experiments of Andrews, Amagat and others have shown that the departure from Boyle's Law becomes more and more marked as the critical point is approached. Amagat's experiments on the compressibility of gases, which extended up to very high pressures, show that when a line is drawn to exhibit the relation of PV to P at a constant temperature, its general form is that illustrated in fig. 86. Instead of being a horizontal straight line, as Boyle's Law would require,

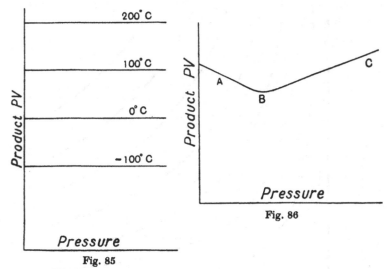

Fig. 85

Fig. 86

Fig. 85. Amagat isothermals for a perfect gas.

Fig. 86. Typical Amagat isothermal for an imperfect gas.

it consists as a rule of two nearly straight parts, A and C, one sloping down and the other sloping up, united by a smooth curve. There is consequently on each isothermal a minimum value of PV at a particular pressure. For pressures less than this $\left(\dfrac{d(PV)}{dP}\right)_T$ is negative; for greater pressures it is positive. The particular pressure at which the minimum of PV is found depends on the temperature. With rising temperature the position of the minimum point B shifts first to the right and then to the left; and if the temperature is high enough it may reach the axis of zero pressure and disappear, with the result that the whole isothermal then consists of an upward-sloping line like BC.

The general features of these isothermals will be apparent from figs. 87, 88 and 89, which are representative examples of Amagat's curves*. The temperature for which each isothermal is drawn is

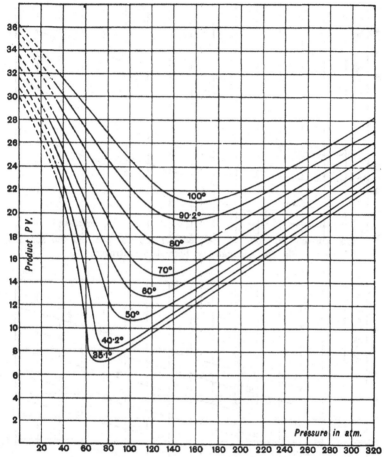

Fig. 87. Amagat's isothermals for carbon dioxide.

marked on it. In fig. 87, which relates to carbon dioxide, the temperatures for which the selected curves are drawn are all above the critical point, but the lowest is not far from it. The left-hand branch of the curve consequently slopes down very fast; if an isothermal were drawn for the critical temperature its direction at the critical

* E. H. Amagat, *Annales de Chimie et de Physique*, vol. XXII, 1881. See also vol. XXIX, 1893.

pressure would become vertical, for at the critical point $\left(\dfrac{d\,(PV)}{dP}\right)_T$ is infinite. At higher temperatures the left-hand branch slopes

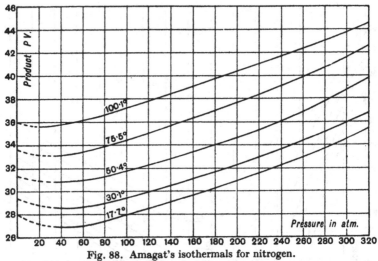

Fig. 88. Amagat's isothermals for nitrogen.

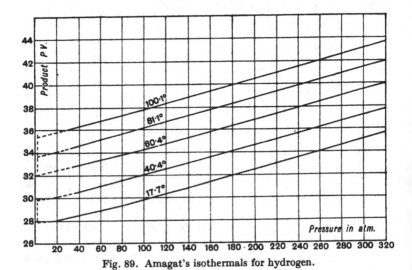

Fig. 89. Amagat's isothermals for hydrogen.

down less steeply, and within the range of this diagram the minimum point moves to the right; but (as further experiments proved) at higher temperatures still it moves to the left. This feature is apparent in the curves of fig. 88, which relate to nitrogen. There the measurements were made at temperatures much more remote from the critical temperature. The downward-sloping branch is short, and becomes shorter when the temperature is raised. Finally, with hydrogen (fig. 89), where the critical point was even more remote, the minimum has disappeared, and each isothermal is a line sloping up along its whole course. At any very low temperature, however, an isothermal for hydrogen would have a branch sloping downwards, for moderate values of P, followed by a minimum of PV and then an upward slope just as in other gases*. If PV is not a constant at constant temperature, then an expression of the form $PV = A + BP + CP^2 + DP^3 \ldots$ can represent the behaviour of a given gas and A, B, C, D, etc. are functions of the temperature called "virial coefficients." For all real gases the second virial coefficient is negative at low temperatures, becomes zero at the so-called "Boyle Temperature," and then becomes positive as the temperature is raised. The higher virial coefficients are much smaller, so that the value of B becomes the same as $\left(\dfrac{\partial (PV)}{\partial P}\right)_{p=0}$ and the slope of the PV curve at $p=0$ gives B the value of the second virial coefficient; when this slope is zero, then the isothermal corresponds to the Boyle temperature. The Boyle temperature for air is about 75° C., for hydrogen about 100° K.

Holborn and Otto have carefully determined the virial coefficients for a number of gases and their variation with the temperature†.

206. Isothermals on the Pressure-Volume Diagram. Another method of exhibiting the departure of real gases from Boyle's Law is to draw isothermal lines on a diagram of the type of fig. 84, the coordinates of which are simply the pressure and the volume. An example of such a diagram was described in Art. 76 and is reproduced here for convenience (fig. 90). In the upper right-hand part of the diagram the substance is in the state of an imperfect gas: the highest isothermal (marked G) does not differ much from a rectangular hyperbola. Below the critical temperature each isothermal is a broken line with three separate parts, namely the part AB, in which the substance is a homogeneous liquid, and the part CD, in which it is a homogeneous vapour. These are joined by the straight line BC which exhibits the change of phase from liquid to

---

* *Phil. Mag.* April, 1896.
† *Zeit. f. phys.* vol. XXXIII, p. 9, 1925.

vapour. During that change the substance is not homogeneous; it consists of a mixture of the two phases, liquid and vapour. The loci of B and of C together constitute the boundary curve, the apex of which is the critical point.

The isothermal for the critical temperature E, fig. 90, touches the boundary curve at the critical point. Its direction at the critical point is horizontal and it has a point of inflection there; consequently at that point

$$\left(\frac{dP}{dV}\right)_T = 0 \text{ and } \left(\frac{d^2P}{dV^2}\right)_T = 0.$$

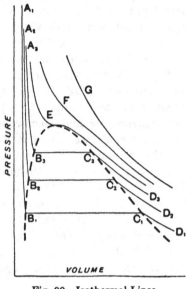

Fig. 90. Isothermal Lines.

207. Continuity of Liquid and Gas. The idea of continuity between the liquid and gaseous states in any substance, which was referred to in Art. 76, received a remarkable development in the speculations of James Thomson[*]. He suggested that we might think of the curves AB and CD as parts of one continuous curve $ABJKLCD$ (fig. 91). The parts BJ and LC correspond to real phenomena of the kind referred to in Art. 79. For in certain circumstances the pressure of a liquid may be reduced below the saturation pressure corresponding to the temperature, without vaporization, and a vapour may be compressed beyond its saturation pressure without condensation. Points between B and J, and between C and L, accordingly represent

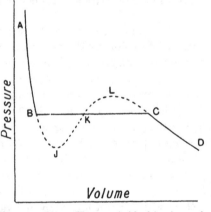

Fig. 91. James Thomson's ideal isothermal.

conditions in which a homogeneous fluid may temporarily exist

[*] *Proc. Roy. Soc.* November, 1871. *Collected Papers*, pp. 276–333.

in metastable states. But points between J and L cannot be realized in a homogeneous fluid: they would be completely unstable, for they would require the pressure and the volume to increase together. Hence the connecting portion of the curve is no more than a mathematical abstraction, but it allows a continuous expression for P in relation to V to be interpreted for isothermals below the critical temperature as well as for isothermals above that temperature. The straight line BC represents the ordinary process of vaporization or condensation at constant pressure. It is interesting to notice that the theoretical connecting curve, which we may call the James Thomson wave, must satisfy this thermodynamic condition, that the area BJK is equal to the area KLC. For we may conceive the fluid to be taken through a complete cycle, from B through JKL to C, and then back to B by the straight line CB. During this cycle its temperature does not change, and therefore, by Carnot's principle, the work done in the cycle as a whole is nil. Accordingly $\int P\,dV$ for the complete cycle must vanish: hence the positive area KLC must be equal to the negative area BJK.

It follows that when we are able to draw for any fluid the theoretical isothermal $AJLD$, from a knowledge of the characteristic equation, we may go on to determine the saturation pressure corresponding to the temperature for which the curve is calculated, since that is the pressure at which the straight line BC must be drawn to make the area BJK equal to the area KLC.

208. Van der Waals' Characteristic Equation. A form of characteristic equation, applicable to any fluid, was devised by Van der Waals* which approximately expresses the relation of P, V and T under all conditions of the fluid through any range of density and temperature, from the state of liquid to that of gas or vapour at low pressures, when the behaviour approaches that of a perfect gas. Although Van der Waals' equation cannot be accepted as exact it gives results which correspond with the broad features that are exhibited by real fluids, in all possible liquid or gaseous states, and throws light on the phenomena of the critical condition and the question of continuity.

Van der Waals' equation was based on the kinetic theory of gases. No more than a rough outline can be given here of the considera-

---

* *The Continuity of the Liquid and Gaseous States of Matter*, published in Dutch in 1878; Eng. Trans. in Physical Society's *Physical Memoirs*, vol. I, part III, 1891.

tions involved in framing it*. The kinetic theory shows that a gas which consists of colliding molecules will conform to the ideal equation $PV = RT$ only if (1) the size of the molecules is indefinitely small compared with the space traversed by them between their encounters, and (2) no appreciable part of the energy of the gas is due to the mutual attraction of the molecules for one another. Neither of these conditions holds in a real gas. In a real gas the volume of the molecules is an appreciable part of the whole volume occupied by the gas, and it is only after making a deduction for it that we have the volume which can be reduced by applying more pressure. Again, during their encounters the molecules attract one another across short distances so that internal work is done in separating them. The result is that this attraction between the molecules assists the pressure exerted by the envelope in preventing the gas from expanding. If the first of these two effects stood alone we should have

$$P\,(V-b) = RT,$$

where b, which is called the "co-volume," represents the deduction due to the volume of the molecules. But in consequence of the second effect we have to add to P a term depending on the attraction between the molecules. Taking any imaginary plane of separation between two portions of the gas, the attraction between molecules across that plane will depend on the number of molecules which are at any moment so near as to be exercising mutual forces: in other words upon the number of encounters that occur on the separating plane per unit of time. But that will depend on the square of the density, for it is proportional to the product of the numbers of molecules per unit of volume on the two sides of the plane. Accordingly Van der Waals takes a/V^2 as the term which is to be added to P to represent the effect of the mutual molecular attractions. He treats a and b as constants for any particular fluid. His characteristic equation therefore takes the form

$$\left(P + \frac{a}{V^2}\right)(V-b) = RT \quad\ldots\ldots\ldots\ldots\ldots\ldots(1).$$

Numerical values of the constants can be found for any fluid by observing experimentally the relations of pressure, volume and temperature in different conditions of the fluid, or they may be inferred from other experimental results. Van der Waals' equation is intended to apply to any homogeneous state, gaseous or liquid. It

---

* Students wishing to pursue the matter should consult Jeans' *Dynamical Theory of Gases*, 3rd edition, Chapter VI.

does in fact reproduce with remarkable comprehensiveness the chief phenomena of both states, and also those of the critical point, though in some particulars it fails to give correct quantitative results.

It may help towards an appreciation of the physical meaning of the term $\frac{a}{V^2}$ if we consider the isothermal expansion of a gas to which Van der Waals' equation applies. When any fluid expands in any manner the heat taken in, dQ, is, by equation (21) of Chapter VIII,

$$dQ = C_v dT + T \left(\frac{dP}{dT}\right)_V dV.$$

If the expansion is isothermal this becomes

$$dQ = T \left(\frac{dP}{dT}\right)_V dV.$$

Now in a Van der Waals gas

$$P = \frac{RT}{V-b} - \frac{a}{V^2} \quad \ldots\ldots\ldots\ldots\ldots\ldots(1\,a),$$

from which

$$\left(\frac{dP}{dT}\right)_V = \frac{R}{V-b} \quad \ldots\ldots\ldots\ldots\ldots\ldots(2).$$

Hence in the isothermal expansion of such a gas

$$dQ = \frac{RT}{V-b} dV = \left(P + \frac{a}{V^2}\right) dV$$

$$= P dV + \frac{a}{V^2} dV \quad \ldots\ldots\ldots\ldots\ldots\ldots(3).$$

But $P dV$ is dW, the external work done during the expansion. Comparing this with the general equation

$$dQ = dW + dE,$$

we see that in a Van der Waals gas there is an increase of internal energy (dE) during isothermal expansion, which is equal to $\frac{a}{V^2} dV$. We may regard this as *internal work* done against a cohesive force $\frac{a}{V^2}$ resisting the expansion, independently of the external pressure P. In a perfect gas there would be no change of E in isothermal expansion (Art. 197).

On assigning various constant values to T the Van der Waals equation gives isothermal curves which have all the general characteristics of those shown in figs. 86 to 91. When the substance is in the gaseous condition and at any very low pressure, V is so large that the terms a/V^2 and b become negligibly small: the gas then approximates to the ideally perfect state and the equation gives nearly the same results as those of the perfect-gas equation $PV = RT$. At higher pressures both of the modifying terms become important.

The equation may be written thus, as a cubic in V,

$$V^3 - \left(b + \frac{RT}{P}\right) V^2 + \frac{aV}{P} - \frac{ab}{P} = 0 \quad \ldots\ldots\ldots\ldots(1\,b).$$

This gives three roots, real or imaginary, for V, corresponding to any assigned value of P, on any isothermal. When the temperature for which the isothermal is drawn is higher than the critical temperature, only one of the three roots is real: that is to say there is only one value of V for each value of P on any isothermal above the one that passes through the critical point. For any temperature below the critical temperature all three roots are real in the mathematical sense. The isothermal curve calculated from the equation then takes the continuous form conceived by James Thomson, and illustrated in fig. 91. One of the three roots corresponds to a point on the curve AJ, one to a point on LD, and the third to a point (not realizable experimentally) on JL.

Van der Waals' equation makes the product PV, for constant T, vary in the manner indicated by Amagat's isothermals, showing a minimum at a particular value of the pressure that depends on the temperature for which the isothermal is drawn. Writing the equation in the form

$$PV = \frac{RTV}{V-b} - \frac{a}{V},$$

and differentiating with respect to P, keeping T constant, we have

$$\left(\frac{d(PV)}{dP}\right)_T = \left[\frac{-RTb}{(V-b)^2} + \frac{a}{V^2}\right]\left(\frac{dV}{dP}\right)_T \quad \ldots\ldots\ldots\ldots(4).$$

Since on any isothermal $\dfrac{d(PV)}{dP}$ is zero at the minimum of PV, the quantity within the square brackets must vanish at that point. Hence, on any isothermal, the minimum of PV is found when the volume is such that

$$\left(\frac{V}{V-b}\right)^2 = \frac{a}{bRT}, \quad \text{or} \quad \left(1 - \frac{b}{V}\right)^2 = \frac{bRT}{a} \quad \ldots\ldots\ldots\ldots(5).$$

This shows that the volume at which the minimum of PV occurs on any isothermal becomes greater as the temperature is raised.

In the particular case when the temperature is, so high that the minimum occurs on the PV axis, where P is zero, V is indefinitely large; $1 - \dfrac{b}{V}$ then becomes equal to 1, and T, the Boyle temperature, is given by the equation $T = \dfrac{a}{bR}$. Hence in a fluid which satisfies Van der Waals' equation an Amagat isothermal for a temperature equal to $\dfrac{a}{bR}$ would slope upwards along its whole course, with increasing P, but an isothermal for any temperature lower than this would first dip towards a minimum of PV and then rise.

209. **Critical Point according to Van der Waals' Equation.** To find the critical point of a fluid which satisfies Van der Waals' equation, we may most conveniently write the equation in the form

$$P = \frac{RT}{V-b} - \frac{a}{V^2},$$

from which

$$\left(\frac{dP}{dV}\right)_T = \frac{-RT}{(V-b)^2} + \frac{2a}{V^3} \quad\ldots\ldots\ldots\ldots\ldots(6),$$

and

$$\left(\frac{d^2P}{dV^2}\right)_T = \frac{2RT}{(V-b)^3} - \frac{6a}{V^4} \quad\ldots\ldots\ldots\ldots(7).$$

At the critical point

$$\left(\frac{dP}{dV}\right)_T = 0 \text{ and } \left(\frac{d^2P}{dV^2}\right)_T = 0.$$

Hence, writing T_c, P_c and V_c for the critical temperature, pressure and volume, we should have, in a Van der Waals fluid,

$$\frac{RT_c}{(V_c-b)^2} = \frac{2a}{V_c^3} \text{ and } \frac{2RT_c}{(V_c-b)^3} = \frac{6a}{V_c^4} \quad\ldots\ldots\ldots(8).$$

This gives

$$\frac{2}{V_c-b} = \frac{3}{V_c},$$

from which

$$V_c = 3b \quad\ldots\ldots\ldots\ldots\ldots\ldots(9).$$

It follows from this result and from (8) above that

$$\frac{RT_c}{(3b-b)^2} = \frac{2a}{27b^3},$$

from which

$$T_c = \frac{8}{27}\frac{a}{Rb} \quad\ldots\ldots\ldots\ldots\ldots(10).$$

Also, from the original equation,

$$P_c = \frac{RT_c}{3b-b} - \frac{a}{9b^2} = \frac{4}{27}\frac{a}{b^2} - \frac{a}{9b^2} = \frac{a}{27b^2} \quad\ldots\ldots\ldots(11).$$

Thus if the constants a and b as well as R were known for a gas which strictly satisfied Van der Waals' equation, the critical volume temperature and pressure might be calculated: or conversely the constants might be inferred from known values of T_c, P_c and V_c.

It follows also that in such a gas the three critical quantities would be connected by the relation

$$P_c V_c = \tfrac{3}{8}RT_c \quad \dots\dots\dots\dots\dots\dots(12),$$

which shows how widely the condition of the fluid then differs from that of a perfect gas.

In applying his equation to carbon dioxide, Van der Waals deduced from the experiments of Regnault and of Andrews these values: $a = 0.00874$, $b = 0.0023$, $R = 0.003685$, the unit of pressure being one atmosphere, and the quantity of gas considered being that which occupies unit of volume at a pressure of one atmosphere and a temperature of $0°$ C. With these constants the calculated critical temperature is $32°$ C., which agrees fairly well with the value observed by Amagat, namely $31.3°$ C. In other particulars the agreement is less good; thus the calculated critical pressure is 61.2 atmospheres, whereas Amagat's observed value is 72.9 atmospheres. This discrepancy is only one illustration of the fact that Van der Waals' equation is inexact and that it fails to represent the behaviour of a gas accurately in the neighbourhood of the critical point.

210. Corresponding States. If we have two or more different fluids to which Van der Waals' equation applies, with different constants for each fluid, an important relation between them can be established by selecting scales of temperature, pressure and volume such that the critical temperatures of the different fluids are expressed by the same number, the critical pressures by the same number, and the critical volumes by the same number. Isothermal curves drawn to these scales for the different fluids will then coincide: in other words a single diagram will show the relation of P to V in all the fluids, when it is read by reference to the appropriate scales. Similarly a single diagram will show the Amagat curves (PV and P) for all. Any point taken in such a diagram, interpreted on the proper scale, marks a definite state for each fluid; and for the different fluids it marks what are called "corresponding states." The critical points for the different fluids furnish an obvious example of corresponding states.

Thus fluids are said to be at corresponding pressures when their pressures bear the same ratio to the respective critical pressures: they are said to be at corresponding volumes when their volumes bear the same ratio to the respective critical volumes, and at corresponding temperatures when their temperatures bear the same ratio to the respective critical temperatures. If substances conform to a characteristic equation of the Van der Waals type all three quantities P, V and T simultaneously have "corresponding" values in the sense here defined. To put this statement in another form, let the unit of temperature chosen for each fluid be its (absolute) critical temperature, the unit of volume its critical volume, and the unit of pressure its critical pressure. Then the same family of curves, either on the pressure-volume diagram or the Amagat diagram, will serve to represent the isothermals for all fluids that conform to a characteristic equation of the Van der Waals type.

That this is true of any fluid to which the Van der Waals equation applies will be seen by reducing the equation to a more general form. Take any such fluid, in any given state, and write its pressure P as $p_r P_c$, where p_r is the number by which the pressure is stated when we use the critical pressure P_c as the unit of pressure. Similarly for V write $v_r V_c$, where v_r is the number by which the volume is stated when we use the critical volume V_c as the unit of volume; and for T write $t_r T_c$, where t_r is the number that expresses the (absolute) temperature when we use the critical temperature T_c as unit of temperature. The quantities p_r, v_r and t_r are called the "reduced" pressure, volume and temperature respectively. Then

$$P = p_r P_c = p_r \frac{a}{27b^2},$$

$$V = v_r V_c = v_r . 3b,$$

$$T = t_r T_c = t_r \frac{8}{27} \frac{a}{bR}.$$

On substituting these values in Van der Waals' equation,

$$\left(P + \frac{a}{V^2}\right)(V - b) = RT,$$

it will be seen that the constants a, b and R cancel out, and the equation becomes

$$\left(p_r + \frac{3}{v_r^2}\right)(v_r - \tfrac{1}{3}) = \tfrac{8}{3} t_r \dots\dots\dots\dots\dots(13).$$

The constants that characterized a particular fluid have disappeared. Accordingly this "reduced" characteristic equation, as it is called, is true of any substance that satisfies a Van der Waals equation, and consequently the forms of the curves connecting p_r, v_r and t_r are the same for all such substances.

In other words, if we compare any two such substances, using say the temperature and pressure as independent variables for the purpose of specifying the state, and choose "corresponding" values of the temperature and pressure for the two, then the volumes will also have "corresponding" values.

This is the theorem of corresponding states, first enunciated by Van der Waals. It was tested by Amagat and found by him to be nearly true of a number of fluids which he examined through a wide range of conditions, and it has been shown to hold approximately in many substances*. The validity of the principle does not depend on the precise form of the characteristic equation: Van der Waals' equation is by no means the only one that would lead to the same conclusion. Any characteristic equation connecting P, V and T with no more than three independent constants (two adjustable constants in addition to R), and giving a critical point, can be brought in like manner to the form of a "reduced" equation in which the constants peculiar to the fluid have disappeared. Hence any such equation serves equally well as a basis for the theorem of corresponding states.

211. Van der Waals' Equation only Approximate. Useful as Van der Waals' equation is in exhibiting broadly the behaviour of a gas even in extreme variations of state, it cannot be brought by any adjustment of the constants into exact agreement with the results of experiment. It appears that the actual properties of a gas are too complex to admit of complete statement by the use of so small a number of constants. The quantities a and b of the equation are not strictly constant; they are to some extent functions of the temperature, or the density, or both. If constants are selected which fit observations of the compressibility, the equation fails to agree with measured values of the critical volume and critical pressure. Further, from equation (12) of Art. 209 we should expect the ratio RT/PV to have at the critical point the value $\frac{8}{3}$ or 2·667, whatever be the values of the constants a and b. But the observations of Young show that this ratio is not the same in all gases at the critical

---

* See S. Young, *Phil. Mag.* February, 1892; also his book on *Stoichiometry*.

point, that in most gases it is about 3·7, but in some it may be less than 3·5 and in others more than 4. The relations of the critical temperature, pressure and volume are in fact less simple than is consistent with the formula of Van der Waals: the critical points in different actual fluids are not strictly "corresponding" states, and there is some departure from the theorem of Art. 210.

Again, taking the Van der Waals equation

$$P = \frac{RT}{V-b} - \frac{a}{V^2},$$

and differentiating with respect to T, keeping V constant, we have

$$\left(\frac{dP}{dT}\right)_V = \frac{R}{V-b} \quad \dots\dots\dots\dots\dots\dots(14),$$

which means that when a Van der Waals fluid is heated at constant volume the increment of pressure per degree of rise in temperature is constant. Hence with such a fluid, whether liquid or gaseous, a thermometer of the constant-volume type would give readings on the thermodynamic scale without correction. In other words the observed pressure coefficient would be independent of the temperature. This is not true of real fluids.

Again, if a substance conformed strictly to the equation of Van der Waals it would follow that C_v, the specific heat at constant volume, would be constant at any one temperature, and would therefore be the same for the liquid as for the gas at any temperature at which both states of aggregation are possible. For taking equation (14) and again differentiating with respect to T, we have

$$\left(\frac{d^2P}{dT^2}\right)_V = 0.$$

Hence by equation (23), Chapter VIII, $\left(\frac{dC_v}{dV}\right)_T = 0$: that is to say C_v would be constant at any one temperature. This, however, is not confirmed by measurements of the specific heat.

Another important particular in which Van der Waals' equation fails to give results that agree with those of experiment is in the cooling effect of throttling. This effect has been measured in various gases and vapours by experiments like the porous-plug experiments of Joule and Thomson. Such experiments show that in any real gas the effect suffers an inversion when the initial temperature of the gas is sufficiently high; that is to say, at high temperatures the effect of throttling is to heat the gas instead of to cool it. The fact of this inversion can be deduced from the Van der Waals equation*, and to that extent the equation is satisfactory. But the amount of the cooling effect in such a gas as carbon dioxide, when calculated from the Van der Waals equation (with constants which suit the form of the isothermal curves) falls much short of the cooling effect that is actually observed; and if the constants of the equation are adjusted to make the observed and calculated cooling effects agree, then the equation does not accord with the observed figures for compressibility†.

212. Other Characteristic Equations: Clausius, Dieterici. Enough has been said to show that Van der Waals' equation cannot be brought into exact agreement with the deviations from Boyle's Law and Joule's Law which are found in an actual gas. The reason has already been indicated—that the "constants" of the equation are not strictly constant. In particular the attraction between the molecules, on which a depends, is probably a function of the temperature, although it is treated in the equation as independent of the temperature. Various attempts have been made to modify the equation so as to bring it into closer accord with the known properties of gases. None of these have been completely successful in giving a formula which will stand all tests throughout a very wide

---

* To show this we may use equation (41 a) of Chapter VIII, which expresses the cooling effect in any fluid as

$$\rho = -\left(\frac{dE}{dP}\right)_T - \left(\frac{d(PV)}{dP}\right)_T.$$

In a Van der Waals fluid, by equation (4) above,

$$\left(\frac{d(PV)}{dP}\right)_T = \left[\frac{-RTb}{(V-b)^2} + \frac{a}{V^2}\right]\left(\frac{dV}{dP}\right)_T.$$

Further $\left(\frac{dE}{dP}\right)_T = \left(\frac{dE}{dV}\right)_T \left(\frac{dV}{dP}\right)_T$, and $\left(\frac{dE}{dV}\right)_T = T\left(\frac{dP}{dT}\right)_V - P$

by equation (35), Chapter VIII. Hence we should have

$$\left(\frac{dE}{dV}\right)_T = \frac{RT}{V-b} - P = \frac{a}{V^2}, \text{ and } \left(\frac{dE}{dP}\right)_T = \frac{a}{V^2}\left(\frac{dV}{dP}\right)_T.$$

Adding the two terms, the whole cooling effect in a fluid which obeys Van der Waals' equation would be

$$\rho = -\left[\frac{2a}{V^2} - \frac{RTb}{(V-b)^2}\right]\left(\frac{dV}{dP}\right)_T \dots\dots\dots\dots\dots(15).$$

By making T sufficiently large the second term within the square brackets exceeds the first, which means an inversion of the effect. When the fluid is a gas at low pressure, and V is consequently very large compared with b, the condition for inversion is that $RTb = 2a$: in other words the inversion temperature in a gas at very low pressure would be $2a/Rb$, or twice the Boyle temperature.
    † See Callendar, *Phil. Mag.* January, 1903, pp. 58–60.

range of states, though in some respects the modified equations approximate better to the observed facts.

Clausius[*] gave a characteristic equation which we may write in the form

$$P = \frac{RT}{V-b} - \frac{a'}{T(V+b')^2} \quad \dots\dots\dots\dots(16),$$

where a' and b', as well as b and R, are constants. On comparing this with equation (1 a), it will be seen to differ from Van der Waals' mainly by the presence of T in the denominator of the last term, which expresses the addition to P that is due to intermolecular attractions. Clausius assumes that these attractions become reduced when the temperature rises; he thereby gets an equation which, while it gives to the isothermals the same general form as is given by the equation of Van der Waals, agrees better with the Joule-Thomson cooling effect. When the same method of finding the critical quantities is applied to it, by writing

$$\left(\frac{dP}{dV}\right)_T = 0 \text{ and } \left(\frac{d^2P}{dV^2}\right)_T = 0,$$

one finds that $V_c = 3b + 2b'$,

$$T_c = \sqrt{\frac{8a'}{27R(b+b')}},$$

$$P_c = \sqrt{\frac{a'R}{216(b+b')^3}}.$$

For carbon dioxide Clausius gives his constants the following values: $R = 0.008688$, $b = 0.000843$, $a' = 2.0935$, $b' = 0.000977$, the unit of pressure being again one atmosphere, and the quantity of gas considered being that which occupies unit volume at one atmosphere and 0° C. With these constants the calculated critical temperature is 31° C. and the calculated critical pressure is 77 atmospheres.

Clausius draws a theoretical isothermal curve of pressure and volume for carbon dioxide at 13·1° C. calculated from his formula. This curve, which is reproduced in fig. 92, shows the form assumed by the James Thomson wave in the Clausius type of characteristic equation. The horizontal straight line BC, which exhibits the process of liquefaction, is so drawn that the crest and hollow of the wave shall have equal areas (Art. 207): this consideration determines its height and therefore fixes the saturation pressure. The

[*] *Phil. Mag.* June, 1880.

dotted portions of the curve exhibit imaginary states, comprised within the characteristic equation, which serve to establish theoretical continuity between the real state of homogeneous liquid AB and the real state of homogeneous vapour CD.

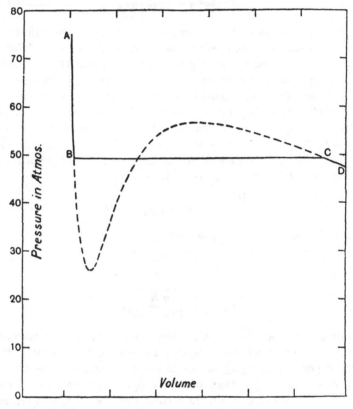

Fig. 92. Theoretical isothermal of CO_2 at $13 \cdot 1°$ C. according to the equation of Clausius.

A modified and more general type of Clausius' equation is obtained by writing

$$P = \frac{RT}{V-b} - \frac{a'f(T)}{(V+b')^2} \quad \dots\dots\dots\dots\dots (17),$$

where $f(T)$ is a function of T such as to diminish with rising temperature. In the original equation of Clausius, $f(T) = \frac{1}{T}$. Van der Waals has suggested that $f(T)$ may be $e^{1-\frac{T}{T_c}}$, where e is $2 \cdot 7188$, the base of the Napierian logarithms, and T_c is the critical temperature. In that case, at the critical temperature $f(T)$ would become equal to 1.

This form of characteristic equation was adopted by Mollier in calculating his tables of the properties of carbon dioxide*.

Still another characteristic equation of the same comprehensive kind is that of Dieterici†, who writes

$$P(V-b) = RTe^{\frac{-a}{RTV}} \quad\text{......................(18)},$$

where e is again the number 2·7183, and a, b and R are constants. Like the others, this formula is founded on the kinetic theory and like them it reproduces the general features of isothermal curves under all conditions and gives a critical point. Since it has only two constants besides R, the principle of corresponding states holds good for the relation it establishes between P, V and T. It makes

the critical temperature $T_c = \dfrac{a}{4bR}$, the critical volume $V_c = 2b$, and

the critical pressure $P_c = \dfrac{a}{4b^2e^2}$. Hence at the critical point the ratio

RT/PV becomes equal to $\frac{1}{2}e^2$ or 3·695, a value which is in much better agreement with observed results than was the value 2·667 calculated from Van der Waals' equation (Art. 209). In respect also of the Joule-Thomson cooling effect and its inversion‡ Dieterici's equation gives a better agreement with experiment than does Van der Waals'.

A more general form of the Dieterici equation is obtained by writing T^n instead of T in the index term, thus introducing one more adjustable constant:

$$P(V-b) = RTe^{\frac{-a}{RT^nV}} \quad\text{..................(19)}.$$

The critical temperature then becomes $\sqrt[n]{\dfrac{a}{4bR}}$. The principle of corresponding states would still apply to any group of substances for which n had the same value, since each substance in the group would still have only two constants individual to itself.

213. Callendar's Equation. Many other forms of characteristic equation have been devised but none of them is completely

---

* Mollier, *Zeitschrift für die gesammte Kälte-Industrie*, vol. ii, 1895 and vol. iii, 1896.

† *Annalen der Physik*, vol. v, p. 51, 1901.

‡ See Porter, *Phil. Mag.* April, 1906 and June, 1910.

successful in representing the behaviour of a fluid in all possible states. But for the practical purpose of enabling tables to be calculated which will show the properties of a fluid *throughout a limited range of variation of state*, it is not impossible to frame a characteristic equation which, by empirical adjustment of the constants, can be made to apply with sufficient accuracy within that range, though it may fail entirely when carried beyond the range. A conspicuous example of this less ambitious type of characteristic equation is the one which Callendar applied to calculate his original tables of the properties of steam*. It served to tabulate the properties of steam within the limits of pressure and temperature which were then usual in steam-engine practice, but it does not apply to the much higher pressures that are now employed.

This equation, which Callendar took as characteristic of any vapour, saturated or superheated, provided the pressure was much lower than the critical pressure, was

$$V = \frac{RT}{P} - c + b \quad \dots\dots\dots\dots\dots\dots(20).$$

Here $\frac{RT}{P}$ is, as usual, the ideal volume of a perfect gas; b is a constant representing the co-volume, as in other characteristic equations; and c is a term which is not constant but is a function of the temperature. Callendar took $c = \frac{C}{T^n}$, where C is a constant and n is a number depending on the nature of the gas. The term c represents the effect of intermolecular forces, but instead of regarding these forces as augmenting the influence of the external pressure (which Van der Waals did by adding the term $\frac{a}{V^2}$ to P), Callendar represented by c their effect in reducing the volume below its ideal value, in consequence of the "co-aggregation" or temporary interlinking of some of the molecules during their encounters. He called c the "co-aggregation volume" and treated it, at the moderate pressures within which the equation was applied, as a function of the temperature only. This assumption would not be true under conditions of high density, but for saturated or superheated steam at what would now be called low pressures it gave results which agreed well with those of experiment.

---

* Callendar, *Proc. Roy. Soc.* vol. LXVII, p. 266, 1900; *Phil. Mag.* January, 1903; *Encyc. Brit.* 11th edition, Articles "Thermodynamics" and "Vaporization."

The fact that Callendar's equation applies only to gases and vapours at low and moderate pressures will be obvious when one considers the form of the isothermal lines which it gives on a diagram of PV and P. We may write it

$$PV = RT - cP + bP \quad \dotfill (20\ a).$$

Since c is a function of T only, and is therefore constant along any one isothermal, this gives

$$\left(\frac{d\,(PV)}{dP}\right)_T = -c + b \quad \dotfill (21).$$

Hence in a gas which obeys Callendar's equation the isothermal lines would be straight, inclined downwards, with increasing P, if c is greater than b, and inclined upwards if b is greater than c. There would be no minimum of PV nor change of inclination along any isothermal line. The equation therefore can apply only under conditions such that the lines are substantially straight, namely at low or moderate pressures. Starting from $P = 0$ the lines are in fact nearly straight for some distance; and, as we saw in Art. 205, they slope down when the temperature is low and slope up when it is high. In any gas, at a sufficiently low temperature c is greater than b, and an isothermal line there will slope down. As the temperature increases for which the isothermal is drawn, c becomes less, since $c = \frac{C}{T^n}$, and a temperature is reached at which the line runs level $(c = b)$. For any higher temperature than this the line slopes up, like the lines for hydrogen in fig. 89.

The temperature at which the sign of the slope changes will be relatively low in a gas which, like hydrogen, has a low critical temperature, and will be relatively high in a gas like carbon dioxide, as might be inferred from the principle of corresponding states.

Callendar's equation has the convenient property that differential expressions deduced from it for the various quantities E, I, ϕ, C_p, C_v and so forth, by applying the general thermodynamic relations of Chapter VIII, take forms such as may be readily integrated. Hence it enabled numerical values of these quantities which were thermodynamically consistent with one another to be calculated to any desired number of figures.

The first publication of steam tables on this basis was an important service, and though the later requirements of engineering have made the original tables obsolete and have compelled modifications of Callendar's procedure his methods still find application in calculating the properties of steam. In the first edition of this book they were described in much detail. A briefer reference will now suffice.

In dealing with a characteristic equation such as Callendar's it would be possible to fix the constants by reference only to experiments on the compressibility of the gas at various temperatures, if sufficiently accurate data of that kind were available. Callendar preferred to fix them by reference mainly to observed values of the Joule-Thomson cooling effect. Their relation to the cooling effect will be apparent from what follows.

214. Deductions from the Callendar Equation. Taking the Callendar equation,

$$V = \frac{RT}{P} - c + b,$$

since

$$c = \frac{C}{T^n},$$

we have

$$\frac{dc}{dT} = -\frac{nC}{T^{n+1}} = -\frac{nc}{T}, \text{ and } \frac{d^2c}{dT^2} = \frac{n(n+1)C}{T^{n+2}} = \frac{n(n+1)c}{T^2}.$$

Also, since

$$\frac{c}{T} = -\frac{1}{n}\frac{dc}{dT}, \quad \frac{d}{dT}\left(\frac{c}{T}\right) = -\frac{(n+1)c}{T^2}.$$

Differentiating the equation with respect to T, keeping P constant,

$$\left(\frac{dV}{dT}\right)_P = \frac{R}{P} - \frac{dc}{dT} = \frac{R}{P} + \frac{nc}{T} \quad\dots\dots\dots\dots(22),$$

$$\left(\frac{d^2V}{dT^2}\right)_P = -\frac{d^2c}{dT^2} = -\frac{n(n+1)c}{T^2} \quad\dots\dots\dots(23).$$

Differentiating with respect to P, keeping T constant,

$$\left(\frac{dV}{dP}\right)_T = -\frac{RT}{P^2} \quad\dots\dots\dots\dots(24).$$

Now by equation (27) of Chapter VIII, in any fluid

$$\left(\frac{dC_p}{dP}\right)_T = -T\left(\frac{d^2V}{dT^2}\right)_P.$$

Hence in a gas or vapour to which Callendar's equation applies,

$$\left(\frac{dC_p}{dP}\right)_T = \frac{n(n+1)c}{T} \quad\dots\dots\dots\dots(25).$$

Integrating, we have

$$C_p = \frac{n(n+1)cP}{T} + C_p' \quad\dots\dots\dots(26),$$

where $C_p{}'$ is the constant of integration. It is the limiting value of the specific heat when $P=0$, at the temperature T. But since any gas in that infinitely rarefied condition may be treated as perfect, Callendar assumes that $C_p{}'$ may be taken as having the same value at all temperatures to which the equation is applied.

It should be noticed that this integration is performed along an isothermal line, and that the constant of integration is not necessarily the same for other temperatures. To treat it as constant when the temperature is varied therefore involves an assumption which is independent of anything in the equation itself.

Again, by equation (41) of Chapter VIII we had for the measure of the Joule-Thomson cooling effect in any fluid

$$\rho = C_p \left(\frac{dT}{dP}\right)_I = T\left(\frac{dV}{dT}\right)_P - V.$$

Hence for a gas to which the Callendar equation applies, the cooling effect is

$$\rho = T\left(\frac{R}{P} + \frac{nc}{T}\right) - V$$
$$= \frac{RT}{P} + nc - V$$
$$= V + c - b + nc - V$$
$$= (n+1)\,c - b \quad\ldots\ldots\ldots\ldots\ldots\ldots(27).$$

As was explained in Art. 189, $\left(\frac{dT}{dP}\right)_I$ is the fall of temperature per unit fall of pressure when the gas passes through a porous plug or any other throttling device, and ρ is the quantity of heat that would serve to maintain the original temperature, if it were supplied from outside during the process. From the above result it follows that Callendar's formula provides for the inversion of the cooling effect which is known to occur in real gases. When $(n+1)\,c$ is greater than b the expression for ρ is positive; the gas is then cooled by throttling. This is the usual case. But when $(n+1)\,c$ is less than b, ρ is negative; the gas is then warmed by throttling, as hydrogen is at ordinary temperatures, and as any gas will be if the initial temperature is sufficiently high. By raising the initial temperature the quantity $(n+1)c$ is reduced, since $c = \frac{C}{T^n}$. Inversion of the Joule-Thomson effect occurs when $(n+1)\,c = b$, or when $nc = -c + b$. But, as we saw above, $-c + b$ is the slope of any isothermal line on the diagram of PV and P, namely $\left(\frac{d\,(PV)}{dP}\right)_T$.

Hence if the isothermal slopes up with a gradient steeper than nc the Joule-Thomson effect will be a heating; it it slopes up less steeply than this, or runs level, or slopes down, the effect will be a cooling. Accordingly, measurements of the cooling effect furnish an important means of settling the values of the constants in Callendar's equation. In calculating his first tables Callendar assumed that the co-volume b was equal to the volume which would be occupied if the gas were all condensed to a liquid, and then found values of n and c from observations of the cooling effect*. In later calculations he assigned a small negative value to b in order to bring the results into better agreement with experiments on high-pressure steam†. For n he took $\frac{10}{3}$.

In fig. 93 isothermals are sketched for a gas obeying Callendar's equation. They are straight lines within the range to which the equation applies. AS is an isothermal drawn for a temperature such that the vapour becomes saturated at a moderate pressure, which is assumed to be within the range of pressure for which the equation holds good. It is straight, up to the saturation point S. The curved line through S is a portion of the boundary curve, below which lies the "wet" region, where the beginning of condensation would be represented by a vertical straight line, SW, P as well as T being then constant. AS slopes downwards, and the effect of throttling, at that temperature, is to cool the gas. $A'S'$ is another isothermal, drawn for a lower temperature, to which the same remarks apply. The effect of throttling is still to cool the gas at the higher temperature for which the horizontal isothermal BB is drawn $(c=b)$, and

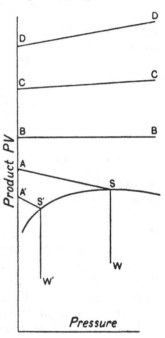

Fig. 93. Amagat isothermals according to Callendar's characteristic equation.

at any temperature up to that of CC, which is the isothermal corresponding to the inversion of the Joule-Thomson effect, namely

---

* *Phil. Mag.* January, 1903, p. 87.
† *Proc. Inst. Mech. Eng.* May, 1929, p. 508.

that for which $(n+1)\,c = b$, the upward gradient of CC being equal to nc. At any higher temperature, such as that for which DD is drawn, the upward gradient is steeper and the effect of throttling is to heat the gas.

215. Total Heat and Entropy of Water.

It is known from the researches of Regnault and others that the specific heat of water is not constant, but increases with rising temperature. Callendar suggests* that this increase may be due to the presence of water-vapour dissolved in the water. He supposes that when water and its vapour are in equilibrium at any temperature a volume of the vapour equal to the volume of the water is contained within the water. Consequently when water is heated its total heat increases more rapidly than it would do if the specific heat were constant, for the heat that is required to form the dissolved vapour becomes a larger proportion of the whole heat. This theory was adopted by Callendar in calculating, for his tables, the total heat and the entropy of water. It has the advantage of allowing these quantities to be expressed in a simple manner.

Let V_s be the volume of 1 lb. of saturated steam at any assigned temperature T, and let V_w be the volume of 1 lb. of water at the same temperature and pressure. Then, according to the theory, 1 lb. of "water" in that state is really 1 lb. of a solution containing dissolved vapour; the water conceals within it a volume of saturated steam equal to V_w. If the remainder were also turned into vapour, under constant pressure, we should have a total volume of vapour equal to $V_w + (V_s - V_w)$ or V_s, and the heat taken in during the process would be the latent heat L. Hence $\dfrac{L V_w}{V_s - V_w}$ represents the heat that is required to produce the vapour initially present in the water before the formation of any separate steam begins. This heat had to be supplied while the water was being warmed up to the temperature of saturation; it constitutes a part of the total heat of water I_w.

The other (and chief) part of the total heat of water is supposed to increase at a uniform rate as the temperature rises: it may therefore be represented by $\kappa\,(T - T_0)$, where κ is a constant and $T - T_0$ is the excess of temperature above $0°$ C., which is taken as the starting point in reckoning the total heat. Hence, adding the two

* Phil. Trans. Roy. Soc. A, vol. cxci, p. 147, 1902.

parts, the total heat of water under saturation pressure at any temperature T is

$$I_w = \kappa \left(T - T_0\right) + \frac{L V_w}{V_s - V_w} - \frac{L_0 V_{w_0}}{V_{s_0} - V_{w_0}} \ldots\ldots\ldots\ldots(28).$$

Here L_0, V_{w_0} and V_{s_0} refer to the state at $0°$ C. The term $L_0 V_{w_0}/(V_{s_0} - V_{w_0})$, which is only about 0·0029 calory, is a constant which has to be subtracted to make $I_w = 0$ when the temperature is $0°$ C.

To find the entropy of water under saturation pressure at any temperature T, we may think of the water as being brought to its actual condition by two stages. Imagine it to be heated to that temperature in an "ideal" manner, namely without the formation of any dissolved steam, and then the dissolved steam to be formed at that temperature. The entropy depends only on the actual condition (Art. 44). Taking, as before, the entropy of water at $0°$ C. to be zero, we therefore have

$$\phi_w = \int_{T_0}^{T} \frac{\kappa dT}{T} + \frac{L V_w}{T \left(V_s - V_w\right)} - \frac{L_0 V_{w_0}}{T_0 \left(V_{s_0} - V_{w0}\right)} \ \ldots\ldots(29),$$

which is the formula used by Callendar for the entropy of water under saturation pressure.

216. Steam Tables (published since 1930). Of the steam tables already mentioned in Art. 59 it is not necessary to add much to what was said there, except to refer to some results of recent research. The pioneer work of Callendar, who himself continued to pursue the subject up to the time of his death, has been followed by investigations which have not only revised his figures but have greatly extended our knowledge of the properties of steam, both saturated and superheated*. Important measurements carried out by the U.S. Bureau of Standards deal with the total heats of water and of saturated steam from $0°$ C. to $270°$ C.† These form to a great

---

* His son, G. S. Callendar, has cooperated with A. C. Egerton in carrying out a long series of precise measurements, mainly at the instance of the British Electrical and Allied Industries Research Association. At the time of writing (1934) their results are not fully published. Broadly speaking, however, it may be said that they agree closer with the Mollier tables of 1932 than with other tables published before that date.

An account of other recent researches will be found in four lectures by Dr Max Jacob, delivered at the University of London in May, 1931. (Reported in *Engineering*, July 31–December 25, 1931.)

† See Osborne, Stimson and Fiock in *Bureau of Standards Journal of Research*, August, 1930. A useful supplementary paper by Fiock follows, reviewing the results in relation to those of other observers.

extent the basis of the Mollier Tables of 1982. Throughout the main part of a range from 0° to the critical temperature of about 874° the properties of saturated steam are now tabulated in satisfactory agreement with the "skeleton" values that were laid down by the International Congress of 1930*. As a result of the 1984 Conference agreed values are available in greater detail and with much narrower tolerances. The values for the state of saturated vapour and saturated liquid are now in very satisfactory agreement up to the critical point. In the region of superheat the range has to be extended further and there are still some differences, but these are being resolved (1984) by the work of Egerton and G. S. Callendar whose results tend on the whole to corroborate experiments by Havliček and to support the figures tabulated by Mollier.

The measurements of the specific volumes of superheated steam carried out at the Massachusetts Institute of Technology under Dr Keyes' direction are also in quite close accord. Mollier, whose tables (the seventh edition, 1982) are the most recent of those that have been cited, founded them largely on the work of the Bureau of Standards and employs a function in the saturation region equal to $\phi - I/T$, or G/T, for which he gives the formula†

$$\log T - \log 273 - t/T$$

as a good approximation available up to 270° C. From that temperature to 810° C. he makes it

$$\log T - \log 273 - t/T - t^2 \times 10^{-8},$$

after which further adjustments are required as the critical temperature is approached.

Unfortunately the compilers of steam tables have not been in strict agreement as to the position of the absolute zero. Knoblauch puts it at $-278\cdot2°$, Mollier at 278°. This accounts in part for differences in the numbers they give for the entropy. At the recent Conference (1984) it was decided to adopt the figure 273·16 pending completion of the work at the Massachusetts Institute designed to increase the accuracy of the determination of the ice point on the thermodynamic scale.

The short tables which will be found in the Appendix are based partly on Mollier's Tables and on Callendar's Tables but mainly on

---

* A comparison of tabulated results is given by Keenan in a Report to the American Society of Mechanical Engineers, December, 1982: see *Mechanical Engineering* for March, 1988.

† Based on Macfarlane Gray, *Proc. Inst. Mech. Eng.* 1889.

the figures of the recent Conference, and consequently the figure −273·16°C. is the absolute zero value involved, which is sufficiently near the value −273·1° C. used throughout the rest of this book to make adjustment unnecessary.

These tables use the Centigrade scale of temperature, which has many advantages. The pressures and volumes are stated in English units*. The numerical figures for quantities of heat, being in lb.-calories per lb., are directly comparable with those in Continental tables, which are in kilo-calories per kilogramme. The numbers which express entropy are non-dimensional, being independent of the units used; they are also directly comparable with the numbers given in Continental tables. Further description will be found along with Tables in the Appendix.

* The conversion factors used are given in the Appendix. The heat unit is the International Steam-Table calory (I.T. cal.), which is equivalent to 4·1860 International Joules, or 4·1873 Joules (absolute units). A table of agreed units and conversion factors is given along with the Report of the 1934 Conference.

CHAPTER X

EFFECTS OF SURFACE TENSION ON CONDENSATION AND EBULLITION

217. Nature of Surface Tension. In Arts. 185–138 it was pointed out that when water-vapour is suddenly expanded it assumes a metastable state, becoming supersaturated owing to what was there called a static retardation in the formation of drops. Wilson's experiments were cited to show that, in the absence of nuclei, a vapour will become much supersaturated before drops will form, and it was mentioned that this is an effect of surface tension in the liquid. In this chapter some account will be given of the nature of surface tension and its thermodynamic influence on the change of phase from liquid to vapour or vapour to liquid; how it retards the formation of drops in a cooled vapour; how on the other hand it may promote condensation in hollow spaces; also how it retards the formation of bubbles within a liquid when the liquid is boiled.

218. Energy Stored in a Surface Layer. In consequence of surface tension the skin of a liquid at any surface separating the liquid from its vapour is the seat of a definite amount of potential energy, and this makes the conditions of equilibrium between the liquid and the vapour depend on the curvature of the surface. It is only when the surface is flat that the liquid and vapour (assumed to be at the same temperature) are in equilibrium when the vapour-pressure has the normal saturation value. In consequence of surface tension a small drop will evaporate into an atmosphere that would be saturated or even supersaturated with respect to large drops or large quantities of the liquid, because the vapour-pressure that is required to prevent evaporation from the highly curved surface of a small drop is greater than the vapour-pressure that will prevent evaporation from a flat surface of the same liquid at the same temperature. In other words, the equilibrium vapour-pressure for a small drop is higher than the normal saturation pressure; and conversely, when vapour is being condensed in a hollow cell with a highly curved internal surface, like the cells of silica gel, the equilibrium vapour-pressure may be much lower than the normal saturation pressure.

The cohesive forces which the molecules of any liquid exert upon one another make the free surface of the liquid behave as if it were a stretched elastic skin. It is to this that the phenomena of capillarity are due—the rise of a liquid column in a tube when the liquid is one that wets it, and the depression of the column when the liquid does not wet the tube. To this also are due the forms assumed by liquid films and by drops. It is the tension of the surface layer that makes a drop take a spherical shape when there are no disturbing forces: the drops of molten metal in a shot-tower, for example, become spheres as they fall freely, and solidify into spherical shot before they reach the bottom. A drop of mercury on a plate, or of dew on a leaf which it does not wet, would be spherical were it not for the upward pressure of the support on which the drop rests; the smaller the drop is the nearer does it come to being a sphere, for the disturbing force due to the weight is relatively unimportant in a small drop. As a result of surface tension, the energy contained in a drop of liquid is greater than the energy contained in an equal quantity when that forms part of a big mass of the same liquid at the same temperature, for energy is stored in the surface layer in much the same way as it would be stored by the stretching of an elastic skin. For the same reason a small drop contains more energy per unit of mass than a large one (of the same liquid at the same temperature), for the surface of the small drop is relatively larger. The energy stored in the surface layer is directly proportional to the area of the layer. To see that this is so we shall consider the amount of work that has to be spent in forming a thin liquid film.

The film that is formed when a soap-bubble is blown, or when a soapy liquid is smeared over a ring or hoop of wire, consists of two surface layers, back to back, with some of the liquid between. When the film is very thin, as, for instance, when it looks black in reflected light just before it breaks, it may be said to consist of two surface layers only; but it can be made a hundred or more times thicker than that and have just the same tension, for the state of tension exists in the surface layers only. The tension of such a film, whether thick or thin, is the tension of two surface layers; in other words, it is twice the surface tension. The tension in a liquid film differs from that of a stretched sheet of india-rubber or other elastic membrane in these important respects: it does not change when the film contracts or is stretched, and it has necessarily the same value in all directions along the surface.

Let a liquid film be formed on a U-shaped frame (fig. 94) by

wetting a wire AB with the liquid, placing it over C, and then drawing it away in the direction of
the arrow. The force that will
have to be applied to draw it
away or to hold it from coming
back is $2Sl$, where l is the
length AB and S is the tension
of the surface layer on each
side of the film per unit of
length. The quantity S so
defined measures the surface
tension of the liquid. In drawing the rod away through any

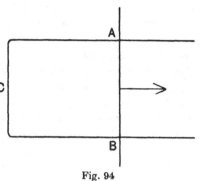

Fig. 94

distance x in the direction of the arrow the work done is $2Slx$, and
this work is stored in the two surface layers of the film, for it is
recoverable by letting the rod slip back. Hence the energy stored
in a single surface layer, in consequence of surface tension, is
numerically equal to S per unit area of surface.

It follows that the surface energy of a spherical drop (that is to
say the potential energy which is due to its surface tension) is
$4\pi r^2S$, where r is the radius of the drop. S is a quantity to be determined by experiment for any liquid; it is a function of the temperature, becoming smaller when the temperature is raised.

The spherical form which a free drop assumes is the form which
will make the surface energy (for a given volume) a minimum. A
drop resting on a support takes such a form as will make its total
potential energy a minimum, namely the sum of the energy of surface tension and the energy of position which the drop has in consequence of the height of its centre of gravity above the level of the
support.

219. Need of a Nucleus. Imagine a drop to be evaporating
under conditions that keep its temperature constant. Energy has
to be supplied in proportion to its loss of mass to provide for the
latent heat of the vapour that is formed. But the drop is losing
surface energy in consequence of its diminution of surface, and to
some extent this reduction of surface energy supplies the latent heat
that is required; only the remainder has to be supplied from outside
the drop. Consequently a drop is more ready to evaporate than the
same liquid in bulk, at the same temperature, and it will continue
to evaporate into an atmosphere which would be saturated with

respect to the same liquid in bulk. There can be no equilibrium between a drop and a surrounding atmosphere of saturated vapour. Moreover, as the drop gets smaller and smaller (if we assume that the reduction of size may go on without altering the nature of surface tension), a stage would be reached when the loss of potential energy due to contraction of the surface would become sufficient to supply all the latent heat of the vapour that is passing off. In that event, no heat would have to be supplied from outside to complete the evaporation of the drop: it would become inherently unstable and would flash into vapour.

For the same reason a drop cannot form except around a nucleus, and the larger the nucleus the more readily it forms. When particles of dust are present in expanding vapour, the first drops to be formed use them for nuclei, as was shown by Aitken (Art. 79), and only a small amount of supersaturation occurs before such drops begin to form. The cloud of particles observed by Wilson when dust-free air containing water-vapour is expanded enough to cause much supersaturation is formed around smaller nuclei which may consist of accidental co-aggregations of the molecules of the gas itself, or of electrically charged molecules, such as are always present in small numbers through ionization of some of the normally uncharged molecules. The presence of an electric charge greatly favours condensation of the vapour upon any nucleus. As an electrified drop evaporates, the charge remains behind; the potential energy due to electrification therefore increases as the drop becomes smaller, for the energy due to a constant electric charge varies inversely as the radius of the sphere that carries it. In this respect the effect of an electric charge is opposite to that of surface tension. Hence when a drop is charged more energy has to be supplied from outside to make it evaporate than would be required if it were uncharged. An electrically charged drop will therefore evaporate less readily than an uncharged drop of the same size, and may grow larger in an atmosphere that is but little supersaturated or even not supersaturated at all. In vapour which is slightly supersaturated it is found that any ionizing action, such as that of an electric spark, or of Röntgen rays or of ultraviolet light, which sets free the ions or particles conveying unbound electric charges, brings about a cloud of condensation, by creating fresh nuclei, or by stimulating the powers of existing nuclei through causing them to acquire an electric charge*.

* See C. T. R. Wilson, *Phil. Trans. R.S.* A, vol. cxcii, 1889.

220. Kelvin's Principle. Confining our attention, however, to drops which are not electrically charged, we shall consider how, as a consequence of surface tension, the equilibrium of a drop of given size depends on the state of supersaturation of the vapour around it. Assume the liquid and the vapour to be at the same temperature. Liquid with a flat surface is in equilibrium with the vapour above it when the vapour is at the pressure of saturation: there is then no tendency on the whole for the liquid to evaporate or for the vapour to condense, any evaporation that occurs being exactly balanced by an equal amount of condensation. Liquid with a convex curved surface is in equilibrium with the surrounding vapour only when the pressure of the vapour exceeds the normal saturation pressure by a definite amount; in other words, only when the vapour is sufficiently supersaturated. The degree of supersaturation necessary for equilibrium depends on the curvature of the surface, in a manner first established by Lord Kelvin*. His reasoning is of fundamental importance in explaining the retarded condensation of expanding steam.

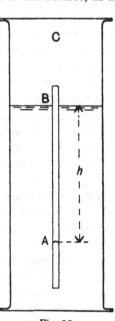

Fig. 95

We may apply Kelvin's general method as follows to find a relation between P_s, the normal pressure of saturation (which is the equilibrium vapour-pressure over the flat surface of a liquid†), and P', the equilibrium vapour-pressure over a curved surface, such as the surface of a small drop. Take for this purpose the curved surface at A, fig. 95, which is formed by holding in the liquid a capillary tube of a material such that the liquid does not wet it. The column of liquid in the tube is accordingly depressed through some distance h, and if the bore is small enough the free surface at A is sensibly part of a sphere. Imagine the liquid to be contained in a closed vessel, and that the space C above it contains nothing but the vapour of the liquid. Let all be at one temperature T. The whole system is in equilibrium.

* *Proc. Roy. Soc. Edin.* vol. VII, 1870; *Popular Lectures and Addresses*, vol. I, p. 64.

† Namely, the saturation pressure for any assigned temperature as given in tables of the properties of saturated steam.

Over the flat surface at B there is vapour whose pressure is P_s: over the curved surface at A there is vapour of a higher pressure P'. The difference $P' - P_s$ is equal to the weight of the column of vapour in the tube (per unit area of cross-section) from the level of A to the level of B. Let σ be the weight of unit volume of the vapour. If this were constant, the weight of the column of vapour in the tube (per unit area of section) would be simply σh. But σ depends upon the pressure P; it is equal to $1/V$ and may therefore be written

$$\sigma = \frac{P}{RT},$$

if we take the equation $PV = RT$ to apply, as it approximately does. The difference in the two vapour-pressures is

$$P' - P_s = \int \sigma dh,$$

integrated between the level at B and the level at A.

Compare next the hydrostatic pressures within the liquid just under the surface at B and just under the surface at A. Just under the flat surface at B the hydrostatic pressure in the liquid is equal to the pressure of the vapour over the surface; it is therefore equal to P_s. As we go down through the liquid to the level of A, the hydrostatic pressure increases by the amount ρh, where ρ is the weight of unit volume of the liquid. Therefore just under the curved surface at A its value is $P_s + \rho h$.

But we may also calculate the hydrostatic pressure under the curved surface at A in another way. The top of the liquid column at A, which has a surface layer in tension, may be treated as a segment of a sphere of radius r. Its surface layer forms a cap whose surface tension S causes it to press down upon the liquid below with a pressure p such that $\pi r^2 p = 2\pi r S$. That this is so will be seen at once by considering the equilibrium of a complete hemisphere of the same curvature and with the same surface tension. Round the circumference $(2\pi r)$ of the horizontal plane forming the base of such a hemisphere there would be a vertical force $2\pi r S$ balancing the resultant force due to the pressure p acting on the area of the base, πr^2. Hence

$$p = \frac{2S}{r},$$

and the hydrostatic pressure just under the curved surface is therefore equal to

$$P' + \frac{2S}{r}.$$

Equating the two expressions for this hydrostatic pressure, we have

$$P' + \frac{2S}{r} = P_s + \rho h,$$

or

$$\frac{2S}{r} = \rho h - (P' - P_s).$$

Hence, since $P' - P_s = \int \sigma dh$,

$$\frac{2S}{r} = \rho h - \int \sigma dh = \int (\rho - \sigma)\, dh.$$

And since $dP = \sigma dh$,

$$\frac{2S}{r} = \int_{P_s}^{P'} \frac{\rho - \sigma}{\sigma}\, dP = \int_{P_s}^{P'} \frac{\rho}{\sigma}\, dP \text{ very nearly,}$$

because σ is small compared with ρ.

On substituting P/RT for σ this approximation gives

$$\frac{2S}{r} = RT\rho \int_{P_s}^{P'} \frac{dP}{P} = RT\rho \log_e \frac{P'}{P_s},$$

or

$$\log_e \frac{P'}{P_s} = \frac{2S}{RT\rho r}.$$

This applies to any liquid surface whose radius of curvature is r. It therefore expresses the relation of the pressure P' in the super-saturated vapour around a spherical drop of radius r to the normal pressure of saturation P_s (over a flat surface) for the same tempera-ture, when the drop is in equilibrium, in the sense that it is neither growing by condensation nor shrinking by evaporation. The ex-pression shows how the degree of supersaturation P'/P_s necessary for the equilibrium of the drop increases when the size of the drop is reduced. For a drop of given radius any increase of P above the value so calculated would cause the drop to grow. The expression also shows what is the least size of drop that can exist in an atmo-sphere with a given degree of supersaturation: any drop for which r is smaller would disappear by evaporation; on the other hand, any drop for which r is larger would grow.

It is only when the drop is very small that the excess of P' over P_s is at all considerable. This is best shown by numerical examples. If we take water-vapour at 10° C. or 283° absolute, and use c.g.s. units, RT (which is treated as equal to PV) is $1\cdot30 \times 10^9$. The surface tension of water at that temperature is about 76 dynes per linear centimetre, and ρ is 1 gramme per cubic centimetre. Hence

$$\log_{10} \frac{P'}{P_s} = \frac{2 \times 76}{1 \cdot 30 \times 10^9 \times r \times 2 \cdot 303*} = \frac{1 \cdot 01}{D},$$

where D is the diameter of the drop in millionths of a millimetre. The formula accordingly gives these results, for drops of water:

Diameter of drop in millionths of a millimetre	Ratio of vapour-pressure in equilibrium with the drop to normal saturation pressure for the same temperature (P'/P_s)
100	1·02
50	1·05
10	1·26
5	1·59
2	3·2
1	10·2

This means, for instance, that a drop of water two millionths of a millimetre in diameter will grow if the ratio of supersaturation in the vapour around it is greater than 3·2, but will evaporate if that ratio is less. Hence when the ratio of supersaturation is 3·2, drops will not form (in the absence of electrification) unless there are nuclei present which are at least big enough to be equivalent to spheres with a diameter of two millionths of a millimetre.

It will be obvious from Kelvin's principle that a drop of water cannot continue to exist in an atmosphere of saturated vapour. When the drop and the atmosphere are at the same temperature, the drop can exist only if the atmosphere around it is supersaturated. For any given degree of supersaturation there is a value of r (determined by the formula) such that a drop of smaller radius will evaporate and a drop of larger radius will grow. Thus the bigger drops in a cloud will tend to grow at the expense of the smaller drops.

221. Ebullition. Similar considerations govern the formation of bubbles in a boiling liquid. We may treat any small bubble as a spherical space of radius r, containing gas, bounded by a spherical envelope in which there is surface tension. Outside of that is the liquid, at a pressure P. The envelope is kept from collapsing by the pressure of the gas within. To maintain the surface tension in the envelope, the pressure inside the bubble P_i must exceed P by the

---

* To convert from Napierian to common logarithms.

amount $2S/r$, where S is the surface tension in the boundary surface of the bubble, making

$$P_i - P = \frac{2S}{r}.$$

When r is very small this implies a great excess of pressure within the bubble. If no particles of air or other nuclei were present to start the formation of bubbles, boiling would not begin until the temperature were raised much above the point corresponding to the external pressure, and would occur with almost explosive violence. Once formed a bubble would be highly unstable, for as the radius increases the tension of the envelope becomes less and less able to balance the excess of pressure within it. This happens, to some extent, when water is boiled after being freed of air in solution: it is then said to boil with bumping.

It follows that a pure liquid may be superheated, that is to say, raised above the temperature of saturation corresponding to the actual pressure. This is an example of a metastable state like the state that is produced when a vapour is supercooled without condensing, or when a liquid is supercooled without solidifying. Water at atmospheric pressure may be heated to 180° C. or more when it has been freed of air and when it is kept from contact with the sides of the vessel by supporting it in oil of its own density, so that the water takes the form of a large globule immersed in oil.

In the ordinary process of boiling, a bubble contains in general some air or other gas besides the vapour of the liquid itself. Without gas in it, the bubble could not exist in stable equilibrium. With gas in it, the bubble will be in stable equilibrium when the partial pressure due to the gas provides the necessary excess of the whole internal pressure P_i over the external pressure P. Any reduction of the bubble's size would then raise the pressure of the gas more than enough to balance the increase of $2S/r$. Let P_v be the vapour-pressure inside the bubble. If we assume that the external pressure and temperature remain constant, the partial pressure due to the gas may be expressed as a/r^3, where a is a constant. Then

$$P_i = P_v + a/r^3,$$

and the equation

$$P_v + \frac{a}{r^3} = P + \frac{2S}{r}, \text{ or } P_v - P = \frac{2S}{r} - \frac{a}{r^3},$$

determines the value of r at which the bubble is in equilibrium. The quantity $P_v - P$ is the excess of the vapour-pressure in the bubble over the pressure in the liquid. Differentiating this with respect to r, to find the limiting condition for stability, we have

$$\frac{d\,(P_v-P)}{dr}=0, \quad \text{when} \quad \frac{a}{r^3}=\frac{1}{3}\frac{2S}{r},$$

and therefore when $\qquad P_v-P=\dfrac{4S}{3r}.$

Hence for stability P_v-P must be less than $4S/3r$. This means that when a liquid containing bubbles of radius r is heated, the temperature will rise until the vapour-pressure within the bubbles exceeds the pressure in the liquid by the amount $\dfrac{4S}{3r}$, but when that point is reached the bubbles will become unstable and ebullition will begin. Callendar* calculates on this basis that with bubbles one millimetre in diameter water (under one atmosphere) will boil at a temperature of $100 \cdot 05°$ C., and that to produce $10°$ of superheat the diameter of the bubbles must not be more than about $\frac{1}{200}$ mm.

222. Action of Silica Gel. The action of silica gel affords another example of the thermodynamic influence of surface tension. Assuming, as we may probably do, that the absorption of moisture by the gel is a purely physical process, it occurs through the deposit, within the minute solid pores, of liquid films which are formed by condensation on their own highly curved inner concave surfaces. The liquid film has a concave curvature not because of pressure from within but because it wets the curved interior of the cell. In consequence of that curvature the vapour-pressure under which the condensation takes place may be much less than the normal vapour-pressure corresponding to the temperature. This explains the efficacy of the gel as a desiccating agent and the possibility of employing it to condense vapour in one of the stages of a refrigerating cycle, in the manner which was described in Art. 118.

* *Enc. Brit.*, Article "Vaporization."

CHAPTER XI

GAS-MIXTURES AND SOLUTIONS

223. Mixture of Gases. We saw in Chapter II that when a vessel contains a mixture of two or more gases in equilibrium, the pressure on the containing walls is equal to the sum of what are called the partial pressures of the constituents. The partial pressure of each constituent gas is the pressure which it would exert on the walls if all other gas were absent. This was experimentally discovered by Dalton and is known as Dalton's Law. It is very nearly true of real gases and vapours at moderate pressures, and is exactly true of the ideal perfect gases of thermodynamic theory (see Art. 62).

Dalton's Law serves to determine the amount of water-vapour that will be present in air, at any assigned temperature, when the atmosphere is "saturated," that is to say when there is equilibrium in respect of evaporation, between the atmosphere and a flat surface of water at the same temperature. The partial pressure of the water-vapour in the air will be equal to the pressure of saturated water-vapour at the same temperature, and the quantity of vapour present, per unit volume of the air, will be equal to the density of saturated water-vapour at that temperature.

The principle, of which Dalton's Law is one manifestation, may be comprehensively stated by saying that in a mixture of perfect gases each constituent behaves as if the others were not there. For any given volume and temperature of the mixture, each constituent quantity that is present contributes to the pressure, to the energy, to the total heat, and to the entropy, just what it would contribute if it alone occupied the given volume at the given temperature.

Imagine two vessels A and B of constant volume, containing two different gases α and β, both at the same temperature, with a partition between them through which an opening can be made, say by having a slide-valve in the partition. When communication is opened a process of diffusion begins which, after a sufficient time, causes both vessels to contain one homogeneous mixture. It is assumed that the gases do not exert any chemical action on one another. If no heat enters or leaves the apparatus during the process, the temperature and the pressure are found to have undergone no change. From this it may be inferred that the gas α originally in A expands into B as if the other gas β were not there, and the gas β expands into A as if the gas α were not there. Each gas behaves like the gas in Joule's experiment (Art. 19); it expands without doing work and neither its temperature nor its internal energy is changed: consequently the mixture keeps the same temperature, and the energy of the mixture is equal to the sum of the energies which the constituents had at first. The partial pressure P_α of one constituent has (by Boyle's Law) changed from the original pressure P to PV_A/V', where $V' = V_A + V_B$, and the partial pressure P_β of the other constituent has changed from P to PV_B/V'. Hence $P = P_\alpha + P_\beta$ as Dalton's Law asserts. Thus the observed fact that when gases become mixed by diffusion there is no change of temperature, provided no heat is taken in or given out and no external work is done, allows Dalton's Law to be anticipated from the other properties of perfect gases.

224. Diffusion an Irreversible Process. Though the process of diffusion does not alter the energy of the system it is an irreversible process, and therefore must be expected to increase the entropy. That it does so is clear from a comparison of the entropy before and after mixture, using the expression for ϕ in Art. 199. Say that there are M_α units of α and M_β units of β in the mixture. The specific volume of the α constituent changes from V_A/M_α to V'/M_α. Before mixture took place its entropy, per unit of mass, was

$$\phi_\alpha = C_v \log_e T + R \log_e \frac{V_A}{M_\alpha} + C.$$

After mixture it is

$$\phi'_\alpha = C_v \log_e T + R \log_e \frac{V'}{M_\alpha} + C.$$

Hence the increase of entropy for the whole constituent α is

$$M_\alpha (\phi'_\alpha - \phi_\alpha) = M_\alpha R (\log_e V' - \log_e V_A).$$

Similarly for the constituent β it is

$$M_\beta (\phi'_\beta - \phi_\beta) = M_\beta R (\log_e V' - \log_e V_B),$$

and adding these terms we have the increase of entropy that results from mixture, for the system as a whole. The calculation may obviously be extended to a mixture of more than two gases.

225. Action of a Semi-permeable Membrane. This increase of entropy implies that energy is dissipated when gases mix. When the gases are separate, in A and B respectively, the availability of the system for doing work is greater than when they are mixed, though there is no change in temperature or pressure or energy. To realize this we have to think of some way by which the system, with separate gases, can be made to do work. Imagine the partition to be made of some porous material but to include what chemists call a *semi-permeable membrane*, such as will allow one of the gases to pass but will hold the other gas from passing. Membranes which have the property of being permeable to one substance and not to another are well known, and their action involves no breach of thermodynamic laws. Assume then, that the partition allows the gas α to pass but not the gas β. The result is some of the gas α passes into B and a difference of pressure is set up in the two vessels. The gas α will continue to diffuse into the gas β against this difference of pressure, until its partial pressure in vessel B is equal to the pressure of what is left of that gas in vessel A. The total pressure in B will then be $P + P_\alpha$, and the pressure in A will be P_α, where P_α is, as before, PV_A/V'. From the system in this state it is obvious that work could be obtained, by allowing the pressures to become equalized through an engine. The change from the original condition of the system took place without any interference from outside: it was thermodynamically self-acting. Hence the system, in its original condition, had an availability for doing work which is not possessed when the gases are completely mixed. If a semi-permeable membrane were fixed between the vessels after complete mixture had taken place it would be without effect, for the partial pressure of the gas capable of passing it would then be the same on both sides, and any diffusion through it would go on equally in both directions.

An example of a semi-permeable partition is furnished by hot palladium. When cold, palladium is impervious to gases, but when strongly heated it allows hydrogen to pass though it still stops a gas with heavier molecules such as nitrogen. Thus if we have two vessels A and B of equal size, separated by a palladium partition, A containing hydrogen and B nitrogen at the same pressure P, and heat the partition, half the hydrogen will pass into B but none of the nitrogen will pass into A. The effect is that when the system is cooled, the pressure in A will be $\frac{1}{2}P$ and the pressure in B will be $\frac{3}{2}P$. The excess of pressure on the nitrogen side corresponds to the "osmotic pressure" which will presently be described in speaking of solutions. The quantity of hydrogen that passes the partition is such as to make the partial pressure of the hydrogen the same on both sides.

226. Planck's Device for Separating Mixed Gases. It will help towards an understanding of the thermodynamics of mixed gases if we consider an imaginary device suggested by Planck for separating mixed gases without taking in heat or doing work, and without change of temperature. The device, which is shown in fig. 96, employs two semi-permeable partitions, one of which is fixed

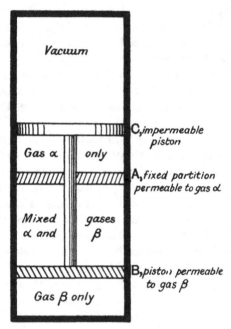

Fig. 96. Planck's imaginary apparatus for separating mixed gases.

and the other is a piston which may sweep through the space containing the mixed gases. At the beginning of the imaginary operation the whole space under the fixed partition A is to be thought of as filled with a mixture of equal volumes of the two gases α and β. The fixed partition A is permeable to gas α only; the piston B is permeable to gas β only. A piston-rod, which passes through the fixed partition A, connects the piston B to an upper piston C which is impermeable to gas. The whole space between the two pistons is equal to the whole space below the fixed partition A. As the piston B begins to rise, from the bottom of the vessel, gas β accumulates in the space below it, and gas α in the space between A and C. The function of the upper piston C is to provide space for the gas α as it passes through the fixed partition A. Suppose the pistons to be slowly raised. Gas α passes into the space above A and gas β into the space below B, and by the time the pistons are fully raised there is complete separation. Each of the gases, separately, then occupies a volume equal to that of the original mixture, so that the two together occupy twice the original volume.

No work is done on the pistons in this operation for the pressures on the moving system are balanced at every stage. The partial pressure of each gas is the same on both sides of the partition to which that gas is permeable. The partial pressure of gas α acts on the under side of the upper piston and that of gas β on the under side of the lower piston, while both partial pressures act on the upper side of the lower piston. Thus the movement occurs under balanced forces. The process is reversible: no heat is taken in or given out, and there is no change of entropy. Taking the equation

$$dE = Td\phi - dW$$

we have T constant, $dE = 0$, $dW = 0$ and $d\phi = 0$.

The system has recovered no availability for doing work: the extent to which the aggregate volume has increased neutralizes the thermodynamic advantage of the separation.

Planck uses this device to establish the proposition that the entropy of a mixture of gases at a given temperature is equal to the sum of the entropies which the constituents would have if, at the same temperature, *each of them separately occupied a volume equal to the volume of the mixture*[*].

227. Solutions. In applying thermodynamic reasoning to the study of solutions much use is made of the notion of hypothetical semi-permeable membranes, and on the experimental side real semi-permeable membranes serve to exhibit fundamental facts and to furnish necessary data. Chemists can cause partitions, otherwise porous, to contain and support a membrane which will allow water to pass freely but will not allow a substance dissolved in the water to pass; and such a partition can be made strong enough to stand, without damage, a large difference of hydrostatic pressure on the

[*] Planck, *Wied. Ann.* 1888. See his *Thermodynamics*, Eng. Trans. of the 7th edition, p. 219.

two sides. Thus it is mechanically possible to have a semi-permeable partition separate a quantity of the solution at one pressure from a quantity of the solvent, or pure liquid, at a lower pressure. As an example of the formation of a membrane of this kind, let a porous pot be filled with a solution of copper sulphate and immersed in a bath of potassium ferrocyanide. These solutions, meeting in the porous wall, will deposit there a layer of copper ferrocyanide which is permeable to water but not to certain substances that may be dissolved in it. A familiar instance is cane sugar, whose molecules are nineteen times as heavy as those of water and are stopped while the molecules of water pass freely.

Imagine now two vessels W and S (fig. 97) separated by a fixed semi-permeable partition. The vessel S contains a quantity of a solution and W contains a quantity of the pure solvent, at the same temperature. The partition is permeable by the solvent, but not by the dissolved substance. It is found that some of the solvent tends to pass through the partition from W into S, weakening the solution. This can only be prevented by increasing the pressure in S by a certain definite amount P_0. We may think of the two vessels as having pistons by means of which pressure may be applied to the liquid

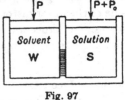

Fig. 97

in each. Whatever be the fluid pressure P on the side of the membrane that faces the pure solvent, there must be a greater fluid pressure $P + P_0$ on the other side if equilibrium is to be maintained. The excess fluid pressure P_0 on the side that faces the solution, when the solution is in equilibrium with the pure solvent on the other side, is called the *Osmotic Pressure*.

228. Osmotic Pressure. The amount of the osmotic pressure depends, for a given dissolved substance and a given solvent, on the "concentration" or quantity of the dissolved substance that is present per unit of volume of the solution. It is increased by increasing the concentration: it is also increased by raising the temperature of the system. We may think of it as a partial pressure due to the presence, in that volume, of the particles of the dissolved substance. This partial pressure is to be added to the partial pressure due to the other component of the solution, namely, the solvent, in determining the total pressure. From this point of view the semi-permeable membrane is exposed on one side to the pressure of **the**

solvent alone, and on the other side to the sum of two partial pressures, one due to the dissolved substance and the other due to the solvent. Hence if the total pressure P were made the same on both sides, that part of it which is due to the solvent would be less on the solution side, and consequently the solvent would tend to flow from W to S in the effort to bring its partial pressure in S up to equality with its pressure in W. This explains why under equilibrium conditions the total pressure in S must be greater, by the amount of the partial pressure of the dissolved substance, which excess constitutes the osmotic pressure.

If the excess pressure actually applied to the solution in the vessel S is less than the osmotic pressure P_0 some of the solvent will flow from W to S. On the other hand, if an excess pressure greater than P_0 be applied, some of the solvent will pass out of the solution into W. These changes will go on until the solution becomes sufficiently less or more concentrated to allow equilibrium to be again attained.

229. Van 't Hoff's Principle. It was pointed out by Van 't Hoff* that in dilute solutions the molecules of the dissolved substance act, in solution, like the molecules of a gas in this sense that the partial pressure which they exert is the same as would be exerted by an equal quantity of the same substance in the gaseous state, occupying the same volume, namely, the volume of the solution, at the same temperature. Thus the osmotic pressure in a dilute solution may be approximately calculated at any temperature and for any (small) concentration by inference from the gas equation $PV = RT$, on the basis that the dissolved substance contributes pressure like a gas whose density is the quantity of dissolved substance divided by the volume of the solution. This applies whether the dissolved substance is itself a gas, a liquid, or a solid; it may, for instance, be a substance that is non-volatile at the given temperature. It follows that the osmotic pressure in weak solutions varies in direct proportion to the absolute temperature. Also that, at any one temperature, the osmotic pressure varies in direct proportion to the quantity of dissolved substance in the solution. Also that when solutions of different substances have the same osmotic pressure at the same temperature they contain the same number of molecules of dissolved substance per unit of volume. These remarkable conclusions of Van 't Hoff are found to be true of very weak solutions, in which the osmotic pres-

* *Phil. Mag.* Aug. 1888.

sure is not so great as to make the deviation from the gas law considerable, provided the molecules of the dissolved substance do not undergo dissociation but retain their chemical character. They are closely true, for example, in dilute solutions of sugar. In solutions of electrolytes, however, there is, as was shown by Arrhenius, much dissociation of the dissolved molecules into their constituent ions, with the result that the salt contributes more than one partial pressure, and the osmotic pressure is consequently greater than it would be if there were no such chemical change. The osmotic pressure serves in fact as a criterion to show whether the molecules of the dissolved substance have undergone dissociation in being dissolved.

It may naturally be asked why, if a substance dissolved in water behaves there like a gas, it does not escape into the atmosphere when the solution lies in an open vessel. The answer is that at the free surface of the solution the effects of surface tension make the free surface virtually act as a semi-permeable membrane, through which molecules of the water may pass while those of the dissolved substance are held back. Similarly, a gas may be absorbed into solution by a non-volatile liquid through a free surface which is exposed to contact with the gas, because the surface is equivalent to a membrane permeable by the gas, and not by the liquid.

230. Vapour-Pressure of a Solution. The vapour given off by a solution of a non-volatile substance is composed entirely of the solvent. At any given temperature its pressure is lower than the vapour-pressure of the pure solvent, to an extent that depends on the osmotic pressure and the relative density of the vapour and the liquid. Let a tall vertical column of homogeneous solution with a free surface (fig. 98) be in equilibrium, through a semi-permeable partition at its base, with a quantity of the pure solvent, the whole being enclosed in a vessel in which the only atmosphere is the vapour of the solvent. The whole system is at a uniform temperature T. Since it is in equilibrium the height h of the column of solution must be such that the hydrostatic pressure on the upper side of the partition exceeds the hydrostatic pressure on the under side by an amount equal to the osmotic pressure P_0; also

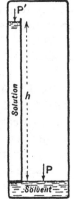

Fig. 98

the vapour-pressure P' at the level of the top of the column must be the saturation vapour-pressure of the solution, and the vapour-pressure P at the free surface of the solvent must be the saturation vapour-pressure of the solvent. Since the vapour has weight, P exceeds P'. Writing σ for the weight of the vapour per unit of volume, we have $P - P' = \sigma h$ if the column is so short that σ may be treated as constant, or $P - P' = \int \sigma dh$ if the variation of density with level is taken into account. The pressure on the upper side of the semi-permeable base is $P' + \rho h$, where ρ is the weight of the solution per unit of volume, and on the under side it is equal to P, if (as is shown in the sketch) the column is held with its base just level with the free surface of the solvent. Hence

$$P' + \rho h - P = P_0.$$

Taking $P - P' = \sigma h$, this gives $h = P_0/(\rho - \sigma)$, and $P - P' = P_0\sigma/(\rho - \sigma)$, or $P_0\sigma/\rho$ nearly, since σ is small compared with ρ. This may also be written $P_0/\rho V_s$, where V_s is the specific volume of the saturated vapour. But unless the solution is exceedingly dilute h is large and $P - P'$ should be taken as $\int \sigma dh$. In this way, an approximate relation is obtained,

$$\log_e \frac{P}{P'} = \frac{P_0}{\rho RT} = \frac{P_0}{\rho P V_s}.$$

231. Boiling Point and Freezing Point. Since the vapour-pressure of the solution is less than that of the pure solvent at the same temperature, its boiling point must be higher, that is to say, a higher temperature is required in the solution than in the solvent to make it give off vapour at the same pressure. To find a relation between the rise of the boiling point and the osmotic pressure in a weak solution, let the curves C and C' of fig. 99 represent curves of vapour-pressure in relation to temperature for solvent and solution respectively. If the pure solvent were

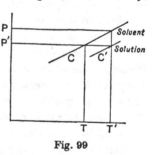

Fig. 99

boiling alone under a vapour-pressure P' its temperature would be T. But when the solution is boiling under the same vapour-pressure P' its temperature is T', a temperature which would correspond to a higher pressure P in vapour coming from the pure solvent. Hence from the geometry of the figure, when the change of boiling point is small, as it is in a very weak solution,

$$\frac{P - P'}{T' - T} = \left(\frac{dP}{dT}\right)_s,$$

where $(dP/dT)_S$ is the slope of the saturation curve C for the pure solvent. But by Clapeyron's equation, $(dP/dT)_S = (L/(V_s - V_w))\,T$. Hence $(P - P')/(T' - T) = L/TV_s$ nearly, since V_w is small compared with V_s. But, as we have seen above, in a very weak solution

$$P - P' = P_0/\rho V_s.$$

Hence in such a solution

$$T' - T = \frac{P_0 T}{\rho L},$$

or the boiling point is raised from T to $T\,(1 + (P_0/\rho L))$ by the presence of the dissolved substance.

It may readily be shown that the freezing point is lowered by a corresponding amount, namely, $P_0 T/\rho\lambda$, where λ is the latent heat of the freezing solvent.

We are here dealing with a solution which when it is cooled begins to freeze by forming crystals of the pure solvent, and when it is heated (under constant pressure) begins to vaporize by giving off vapour of the pure solvent. The vapour is given off at a definitely higher temperature, and the crystals begin to form at a definitely lower temperature, than if the liquid were pure. In either case the difference $T' - T$ depends on the concentration. The vapour of the boiling solution is superheated to that extent, at the moment it comes off. The efficacy of a "freezing mixture," such as a mixture of ice and salt, depends on this lowering of the freezing point: equilibrium is reached only when some of the ice has melted to form a solution, and the temperature of the whole, including the unmelted ice, has assumed the lower value which corresponds to the freezing point of the solution. Conversely, when saturated steam at temperature T is passed into a solution of salt in water it will condense until equilibrium is reached, namely, until the temperature of the solution rises to the higher value T', for steam can only be given off by the solution at that higher temperature. The apparent anomaly of steam being condensed in a liquid warmer than itself is no more remarkable than what occurs during the approach to equilibrium on the part of a freezing mixture, when ice melts in a liquid that is colder than itself: both are effects of osmotic pressure.

232. **Solution of a Gas in a Liquid.** To illustrate the thermodynamic reasoning by which Van 't Hoff's principle is established, we may consider the solution of a gas in a liquid. According to that principle the osmotic pressure, at any temperature, should be equal to the pressure which the gas would have, at the same temperature,

if it alone occupied a space equal to the volume of the solution. To prove that the osmotic pressure actually has that value, imagine a very long cylinder (fig. 100) with a fixed partition a and two movable pistons b and c. Both a and c are semi-permeable: a is permeable only to the gas, and c only to the solvent. Behind c there is pure sol-

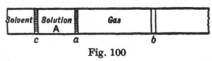

Fig. 100

vent; in the space between a and b there is gas; in the space between c and a is the liquid which dissolves the gas as the operation proceeds. Suppose the whole system to be at one temperature T and to be kept at that temperature (say by a water-jacket). At the beginning suppose c to be fixed and b to have been drawn so far away to the right that the pressure of the gas is negligibly small. Then equilibrium requires that the liquid in the space A shall contain no gas, or, to be exact, a negligible quantity of gas, for it is known as an experimental result (called Henry's Law) that the quantity of gas which a liquid will dissolve is directly proportional to the pressure. We begin therefore with practically pure solvent in the space A, whose volume we shall call V_A. Now imagine b to be slowly pressed in, compressing the gas isothermally and causing it to be gradually absorbed by the liquid in A. This is a reversible process: if, at any stage, b were stopped and slowly moved out again the action would be exactly reversed. When b reaches a all the gas is dissolved. The work spent in forcing the piston home is $\int_0^{P_1} P\,dV$, where P_1 is the pressure that has to be applied at the finish, under which the last part of the gas passes the partition a into the liquid. Now, keeping b with an external pressure P_1 still applied to it, suppose c to be forced slowly towards a. To do this will require that a pressure equal to the osmotic pressure P_0 be applied to c. The solution will thereby be separated into its components, the solvent passing behind c, and the gas passing through a and pushing out the piston b with the constant pressure P_1. P_1 does not change, for, as c advances, there is no change in the concentration of the remaining part of the solution. When c reaches a all the gas has left the solution, and is now behind the piston b, still at pressure P_1 and occupying a volume which we shall call V_1. The work done by the gas on b is $P_1 V_1$, and the work that has been spent in forcing in the semi-permeable piston c is $P_0 V_A$. Now let the gas expand isothermally till the pressure of the gas is again negligibly small: the work done in that expansion is the same as was originally spent, namely, $\int_0^{P_1} P dV$. To complete a cycle of operations we have only to withdraw c to its original position, which requires no work to be done, for it now has pure solvent on both sides. Since the cycle is isothermal the work done must be equal to the work spent; hence

$$P_0 V_A = P_1 V_1,$$

which proves the osmotic pressure P_0 to be equal to the pressure the gas would have if it alone occupied the space V_A, as Van 't Hoff's principle requires.

CHAPTER XII

ELECTROLYTIC TRANSFORMATIONS AND THERMOELECTRIC EFFECTS

233. Electrolytic Transformations. It has been pointed out that when the fundamental equation of energy

$$dW = dQ - dE$$

is applied to a complex system, dQ being the heat taken in from outside, and $-dE$ the decrease of internal energy, the external work dW may be done in other ways than by expansion of volume. In an electrolytic system, such as a galvanic cell, the transformation which goes on within the system results in the doing of electric external work, the measure of which (in a small transformation) is Fde, where F is the electromotive force and de is the quantity of electricity generated. In many electrolytic actions the amount of mechanical work due to change of volume, or to alterations of level of substances within the cell, is negligibly small. This is the case when electric energy is produced by a battery such as Daniell's, or when it is stored and restored by the chemical action on the lead plates of a storage battery. In what follows regarding electrolytic action we shall confine our attention to those cases in which, sensibly, all the external work is electrical. The action may take place under reversible conditions: the deposit of copper from a copper sulphate solution, for example, such as occurs in a Daniell cell, is exactly reversed when a reversed current is caused to pass through the cell. The energy equation for a reversible electrolytic action, involving no appreciable change of volume, may accordingly be written

$$Fde = dQ - dE.$$

Here dQ represents heat taken in reversibly from outside of the cell, and dE, the change of internal energy, represents what chemists call the heat of reaction in the chemical changes which are associated with the passage of the current. The heat of reaction is the quantity of heat that would be generated (say in a calorimeter) if the same chemical action were to take place without giving out electrical energy. If the electrical energy given out by a galvanic cell were dissipated within the cell itself, instead of being employed to do work outside of it, dE is the quantity of heat which would appear. In a Daniell cell, for example, dE is the quantity of heat which would appear if zinc were consumed to form zinc sulphate in solution, less the quantity of heat which would appear if an equivalent amount of copper were consumed to form copper sulphate in solution, without the production of any external electrical effect. It is the difference between these quantities that measures the "heat of reaction" in the Daniell cell as a whole, and this is numerically equal to the loss of internal energy that occurs when the cell is employed to do external work by producing electrical energy reversibly.

234. Action of a Galvanic Cell. In this action heat may or may not be taken in from outside. Suppose a galvanic cell to be placed in a bath of water or other isothermal enclosure so that its temperature is kept uniform. Experiment shows that the quantity of heat taken in during its action, namely dQ, may be either positive or negative. In other words, the reversible chemical action which goes on within the cell may tend either to make it colder or to make it warmer. In the former case some heat, dQ, will be taken in from the isothermal enclosure in which we have imagined the cell to be placed; in the latter case some heat will be given out to the enclosure.

A Daniell cell working reversibly, and therefore with an internal resistance so low that no sensible amount of the electric energy which it produces is dissipated within the cell by the heating effect of the current, must take in a small quantity of heat from outside if its temperature is not to fall. In the Daniell cell dQ has a positive value amounting to rather less than 1 per cent. of the output of electrical energy. The ordinary storage battery also requires a small addition of heat to maintain its temperature constant while it discharges. In the Clark cell, on the other hand, dQ is negative and its numerical amount is greater. If dQ were zero, which is nearly true of the Daniell cell, we should have $Fde = -dE$, which would furnish a very simple means of calculating the electromotive force when the heat of reaction is known. By Faraday's Laws one and the same quantity of electricity (about 96,540 coulombs) is required for the deposition of one gramme-equivalent of any substance. From the known heats of reaction of the active substances in a given cell it is therefore easy to calculate the change of the internal energy E which occurs in the passage of one unit of electricity, at constant temperature, and the value so obtained would be numerically equal to F if no heat were taken in. On this basis Kelvin in 1851 calculated the electromotive force of a Daniell cell, obtaining a number which is a little short of the actual value as determined by experiment. When account is taken of the dQ term a correct value is deduced.

Direct measurements of the quantity of heat which is reversibly taken in or given out during the action of a cell are difficult, for the effect is inevitably mixed up, in any experiment, with the irreversible development of heat within the cell that arises from its electrical resistance. But the quantitative influence of the dQ term may be inferred, without direct measurement of the heat taken in or given out, from observations of the extent to which the electromotive force of the cell is affected by changing the constant temperature at which the cell works. This was shown independently by Willard Gibbs and Helmholtz, who thereby applied the necessary correction to the original calculation of Kelvin.

235. Calculation of the Electromotive Force. Gibbs-Helmholtz Equation. For the purpose of calculating the electromotive force it is convenient to make use of the function ψ, or $E - T\phi$, for the system as a whole. We saw that in any system undergoing a reversible transformation at constant temperature the decrement of ψ measures the external work done. Let the amount by which ψ is diminished while one unit of electricity is generated be represented by ψ_1: for e units the amount will be $e\psi_1$, and this is equal to the external work. The amount of electrical work done while e units of electricity are generated is eF. Hence if all the external work is electrical (a condition substantially satisfied in a cell where volumetric or gravitational work is negligible),

$$eF = e\psi_1, \text{ or } F = \psi_1.$$

Then from the equation (Art. 195)

$$\psi = E + T \left(\frac{d\psi}{dT}\right)_W$$

we have at once

$$F = E_1 + T \frac{dF}{dT},$$

where E_1 is the "heat of reaction" corresponding to the passage of one unit of electricity, and dF/dT is the rate of change of the electromotive force with temperature as observed when the cell is on "open circuit," doing no work. The term $T (dF/dT)$ corresponds to Q_1, the amount of heat that is taken in, reversibly, during the passage of one unit of electricity, when the cell works at constant temperature T. This important expression for the electromotive force of a cell is known as the Gibbs-Helmholtz equation. We may illustrate it by numerical values for the Daniell cell. When zinc replaces copper in a (moderately strong) solution of the sulphate the "heat of reaction" is found, by measurements in a calorimeter, to be 104,900 joules per gramme-equivalent of either metal. The corresponding quantity of electricity is 96,540 coulombs. Thus E_1 is $\frac{104900}{96540}$ or 1·0866 joules per coulomb. The electromotive force of a Daniell cell would therefore be equal to 1·0866 volts if the temperature coefficient were zero. But dF/dT is observed to be positive, and equal to 0·000034 volt per degree. Hence at, say, 15° C. the term $T\,(dF/dT)$ or Q, namely the heat which the cell takes in to keep its temperature from falling, is $288 \times 0·000034$ or 0·0098 joule per coulomb. With these data the calculated electromotive force of the Daniell cell is accordingly

$$F = 1·0866 + 0·0098 = 1·0964 \text{ volts.}$$

The Gibbs-Helmholtz equation may obviously be applied in the converse manner, to calculate the aggregate heat of reaction for the chemical changes which go on in a reversible cell, from observations of the electromotive force and its temperature coefficient.

236. Alternative Method of establishing the Gibbs-Helmholtz Equation. Readers unfamiliar with the use of the function ψ may find it more satisfying to have the Gibbs-Helmholtz equation established by another method, namely, by considering a cyclic process of four operations in which the cell does electrical work during one part of the cycle and has electrical work spent upon it during another part. We shall assume, as before, that the action of the cell is reversible, and that it is surrounded by an isothermal jacket containing a fluid which will serve as source or receiver of heat. We shall further imagine that the temperature of the jacket, and therefore of the system as a whole, can be reversibly altered through some small amount δT, say by means of adiabatic expansion, so that a part of the cycle of the cell's action can be performed at temperature T and another part at temperature $T - \delta T$, the heat taken from the system in lowering its temperature from T to $T - \delta T$ being returnable to the system without loss, with the effect of restoring the temperature of the system to T. Suppose that the cell first produces electric energy at temperature T; then has its temperature lowered to $T - \delta T$; then has enough electric energy spent upon it at that lower temperature to cause the chemical changes which took place in the first operation to be exactly reversed in this third operation; and finally has its temperature restored to T. By Faraday's Laws the same quantity of electricity must pass through the cell in the third operation as in the first, in the reverse direction. But the electromotive force depends on the temperature: call it F in the first operation and $F - \delta F$ in the third. We assume that each of the four operations is reversible in the thermodynamic sense, and also that no appreciable amount of work is done by or upon the cell except the electric work. There is no chemical action, and no passage of electricity, in the second operation or in the fourth. Let e represent the quantity of electricity that passes in each of the first and third operations, and, as before, let E_1 represent the "heat of reaction" for the chemical changes that are associated with the passage of one unit of electricity. Let Q be the heat taken in (reversibly) from the jacket during the first operation, per unit of electricity that passes, and let $Q_1 - \delta Q$ be the quantity of heat returned to the jacket during the third operation, also per unit of electricity. Then the energy equation for the first operation is

$$eF = eE_1 + eQ_1,$$

and for the third operation it is

$$e\,(F - \delta F) = eE_1 + e\,(Q_1 - \delta Q).$$

The quantity of heat reversibly abstracted from the jacket in lowering its temperature in the second operation is returned to it in the fourth, and may therefore be omitted in summing the energies for the cycle as a whole. The cell is now restored exactly to its original state, and for the cycle as a whole, by adding the above expressions, we have

$$e\delta F = e\delta Q,$$

where $e\delta F$ is the net amount of electrical work done by the cell, and $e\delta Q$ is the net amount of heat taken in from the isothermal jacket. The energy of the cell is the same as at first. The result of the cycle as a whole is to convert an amount of heat $e\delta Q$ into electrical work $e\delta F$, and this conversion has been effected in a reversible process, by taking in heat eQ at temperature T and rejecting heat at the lower temperature $T - \delta T$. Hence by the Second Law the work done is equal to $\delta T/T$ times the heat taken in, or

$$e\delta F = \frac{eQ_1\delta T}{T},$$

from which $$Q_1 = T\frac{\delta F}{\delta T} = T\frac{dF}{dT}.$$

Substituting this in the energy equation $F = E_1 + Q_1$ we have the Gibbs-Helmholtz result

$$F = E_1 + T\frac{dF}{dT}.$$

It should be added that to simplify this argument no account was taken (1) of any difference there may be in the heat of reaction E at T and at $T - \delta T$, also (2) of any difference there may be in the thermal capacity of the substances in the cell before and after the reaction. These two small quantities in fact cancel out. For in a calorimetric experiment in which the reaction was allowed to develop heat, it would be a matter of indifference whether the reaction took place at a temperature T and the products were then cooled to $T - \delta T$, or whether the substances were first cooled to $T - \delta T$ and the reaction were then to take place. In both cases the same total quantity of heat would be given out. Thus in ignoring both (1) and (2) we ignored quantities whose sum amounts to zero in the cyclic operation as a whole, and the validity of the argument was not affected.

237. Thermoelectric Circuits. Peltier Effect and Thomson Effect. When a circuit is made up of two different metallic conductors joined at their ends, and the junctions are kept at different temperatures, electricity is continuously generated. The circuit is a heat-engine which converts heat into electrical energy. Its thermodynamic action depends on the following facts:

(1) When two metals are in contact and electricity passes from one to the other, the junction being kept at any constant temperature, heat is taken in or given out at the junction. This action, which was discovered by Peltier soon after thermoelectric phenomena were first observed, and is called the *Peltier Effect*, is thermodynamically reversible. If a quantity of electricity passes through the junction in one direction heat is taken in; if the same quantity of electricity passes in the opposite direction, an equal quantity of heat is given out. The reversible character of the Peltier effect distinguishes it at once from any generation of heat through imperfect electric conductivity. The quantity of heat taken in or given out depends on the nature of the metals in contact and on the temperature: for a given junction at a given temperature it is directly proportional to the quantity of electricity that passes. The Peltier

coefficient, which is represented by Π, may accordingly be expressed in joules per coulomb, and the quantity of heat that is reversibly taken in or given out at a junction when e units of electricity pass is $e\Pi$.

(2) When electricity passes along a metallic conductor of uniform quality along which a gradient of temperature is maintained, heat is taken in or given out. This again is a reversible action, quite distinct from any heating of the conductor through imperfect conductivity. It was predicted by W. Thomson (Lord Kelvin) in 1854 as a thermodynamic consequence of the observed properties of circuits in which electric currents flowed from hot parts to cold parts of a conductor, and is called the *Thomson Effect*. In a given metal, electricity passing from hot to cold will cause heat to be given out in any part along which there is a gradient of temperature; the same quantity of electricity passing in the reverse direction will cause an equal amount of heat to be absorbed, in the same part of the conductor. The Thomson effect is proportional to the quantity of electricity that passes: it may be expressed as $e\sigma\delta T$ for any length of a conductor between points whose temperatures differ by δT, σ being a coefficient which depends on the nature of the conductor and differs in amount and even in sign in different metals. In the conductor as a whole the Thomson effect per coulomb is therefore equal to $\int_{T_2}^{T_1} \sigma dT$, where T_1 and T_2 are the temperatures of the ends.

238. Calculation of the Peltier and Thomson Effects. Thermo-electric Power of a Pair. Thermoelectric Diagram. Imagine now an ideal circuit of two metallic conductors a and b—ideal in the sense that the conductors have no appreciable electrical resistance, so that we may omit consideration of irreversible heating effects. Let one junction be kept at a temperature T_1 and the other at a lower temperature T_2. If the metals are different, there will in general be a Peltier effect, as well as a Thomson effect, and a current will pass. Assume that in the circuit there is a perfectly efficient electric motor by which the electrical energy which is produced by the agency of heat is utilized as external work. Calling F the electromotive force produced in the circuit and employed to drive the motor, the work done by e units of electricity is eF, and the energy equation is

$$eF = e\Pi_1 - e\Pi_2 + e\int_{T_2}^{T_1} \sigma_a dT - e\int_{T_2}^{T_1} \sigma_b dT.$$

Here $e\Pi_1$ is the Peltier effect at the hot junction, namely, the heat taken in there, $e\Pi_2$, is the heat given out at the cold junction; and the two other terms are the Thomson effects in the two conductors, a taking in of heat in conductor a and a giving out of heat in conductor b. On dividing by e we have

$$F = \Pi_1 - \Pi_2 + \int_{T_2}^{T_1} (\sigma_a - \sigma_b)\, dT \dots\dots\dots\dots(1).$$

The whole system forms a reversible heat-engine in which the heat taken in does work as in a Carnot engine, by being let down in temperature. Hence by the Second Law $\Sigma\,(\delta Q/T) = 0$ for the action as a whole, and

$$\frac{\Pi_1}{T_1} - \frac{\Pi_2}{T_2} + \int_{T_2}^{T_1} \frac{(\sigma_a - \sigma_b)}{T}\, dT = 0.$$

Differentiating this with respect to T,

$$\frac{d}{dT}\frac{\Pi}{T} + \frac{\sigma_a + \sigma_b}{T} = 0,$$

or

$$\sigma_a - \sigma_b = \frac{\Pi}{T} - \frac{d\Pi}{dT} \dots\dots\dots\dots\dots(2).$$

On substituting this expression for $\sigma_a - \sigma_b$ in equation (1) we obtain

$$F = \int_{T_2}^{T_1} \frac{\Pi}{T}\, dT \dots\dots\dots\dots\dots(3).$$

Apply these equations to a circuit in which the temperatures of the junctions differ by only an infinitesimal quantity δT, and write δF for the corresponding electromotive force, and $\delta\Pi$ for the difference between the two Peltier coefficients. Then the equation

$$F = \Pi_1 - \Pi_2 + \int_{T_2}^{T_1} (\sigma_a - \sigma_b)\, dT$$

becomes

$$\delta F = \delta\Pi + (\sigma_a - \sigma_b)\,\delta T,$$

or, since

$$\delta\Pi = \frac{d\Pi}{dT}\,\delta T,$$

$$\delta F = \left\{ \frac{d\Pi}{dT} + (\sigma_a - \sigma_b) \right\} \delta T \dots\dots\dots\dots(4).$$

By equation (2), this gives

$$\delta F = \frac{\Pi}{T}\,\delta T,$$

from which

$$\Pi = T\frac{dF}{dT} \dots\dots\dots\dots\dots(5),$$

a result which might have been got more shortly by differentiating equation (3). On substituting this expression for Π in equation (2) we have

$$\sigma_a - \sigma_b = \frac{dF}{dT} - \frac{d}{dT}\left(T\frac{dF}{dT}\right) = -T\frac{d^2F}{dT^2} \quad \ldots\ldots\ldots\ldots(6).$$

Equation (5) allows the Peltier effect for any given pair of metals to be calculated from the observed value of what is called the "thermoelectric power" of the pair, namely dF/dT or the ratio of δF, the observed electromotive force for a small difference of temperature between the junctions, to δT, the amount of that difference. Equation (6) allows the difference of Thomson effects for the two metals to be calculated when the relation of the thermoelectric power to the temperature is ascertained. The Thomson effect is positive in some metals and negative in others; it is sensibly nil in lead over a wide range of temperature. Consequently, in tabulating values of the Peltier and Thomson effects, lead is usually taken as one metal of the pair.

The thermoelectric power dF/dT of any pair is a function of the temperature: in most metals its rate of variation with temperature is constant or nearly constant*, so that a line drawn on what is called a thermoelectric diagram to exhibit the values of the thermoelectric power of a given metal with respect to lead, in relation to the temperature, is straight or nearly straight. The line for lead is taken as a horizontal straight line. When the lines for two metals cross one another, it means that the thermoelectric power of that pair vanishes at the corresponding temperature, and has opposite signs at temperatures above and below that "neutral point." The thermoelectric power of a given pair was in fact discovered very early by Cumming to suffer inversion of sign at a particular temperature. Thus when the temperature of one junction is fixed a maximum of electromotive force is obtained by bringing the temperature of the other junction to the temperature of inversion. With a copper-iron pair, for example, when the hot junction is raised to about 275° C., the thermoelectric power vanishes and the electromotive force of the circuit is a maximum. It was this fact of inversion that led W. Thomson to the discovery of the Thomson effect. If the Peltier effect were the only reversible thermal phe-

---

* Unless the metal—like iron or nickel—undergoes allotropic modification (change of phase), in which case there is a sharp bend in the region of temperature where the change occurs.

nomenon in the action of a thermoelectric circuit no inversion could occur. The circuit would then be a very simple reversible heat-engine taking in heat only at T_1 and rejecting heat only at T_2. With any assigned value of T_1 the amount of heat converted into electrical work would necessarily, by the Second Law, be proportional to $T_1 - T_2$. Hence the electromotive force would also be proportional to $T_1 - T_2$, for all values of T_2, and there would be no inversion. Thomson inferred that the passage of the current must cause, in addition, some other kind of reversible thermal effect, and that it could only occur in the conductors as a consequence of the fact that along each of them there was a temperature gradient.

APPENDIX

TABLES OF THE PROPERTIES OF STEAM

THESE Steam Tables contain some representative numbers but reference should be made to tables such as those mentioned in Art. 59 for more complete data. The figures in the tables which follow, however, are based on the latest figures agreed internationally (International Steam-Table Conference, New York, 1934). Mr G. S. Callendar has carried out the computations for the tables which follow.

Table A. Properties of Saturated Steam in relation to the Temperature, 0–374° C.

A′. Properties of Water at Saturation Pressure, 0–374° C.

B. Properties of Saturated Steam, in relation to the Pressure, 0·1 lb. to 3200 lbs.

C. Volume of Steam in the Dry State, from 20 lbs. to 3200 lbs. and to 550° C.

D. Total Heat of Steam in the Dry State, from 20 lbs. to 3200 lbs. and to 550° C.

E. Entropy of Steam in the Dry State. 20 lbs. to 400 lbs. and to 400° C.

Tables A and B relate to the special case of saturated steam. When steam is saturated a single property, such as either the temperature or the pressure, fixes its state. In Table A, the property which is assumed to be known is the temperature, and the table gives corresponding values of the pressure, volume, total heat and entropy—all for the saturated state. It also gives the latent heat and internal energy. (These values are based on the International Steam-Table Conference values, the figures for the entropy were obtained by noting the departure of the total heat value from Mollier's (1932) tabulated values and so correcting the Mollier figures for entropy.) The internal energy is given by $I_s - APV$.

Table A′ contains the volume, total heat and entropy of water at saturation pressure. The function $G = T\phi_s - i_s$, where i_s is the total heat of water. In Table B the property which is assumed to be known is the pressure and corresponding values are given of the other properties in the saturated state, namely the temperature, volume, total heat, entropy, latent heat and the function G. (The values were obtained from the departure of the Callendar Table (1931) figures at the International Steam-Table Conference points.)

Tables C, D and E relate to steam in a dry state and superheated. A knowledge of two properties is then required to specify the state: the two that are selected as independent variables are the temperature and the pressure. Table C gives the volume, Table D the total heat and Table E the entropy in relation to these variables. (The departures of the latest agreed values from the Callendar (1915) Table were used to derive the tabulated figures for specific volume up to 800 lbs., and from the Mollier (1932) Table, from 700 to 3200 lbs. per sq. in. Similarly Mollier values from 250 to 3200 lbs. per sq. in. were used to derive the total heat figures from the latest agreed values in a similar manner. Values near saturation are more uncertain. For the construction of the entropy Table E, the departures from the Callendar (1915) values for each 10° increase of superheat were employed.)

From Table D it is easy to find the heat of formation, under constant pressure, of steam in any condition of superheat. The total heat at the given temperature and pressure is obtained from the table, and the heat of formation is found by subtracting from the total heat of water at the same pressure and at the temperature of the feed. Again from Table D values may be inferred of the specific heat (C_p) of steam at temperatures and pressures within the range of the table. C_p for any condition of steam is equal to the amount by which I increases per degree of rise in the temperature at constant pressure, and the mean values of C_p over 10° intervals, which are practically the same as C_p at the middle temperature, can be obtained from the table.

The following agreed conversion factors have been employed in computing the tables:

Pressure.   1 lb. per sq. in. at $g = 980 \cdot 665$ cm./sec.$^2 = 0 \cdot 070307$ kg./cm.^{2}

(1 lb. per sq. in. at London $= 1 \cdot 00053$ normal lbs. per sq. in.)

Volume.   1 ft.^{3}/lb. $= 0 \cdot 062428$ m.^{3}/kg.

Heat Energy.   1 I.T. Cal. $= 4 \cdot 18730$ Joules
$$= 4 \cdot 18605 \text{ Int. Watt secs.}$$
$$= \tfrac{1}{252} \text{ B.T.U.}$$
$$= 3 \cdot 0884 \text{ ft. lbs. (normal).}$$

Temperature. International Scale. Ice point $273 \cdot 16°$ K.

Table A. *Properties of Saturated Steam.*

Temp. °C.	Pressure lbs./in.²	Volume ft.³/lb.	Total heat Calories	Entropy	Latent heat Calories	Int. energy Calories
0	0·0886	3305·1	597·3	2·1864	597·26	567·2
10	0·1780	1704·3	601·6	2·1264	591·54	570·4
20	0·3390	926·22	605·9	2·0704	585·86	573·6
30	0·6153	527·40	610·2	2·0191	580·20	576·9
40	1·0697	313·09	614·5	1·9722	574·56	580·1
50	1·7891	192·94	618·9	1·9292	568·91	583·4
60	2·8892	122·98	623·1	1·8898	563·16	586·6
70	4·5200	80·83	627·4	1·8534	557·48	589·9
80	6·8690	54·61	631·4	1·8188	551·40	592·8
90	10·168	37·83	635·3	1·7871	545·32	595·7
100	14·696	26·80	639·1	1·7572	539·06	598·6
110	20·78	19·38	642·7	1·7293	532·6	601·3
120	28·80	14·28	646·2	1·7032	526·0	603·9
130	39·20	10·70	649·6	1·6785	519·2	606·5
140	52·41	8·145	652·7	1·6553	512·1	608·8
150	69·03	6·287	655·7	1·6332	504·8	611·1
160	89·63	4·914	658·5	1·6121	497·2	613·2
170	114·85	3·885	661·0	1·5921	489·3	615·1
180	145·43	3·104	663·3	1·5727	481·1	616·9
190	182·04	2·504	665·2	1·5540	472·4	618·3
200	225·51	2·037	666·8	1·5360	463·4	619·6
210	276·73	1·670	668·0	1·5184	453·7	620·5
220	336·50	1·379	669·0	1·5014	443·7	621·3
230	405·81	1·145	669·4	1·4843	433·0	621·6
240	485·61	·9561	669·4	1·4675	421·7	621·7
250	576·92	·8019	668·9	1·4504	409·6	621·3
260	680·80	·6752	667·8	1·4334	396·8	620·5
270	798·46	·5702	666·0	1·4162	383·0	619·2
280	931·00	·4825	663·6	1·3989	368·3	617·4
290	1079·6	·4088	660·4	1·3808	352·4	615·0
300	1245·9	·3464	656·1	1·3619	335·1	611·7
310	1431·3	·2932	650·8	1·3424	316·2	607·6
320	1637·3	·2473	644·2	1·3217	295·2	602·6
330	1865·5	·2075	636·0	1·2989	271·8	596·2
340	2118·7	·1724	625·6	1·2736	244·9	588·1
350	2398·5	·1410	611·9	1·2442	213·0	577·1
360	2708·3	·1119	592·9	1·2077	172·1	561·7
370	3053·6	·08005	559·3	1·1502	107·0	534·2
374	3203·5	·05842	523·3	1·0928	35·3	504·2

Table A'. *Properties of Water at Saturation Pressure.*

Temp. ° C.	Pressure lbs./in.²	Volume ft.³/lb.	Total heat Calories	Entropy	Function G Calories
0	0·0886	0·01602	0	0	0
10	0·1780	·01602	10·04	0·0360	0·15
20	0·3390	·01605	20·03	·0706	0·67
30	0·6153	·01609	30·00	·1041	1·55
40	1·0697	·01614	39·98	·1365	2·76
50	1·7891	·01621	49·95	·1680	4·31
60	2·8892	·01629	59·94	·1984	6·17
70	4·5200	·01639	69·93	·2281	8·32
80	6·8690	·01648	79·95	·2568	10·75
90	10·168	·01659	89·98	·2849	13·43
100	14·696	·01672	100·04	·3120	16·38
110	20·78	·01684	110·12	·3388	19·65
120	28·80	·01699	120·25	·3648	23·17
130	39·20	·01714	130·42	·3903	26·94
140	52·41	·01730	140·64	·4153	30·95
150	69·03	·01745	150·92	·4397	35·20
160	89·63	·01765	161·26	·4639	39·70
170	114·85	·01785	171·68	·4875	44·41
180	145·43	·01806	182·18	·5109	49·34
190	182·04	·01829	192·78	·5339	54·52
200	225·51	·01853	203·49	·5567	59·90
210	276·73	·01878	214·32	·5792	65·50
220	336·50	·01906	225·29	·6014	71·30
230	405·81	·01936	236·41	·6234	77·30
240	485·61	·01969	247·72	·6454	83·50
250	576·92	·02004	259·23	·6671	89·80
260	680·80	·02043	270·97	·6889	96·50
270	798·46	·02086	282·98	·7108	103·1
280	931·00	·02134	295·30	·7328	110·0
290	1079·6	·02188	307·99	·7549	117·1
300	1245·9	·02249	320·98	·7771	124·4
310	1431·3	·02320	334·63	·8000	131·8
320	1637·3	·02403	349·00	·8237	139·5
330	1865·5	·02503	364·23	·8480	147·3
340	2118·7	·02628	380·69	·8742	155·3
350	2398·5	·02798	398·90	·9022	163·4
360	2708·3	·03054	420·8	·9359	171·7
370	3053·6	·03574	452·3	·9838	180·4
374	3203·5	·04470	488·0	1·0383	184·0

Table B. *Properties of Saturated Steam. Pressure Basis.*

Pressure lbs./in.²	Temp. ° C.	Volume ft.³/lb.	Total heat Calories	Entropy	Latent heat Calories	Function G Calories
0·1	1·68	2968	597·9	2·1768	596·3	·004
0·2	11·76	1533	602·4	2·1168	590·7	·22
0·3	18·06	1043	605·1	2·0813	587·0	·54
0·4	22·71	793·8	607·1	2·0561	584·5	·87
0·5	26·45	643·3	608·5	2·0371	582·1	1·20
1	38·75	334·1	614·5	1·9779	575·2	2·59
2	52·27	173·9	619·8	1·9202	567·6	4·70
3	60·82	118·9	623·4	1·8872	562·6	6·34
4	67·20	90·82	626·1	1·8636	558·9	7·69
5	72·35	73·64	628·0	1·8455	555·8	8·87
6	76·69	62·09	629·8	1·8307	553·1	9·93
7	80·46	53·74	631·2	1·8184	550·9	10·88
8	83·81	47·43	632·7	1·8075	548·8	11·75
9	86·82	42·48	633·8	1·7979	547·0	12·56
10	89·55	38·49	635·0	1·7895	545·4	13·31
12	94·42	32·46	636·9	1·7748	542·4	14·71
14	98·65	28·10	638·6	1·7624	539·9	15·97
16	102·40	24·80	640·0	1·7515	537·5	17·14
18	105·78	22·22	641·3	1·7418	535·4	18·23
20	108·86	20·13	642·5	1·7334	533·6	19·26
22	111·70	18·42	643·5	1·7255	531·7	20·23
24	114·35	16·97	644·4	1·7182	529·9	21·15
26	116·80	15·75	645·2	1·7116	528·3	22·01
28	119·11	14·69	646·1	1·7054	526·8	22·85
30	121·30	13·75	646·9	1·6998	525·4	23·63
32	123·36	12·95	647·7	1·6945	524·1	24·40
34	125·32	12·23	648·3	1·6897	522·7	25·14
36	127·19	11·60	649·0	1·6852	521·5	25·85
38	128·98	11·03	649·6	1·6810	520·3	26·53
40	130·70	10·51	650·2	1·6769	519·1	27·20
42	132·34	10·04	650·7	1·6730	517·9	27·84
44	133·92	9·607	651·3	1·6693	517·0	28·48
46	135·45	9·215	651·8	1·6657	515·8	29·09
48	136·92	8·855	652·1	1·6623	514·7	29·68
50	138·35	8·523	652·6	1·6590	513·7	30·25
60	144·85	7·180	654·6	1·6444	509·0	32·95
70	150·51	6·212	656·3	1·6321	504·9	35·40
80	155·55	5·480	657·6	1·6215	501·0	37·65
90	160·15	4·902	658·8	1·6120	497·4	39·73
100	164·35	4·438	659·8	1·6035	494·0	41·70
110	167·21	4·049	660·7	1·5958	490·9	43·52
120	171·81	3·731	661·6	1·5887	488·1	45·25
130	175·18	3·458	662·2	1·5822	485·2	46·90
140	178·35	3·222	662·9	1·5761	482·4	48·49
150	181·35	3·016	663·6	1·5705	480·0	50·01

Table B (*continued*)

Pressure lbs./in.²	Temp. ° C.	Volume ft.³/lb.	Total heat Calories	Entropy	Latent heat Calories	Function G Calories
160	184·19	2·835	664·2	1·5651	477·6	51·47
170	186·90	2·676	664·8	1·5600	475·4	52·88
180	189·48	2·532	665·2	1·5551	473·0	54·24
190	191·95	2·404	665·5	1·5507	470·8	55·56
200	194·33	2·288	665·9	1·5464	468·6	56·83
210	196·61	2·184	666·3	1·5424	466·6	58·08
220	198·81	2·087	666·7	1·5384	464·6	59·28
230	200·94	2·000	667·0	1·5346	462·5	60·46
240	202·99	1·920	667·3	1·5310	460·6	61·60
250	204·92	1·846	667·7	1·5275	458·7	62·74
260	206·90	1·777	667·9	1·5242	456·9	63·82
270	208·77	1·713	668·1	1·5210	455·1	64·89
280	210·59	1·654	668·3	1·5177	453·4	65·94
290	212·35	1·597	668·4	1·5146	451·7	66·96
300	214·07	1·545	668·6	1·5116	449·9	67·97
350	222·06	1·328	669·2	1·4977	441·7	72·75
400	229·21	1·163	669·4	1·4852	433·8	77·15
450	235·71	1·034	669·4	1·4737	426·5	81·26
500	241·66	·930	669·3	1·4640	419·5	85·12
550	247·19	·845	669·1	1·4548	413·0	88·30
600	252·33	·773	668·6	1·4461	406·6	91·40
650	257·16	·711	668·1	1·4380	400·5	94·50
700	261·71	·657	667·5	1·4302	394·4	97·40
750	266·02	·611	666·7	1·4231	388·4	100·3
800	270·12	·569	665·9	1·4162	382·6	103·2
850	274·02	·533	665·0	1·4095	377·0	105·9
900	277·76	·501	664·1	1·4031	371·5	108·4
950	281·34	·472	663·2	1·3968	366·0	110·8
1000	284·77	·445	662·3	1·3908	360·3	113·2
1100	291·27	·399	660·1	1·3792	350·2	117·9
1200	298·33	·361	657·6	1·3679	339·8	122·2
1300	303·01	·329	654·9	1·3571	329·6	126·3
1400	308·37	·301	652·1	1·3467	319·4	130·4
1500	313·44	·277	649·3	1·3364	309·4	134·2
1600	318·25	·254	646·2	1·3264	299·4	137·8
1700	322·85	·236	643·0	1·3164	289·3	141·3
1800	327·22	·219	639·5	1·3064	279·3	144·8
1900	331·41	·203	635·8	1·2964	269·1	148·3
2000	335·43	·189	631·9	1·2863	258·7	151·5
2200	343·01	·164	622·8	1·2658	237·0	157·5
2400	350·05	·142	612·6	1·2438	213·8	163·4
2600	356·61	·122	600·9	1·2206	187·9	168·4
2800	362·76	·103	586·4	1·1933	156·8	173·0
3000	368·52	·083	567·9	1·1548	117·3	—
3200	373·91	·058	518·2	1·0960	36·0	—

Table C. *Volume of Superheated Steam.*

Pressure, lbs. per sq. in.

T. °C.	20	40	60	80	100	120	140	160	180	200	250	300	350	400	450
110	20·14														
120	20·71														
130	21·28														
140	21·85	10·77													
150	22·41	11·06	7·289												
160	22·96	11·35	7·488	5·556											
170	23·52	11·64	7·685	5·708	4·517										
180	24·07	11·92	7·880	5·859	4·641	3·825	3·242								
190	24·62	12·21	8·074	6·008	4·764	3·930	3·335	2·892	2·545						
200	25·17	12·49	8·267	6·156	4·885	4·033	3·427	2·974	2·619	2·338					
210	25·72	12·77	8·457	6·302	5·004	4·136	3·517	3·055	2·691	2·404	1·887				
220	26·26	13·05	8·647	6·447	5·121	4·239	3·606	3·154	2·763	2·469	1·941	1·584			
230	26·80	13·33	8·836	6·590	5·239	4·339	3·695	3·213	2·833	2·534	1·995	1·631	1·371	1·177	
240	27·35	13·60	9·024	6·733	5·356	4·438	3·782	3·291	2·902	2·597	2·048	1·676	1·413	1·214	1·056
250	27·89	13·88	9·210	6·875	5·472	4·536	3·868	3·368	2·971	2·660	2·100	1·721	1·454	1·250	1·092
260	28·44	14·16	9·396	7·016	5·587	4·634	3·952	3·443	3·039	2·722	2·151	1·765	1·494	1·286	1·126
270	28·98	14·43	9·582	7·157	5·701	4·731	4·037	3·517	3·106	2·783	2·201	1·809	1·533	1·321	1·159
280	29·52	14·71	9·767	7·297	5·815	4·827	4·120	3·592	3·172	2·843	2·251	1·852	1·570	1·356	1·191
290	30·06	14·98	9·951	7·438	5·929	4·923	4·203	3·665	3·238	2·903	2·301	1·895	1·608	1·390	1·223
300	30·60	15·25	10·13	7·576	6·042	5·018	4·286	3·738	3·304	2·963	2·350	1·938	1·645	1·424	1·253
350	33·33	16·62	11·05	8·268	6·600	5·487	4·689	4·094	3·626	3·255	2·588	2·143	1·824	1·585	1·399
400	36·01	17·97	11·95	8·955	7·146	5·943	5·086	4·443	3·940	3·540	2·819	2·339	1·995	1·736	1·536
450	38·64	19·30	12·83	9·632	7·682	6·389	5·473	4·782	4·244	3·818	3·043	2·529	2·159	1·882	1·668
500	41·26	20·61	13·70	10·305	8·212	6·829	5·853	5·112	4·540	4·092	3·262	2·713	2·317	2·026	1·796
550	43·80	21·91	14·55	10·976	8·738	7·264	6·227	5·437	4·829	4·362	3·477	2·893	2·471	2·168	1·921

Table C (*continued*)

Pressure, lbs. per sq. in.

T. °C.	500	600	700	800	900	1000	1200	1400	1600	1800	2000	2400	2800	3200
240	·928													
250	·963	·768												
260	·995	·799	·656											
270	1·026	·827	·683	·570										
280	1·056	·854	·708	·596										
290	1·085	·881	·733	·621	·532									
300	1·113	·906	·757	·645	·555	·482	·368							
350	1·248	1·024	·863	·746	·652	·575	·461	·377	·312	·262	·218			
400	1·374	1·132	·960	·834	·733	·651	·529	·442	·378	·324	·282	·219	·170	·130
450	1·494	1·235	1·050	·914	·806	·719	·589	·496	·425	·372	·328	·262	·214	·177
500	1·612	1·334	1·137	·991	·876	·783	·644	·546	·469	·414	·366	·297	·247	·209
550	1·726	1·428	1·217	1·066	·943	·844	·697	·592	·513	·456	·402	·328	·276	·236

Table D. *Total Heat of Superheated Steam.*

Pressure, lbs. per sq. in.

T. °C.	20	40	60	80	100	120	140	160	180	200	250	300	350	400	450
110	643·0														
120	647·8														
130	652·6														
140	657·4	655·0													
150	662·2	660·0	657·6												
160	667·0	665·0	662·7	660·6											
170	671·8	669·9	667·7	665·8											
180	676·5	674·8	672·7	670·9	668·7	666·2	664·2								
190	681·0	679·7	677·7	676·1	674·1	671·9	670·0	667·6							
200	686·0	684·5	682·7	681·2	679·4	677·5	675·7	673·6	671·4	669·3					
210	690·7	689·3	687·6	686·3	684·6	682·8	681·2	679·4	677·4	675·5	670·8				
220	695·4	694·1	692·6	691·3	689·8	688·1	686·6	685·0	683·2	681·5	677·0	672·7			
230	700·1	698·9	697·6	696·3	694·9	693·3	691·9	690·5	688·8	687·3	683·3	679·4	674·6	670·0	
240	704·9	703·8	702·5	701·3	700·0	698·4	697·1	695·8	694·4	693·0	689·3	685·8	681·7	677·4	
250	709·6	708·6	707·3	706·2	705·1	703·6	702·4	701·2	699·9	698·6	695·2	691·9	688·3	684·5	
260	714·4	713·4	712·2	711·2	710·1	708·7	707·6	706·5	705·3	704·1	700·9	697·9	694·6	691·2	687·0
270	719·2	718·2	717·1	716·1	715·1	713·8	712·8	711·7	710·6	709·5	706·5	703·7	700·7	697·6	694·0
280	724·1	723·0	722·0	721·1	720·1	718·9	717·9	716·9	715·8	714·8	712·1	709·5	706·7	703·9	700·6
290	728·9	727·9	726·9	726·0	725·1	724·0	723·0	722·1	721·1	720·1	717·6	715·1	712·5	710·0	707·0
300	733·6	732·7	731·8	731·0	730·0	729·1	728·1	727·3	726·3	725·4	723·1	720·7	718·3	716·0	713·3
350	758·0	757·4	756·5	755·8	755·1	754·4	753·6	752·8	752·2	751·5	749·7	748·2	746·5	744·7	742·5
400	782·2	781·9	781·2	780·9	780·3	779·6	779·0	778·5	778·0	777·4	776·2	774·8	773·5	772·1	770·5
450	807·0	806·3	806·3	806·0	805·6	805·1	804·6	804·2	803·8	803·3	802·5	801·2	800·1	799·0	779·7
500	832·2	831·8	831·5	831·1	830·9	830·7	830·3	830·0	829·6	829·1	828·5	827·5	826·5	825·5	824·3
550	857·8	857·3	857·0	856·8	856·6	856·4	856·1	855·8	855·5	855·0	854·5	853·5	852·7	851·8	850·8

Table D (*continued*)

Pressure, lbs. per sq. in.

T. °C.	500	600	700	800	900	1000	1200	1400	1600	1800	2000	2400	2800	3200
250	676·0													
260	683·7	674·6												
270	690·8	682·9	673·5											
280	697·7	690·8	682·5	675·3	661·8									
290	704·4	698·3	690·8	684·3	672·3	668·0								
300	710·9	705·3	698·7	693·0	681·9	678·0	660·0							
350	741·0	737·2	733·0	729·4	722·4	720·7	710·6	699·3	687·9	674·3	658·8			
400	769·2	766·4	763·2	760·4	755·9	754·2	747·8	740·8	733·7	726·3	718·0	700·0	679·2	651·3
450	796·5	794·1	791·8	789·6	786·0	784·9	780·1	775·2	769·9	764·6	759·2	747·3	735·0	721·2
500	823·6	821·3	820·0	818·0	815·2	814·3	810·5	806·6	802·7	798·7	794·7	785·8	777·0	767·7
550	850·5	848·0	847·5	846·0	843·5	842·8	839·5	836·6	833·3	829·9	826·5	819·7	813·0	805·6

Table E. *Entropy of Superheated Steam.*

Pressure, lbs. per sq. in.

T. °C.	20	40	60	80	100	120	140	160	180	200	250	300	350	400
110	1·7346													
120	1·7470													
130	1·7591													
140	1·7708	1·6888												
150	1·7823	1·7007	1·6511											
160	1·7935	1·7124	1·6635	1·6279										
170	1·8045	1·7236	1·6750	1·6403	1·6110									
180	1·8152	1·7345	1·6861	1·6516	1·6234	1·5991	1·5791							
190	1·8256	1·7452	1·6971	1·6630	1·6352	1·6115	1·5917	1·5727						
200	1·8358	1·7555	1·7078	1·6739	1·6466	1·6235	1·6039	1·5853	1·5686	1·5537				
210	1·8456	1·7655	1·7180	1·6847	1·6575	1·6345	1·6154	1·5974	1·5811	1·5667	1·5332			
220	1·8552	1·7755	1·7282	1·6950	1·6681	1·6454	1·6264	1·6089	1·5930	1·5794	1·5468	1·5200		
230	1·8648	1·7850	1·7382	1·7048	1·6784	1·6559	1·6372	1·6200	1·6046	1·5906	1·5594	1·5334	1·5087	1·4864
240	1·8741	1·7947	1·7480	1·7147	1·6884	1·6661	1·6475	1·6305	1·6153	1·6019	1·5712	1·5480	1·5226	1·5009
250	1·8834	1·8040	1·7573	1·7244	1·6981	1·6759	1·6576	1·6408	1·6259	1·6126	1·5826	1·5577	1·5352	1·5146
260	1·8923	1·8131	1·7666	1·7337	1·7077	1·6855	1·6674	1·6508	1·6361	1·6231	1·5934	1·5691	1·5472	1·5273
270	1·9014	1·8220	1·7757	1·7428	1·7170	1·6951	1·6771	1·6605	1·6459	1·6331	1·6039	1·5800	1·5585	1·5393
280	1·9101	1·8308	1·7847	1·7518	1·7261	1·7044	1·6864	1·6700	1·6555	1·6428	1·6140	1·5904	1·5694	1·5507
290	1·9187	1·8395	1·7935	1·7607	1·7350	1·7135	1·6956	1·6794	1·6649	1·6523	1·6239	1·6005	1·5797	1·5616
300	1·9270	1·8480	1·8020	1·7695	1·7437	1·7225	1·7045	1·6885	1·6741	1·6616	1·6335	1·6103	1·5901	1·5721
350	1·9678	1·8893	1·8435	1·8110	1·7857	1·7648	1·7472	1·7312	1·7175	1·7053	1·6781	1·6564	1·6373	1·6202
400	2·0051	1·9272	1·8815	1·8500	1·8246	1·8037	1·7864	1·7708	1·7573	1·7454	1·7190	1·6975	1·6790	1·6625

INDEX

Printed in the United States
By Bookmasters